Engineering Design: A Practical Guide

Madara M. Ogot

School of Engineering Design, Technology
and Professional Programs
Department of Mechanical and Nuclear Engineering
The Pennsylvania State University

Gül Kremer

School of Engineering Design, Technology
and Professional Programs
The Pennsylvania State University

Printed in Victoria, Canada

Cover Design: Nicole Drey

Books purchased for course use come bundled with an academic version of **Solidworks** through Togo Press, LLC at no additional cost. Academic institution orders and desk copy requests should be directed to:

> Togo Press, LLC
> 320 Anthon Drive
> Pittsburgh, PA 15235
> Tel: (412) 613-3497 Fax: (412) 798-9428
> Email: orders@togopressllc.com

Note for Librarians: a cataloguing record for this book that includes Dewey Classification and US Library of Congress numbers is available from the National Library of Canada. The complete cataloguing record can be obtained from the National Library's online database at: www.nlc-bnc.ca/amicus/index-e.html

ISBN 1-4120-3850-2

TRAFFORD

This book was published on-demand in cooperation with Trafford Publishing.
On-demand publishing is a unique process and service of making a book available for retail sale to the public taking advantage of on-demand manufacturing and Internet marketing. On-demand publishing includes promotions, retail sales, manufacturing, order fulfilment, accounting and collecting royalties on behalf of the author.

Suite 6E, 2333 Government St., Victoria, B.C. V8T 4P4, CANADA
Phone   250-383-6864   Toll-free 1-888-232-4444 (Canada & US)
Fax   250-383-6804   E-mail   sales@trafford.com
Web site www.trafford.com   TRAFFORD PUBLISHING IS A DIVISION OF TRAFFORD HOLDINGS LTD.
Trafford Catalogue #04-1658   www.trafford.com/robots/04-1658.html
10 9   8   7   6   5   4   3   2   1

Dedication

To my parents, Prof. Bethwell A. and Mrs. Grace A. Ogot, to my wife Artie for her patience and support, and to my daughters Akinyi and Ger. – *Madara*

To my parents Muharrem and Hamdiye, my brother Salih, my sister Aylin, and my husband Paul with love. – *Gül*

Contents

V Appendices   341

A Creation of the PMWs in Excel   343

B Illustrator 10 Tutorial   357

Preface

Engineering Design: *A Practical Guide*, the secondary title set the theme for the way we approached writing this introductory text on engineering design. We strongly felt that in addition to providing readers with a conceptual understanding of the early stages of the design process, the text should also provide instruction on practical tools that readers can apply to aid in the creation of better designs. For example in Chapter 2, we provide an extensive discussion on project management. Realizing that project management software may not be available to all readers, we have developed and incorporated into the discussion the use of a powerful **Excel-based** automated project management tool, **Project Management Workbooks**.

Since the early 1980s companies have increasingly adopted the concurrent engineering approach to design. This is where teams from diverse disciplines (e.g., marketing, industrial design and engineering) work together from the beginning of the design process through production. This is not a simple task due to both individual and discipline differences amongst team members. Chapter 3 provides a discussion on collaborative design along with tools that teams can use to monitor team health and manage team conflict.

Documentation and communication of ideas is an essential component of the design process. We dedicate two chapters to engineering communication. Chapter 4 provides extensive guidelines for writing technical reports, giving oral presentations, as well as presentations through poster sessions. Chapter 5 provides a discussion on the use of graphics – hand sketches, computer-generated diagrams, and solid modeling – as a means to communicate design ideas amongst team members and to other stakeholders. Computer software should be used for the creation of diagrams and schematics used in formal presentations or in final documentation of the design. As most novice designers have little exposure to computer illustration software, a project-based tutorial on **Adobe Illustrator** is included in the appendix. On completion of the tutorial, the reader should have an intermediate level of proficiency with the software.

Engineering design can be viewed as a decision making process. Decisions need to be made on concepts to pursue further, criteria to use for evaluating concepts, materials to use, etc. Chapter 6 discusses several decision making tools including decision matrices, pairwise comparison charts, the analytic hierarchy process and Pugh charts (introduced in Chapter 9). All methods are discussed in the context of implementation in a spreadsheet program, such as **Excel**. Use of spreadsheets for routine calculations, frees up more time for the creative part of the design process.

Finally, in addition to discussing traditional concept generation methods in Chapter 9, an entire chapter is devoted to an introductory discussion on the *theory of inventive problem solving*, TRIZ. One of the barriers to wider use of TRIZ by the engineering design community has been TRIZ's unique terminology and modeling methods. The introduction of TRIZ in Chapter 10 incorporates terminology commonly used in engineering design. In addition, the black-box modeling technique common in engineering design has been adapted and incorporated into the TRIZ discussion in order not to require the reader learn a new modeling method.

Acknowledgments

A text such as this could not be put together without the assistance of a large number of individuals. First and foremost we would like to thank our students some of whose designs from class projects we have used as examples throughout the text. We are grateful to Sven Bileń, Richard Devon and Andrew Lau for their numerous constructive criticisms and comments. We would like to acknowledge the tremendous technical writing assistance we received from Andrew Alexander. We thank Nicole Drey for assisting with editorial changes, generation of the technical drawing exercises and designing the cover. Finally, we would like to thank our families for their patience, support and encouragement.

Madara Ogot
Gül Kremer
June 2004

Acronyms

AHP	Analytic Hierarchy Process
ANSI	American National Standards Institute
BOM	Bill of Materials
CAD	Computer-Aided Design
CPM	Critical Path Method
CPT	Critical Path Time
CYMK	Cyan Yellow Magenta Black color model
DSM	Design Structure Matrix
EF	Ecological Footprint
EMS	Energy-Materials-Signals model
FDM	Fused Deposition Modeling
FFM	Five Factor Model
HTML	Hypertext Markup Language
LCA	Life Cycle Assessment
PCC	Pairwise Comparison Charts
PMW	Project Management Workbook
RGB	Red Green Blue color model
SFA	Substance-Field Analysis
SLA	Stereolithography
TRIZ	Theory of Inventive Problem Solving
WBS	Work Breakdown Structure

Chapter 1

Engineering Design

1.1 Design

Numerous definitions have been proposed for the word *design*. A dictionary definition states that design is "..to map out in the mind; plan; invent" . Bertoline and Wiebe (2002) describe design as "..the process of conceiving or inventing ideas [...] and communicating those ideas to others." Design can either be artistic or technical (Figure 1.1). Artistic design is based on personal expression and includes paintings, sculptures (Figure 1.2), landscapes, housing interiors, etc. Technical design on the other hand focuses on the development of a product or process, like those shown in Figure 1.3. Technical design can further be sub-divided into two broad categories:

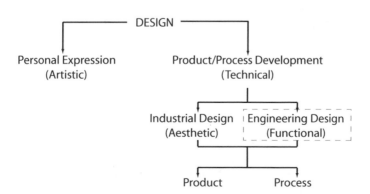

Figure 1.1 – *Engineering design in the overall design framework. Adapted from Bertoline et al., (2002).*

1

Figure 1.2 – *Mask from the Dogon peoples of Mali. The mask represents the ancient ancestress named Yasigi and is an example of artistic design (Courtesy University of Virginia).*

1. Industrial Design - concentrates on the aesthetics (appearance) and the ergonomics (human factors) associated with the product or process.
2. Engineering Design - focuses on the function, technology, and innovation behind the product or process.

This text provides an introduction to the process of *engineering design* and to tools and methods that will help designers to complete design projects.

1.2   Engineering Design

Formal definitions of engineering design abound in the literature. For example, Blumrich (1970) defines engineering design as

'[a process that] establishes and defines solutions to pertinent structure for problems not solved before, or new solutions to problems which have previously been solved in a different way.'

Figure 1.3 – *Collage of images that represent examples of technical design (Courtesy of Dell Computer, Ford Motor Co., and Boeing Co.)*

The US Accreditation Board for Engineering and Technology (ABET, 1995) identifies engineering design as

> 'the process of deriving a system, component, or process to meet desired needs. It is a decision-making process (often iterative), in which the basic sciences, mathematics, and engineering sciences are applied to convert resources optimally to meet a stated objective. Among the fundamental elements of the design process are the establishment of objectives and criteria, synthesis and analysis, construction, testing and evaluation. ... Further, [engineering design] is entailed to include a variety of ... constraints such as economic factors, safety and reliability, aesthetics, ethics and social impact.'

Successful engineering design requires a strong understanding of fundamental concepts in the basic sciences (physics[1], chemistry, and biology) combined with a solid background in mathematics. Engineering design is (1) the application of this technical knowledge with knowledge of from non-technical disciplines (for example, economics or aesthetics) and (2) the use of design and analysis tools

[1]Most concepts in engineering can be viewed as applied physics.

to synthesize a product or a system that solves a particular problem or meets a specific need.

Dym and Little (2003) describe engineering design problems as *open-ended* and *ill-structured*. Open-ended in that they typically have several acceptable solutions, and ill-structured in that a solution cannot be found by simply applying mathematical formulas in a routine or structured way. In addition, time and financial constraints further bound the entire design process, i.e. projects must be completed within a certain time-frame and budget. As a result, the final design is rarely the 'optimal' technical design but rather a 'good design' attainable within the overall project constraints.

This text provides an introduction to design tools that, when integrated with science and mathematics and applied in a systematic way form the engineering design process. Engineering design typically involves either the design of systems (for example highways, chemical plants, assembly lines) or products (for example, cars, computers, planes). The design processes for both, however, are quite similar. Where differences exist, they will be pointed out in the text.

In the design of products, especially consumer products, three distinct disciplines are involved (Cagan, 2002):

1. **Marketing** - focuses on the customer from a business perspective. It aims to answer the following questions: Who will buy the product? How much would they be willing to pay? What will it cost to introduce it to market? What is the potential market size?

2. **Industrial Design** - views the product from an aesthetic and ergonomics point of view. It aims to answer questions such as the following: What should the product look like? How should it be used?

3. **Engineering Design** - focuses on the function, technology and innovation behind the product, addressing issues such as the following: How will the product work? How will it be manufactured?

Considering everyday products and systems, one can see the interaction of all three disciplines. For example, the design of a building, involves (1) marketing who articulates what the building owners wants and how much they are willing to pay, (2) architects determine the look and layout of the building, thereby taking on the role of industrial designers, and (3) mechanical, electrical and structural engineers, the engineering designers. The design of an automobile also involves (1) marketing who determine what is the latest trend, how much are car buyers willing to pay, and what features would the customers would like to see in the vehicle, (2) industrial designers who determine what the car will look like, how the driver will interact with the car, etc., and (3) mechanical, electrical and industrial engineers who define the technology that goes into the car, determine the appropriate materials and manufacturing processes, and define how the various systems in the vehicle work.

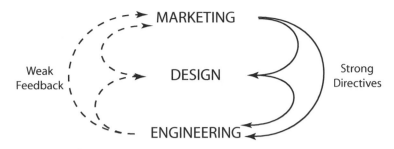

Figure 1.4 – *Traditional product development model.*

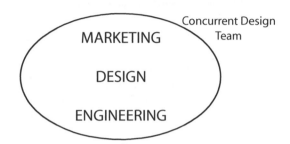

Figure 1.5 – *Concurrent engineering product development model*

Concurrent Engineering Design

Traditionally, each discipline involved in the design process has operated separately, with marketing first defining the product and then designers and engineers iterating between themselves to develop the product (Figure 1.4). Since the early 1980s, however, there has been a push for all three disciplines to work together on a product from the beginning of the design process through production. In this *Concurrent Engineering* model[2] (Figure 1.5), differing points of view from the three disciplines are always represented at the table when design decisions are being made, often resulting in better products that reach the market in a shorter time.

For concurrent engineering to be successful, teams of very diverse individuals must be able to work together as a single unit. This is not a simple task due to both individual and discipline differences amongst the team members. Often, team members are in different geographic locations or time-zones. In general,

---

[2]The concept of having the different disciplines interact more closely during the design of a product, when first introduced in the early 1980s, was referred to as *simultaneous engineering*. The concept later evolved in the late 1980s to include more disciplines and was referred to as *concurrent engineering*. Finally, in the 1990s the concept was referred to as *integrated product and process design* (Ullman, 2003). As the three terms are essentially synonymous, *concurrent engineering* will be used throughout the text.

teams that are able to use the differences as a strength rather than a weakness experience more success. Throughout the text, therefore, emphasis will be placed on teamwork and the roles that designers can play to create high performing design teams. A discussion on teamwork is presented in Chapter 3.

Technical communication is essential to the success of any design endeavor. This includes communication within the design team, between the design team and the customer, and between the design and production teams. Communication can be in the form of

1. Oral presentations, such as formal presentations at meetings.
2. Written communication, such as design memos, design proposals, technical reports, and fabrication specifications.
3. Graphical communication, such as concept illustrations, data plots, and technical drawings.

A formal discussion of technical communication is presented in Chapters 4 and 5.

User-Centered Design

User-centered design places the final user of the product, process or system at the heart of the design process (Figure 1.6). The basic premise is that by including the user throughout the design process a design team is more likely to produce a product that meets the user's expectations. If the user has a good experience with the product and if her expectations are met, she will buy it and recommend it to others. On the other hand, if the experience is poor, the user will feel frustrated and have a negative view of the product (Cagan and Vogel, 2002). These feelings will be reflected in negative recommendations to others and poor sales.

User-centered design (UCD) begins by identifying a target user group and recruiting representative users to work with members of the design team (drawn from marketing, engineering and industrial design) through each step of the design process. After assembling the entire team, the UCD process goes through three broad phases (Cagan and Vogel, 2002):

1. **Expectation:** What expectations do users have when they interact with the product? An expectation has three main attributes:

   a) Aesthetics - Is the product pleasing to the eye? Does it 'feel good' to use? Does it fit in with the image or lifestyle of the user?
   b) Performance - Does the product perform as expected? What does the user expect the product to do? How are the tasks currently achieved? What does the user like or dislike about the current method to achieve the task(s)? The latter two questions begin the process of

Table 1.1 – *Buckminister 'Bucky' Fuller and the Geodisic Dome*

Bucky Fuller

Image of a geodesic dome

Buckminster 'Bucky' Fuller was an avid designer and a humanitarian. Over his lifetime he was awarded 25 US patents, received 47 honorary doctorates in the arts, science, engineering and the humanities, and received dozens of major architectural and design awards. Fuller is best known for the invention of the geodesic dome, touted as the the lightest, strongest, and most cost-effective structure ever devised. The geodesic dome is able to cover more space without internal supports than any other enclosure. The dome can cover extremely large surfaces. For example, Fuller designed a 20-story dome housing the US pavilion at Montreal's Expo '67.

The geodesic dome's strength comes from the creation of a spherical structure based on a series of triangle. Compared to other shapes triangles form rigid structures and can withstand large forces without collapsing. In addition, triangles can be interconnected to create a continuous spherical surface resulting in a structure with very high strength. The domes also provide an energy efficient atmosphere for human habitation as the air and energy are allowed to naturally flow within the domes without obstruction. Over 300,000 major geodesic structures have been built worldwide. More information about Bucky Fuller can be found on the Buckminster Fuller Institute website (www.bfi.org).

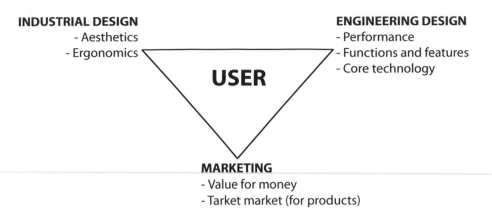

Figure 1.6 – *Illustration of how the user is placed at the core of the design process in user-centered design*

        benchmarking competitors' products and ensure that the final design is competitive in the market place.

c) Value for the money - Every target market has a spending profile that reflects the product value that people are willing to pay for. Is the profile understood, and is the price of the product within the profile?

It is important that the three main groups in the design process (marketing, industrial and engineering design) collectively understand these expectations or needs and translate them into product features. A more detailed description of how these expectations are understood from the engineering designer's perspective is given in Chapter 7 *Determination of Need/Problem Definition*.

2. **Attributes:** User expectations must then be translated into a set of product attributes. For example, the performance expectations are translated into product features, or the value for money perception is achieved by focusing the product to a particular target market.

3. **Manifestation:** Finally the attributes are manifested in the product by technology (enabling features), style (aesthetics and ergonomics), and pricing and branding strategy (suitable for target market).

    During the UCD process, several prototypes of increasing complexity incorporating successively more of the desired attributes are created and tested by participants recruited from the target audience. The participants' task performance, reactions and comments provide invaluable information that helps the design team decide what aspects of the current design to keep and what to change. A large number of iterations occur between re-design and re-testing until functional requirements and usability criteria are met. It is only then that the product is launched into the market.

User input, however, continues to be collected through participation in benchmark assessments where the product is rated against the users' requirements and competitive products. In addition, customer service keeps track of user-reported problems. This information is passed back to the design teams to let them know what to improve in the next generation of the product.

Numerous corporations have adopted the UCD philosophy of design. IBM, for example, is a strong advocate of UCD and has devised a set of UCD principles that clearly defines and communicates the UCD process (IBM, 2004):

1. **Set business goals:** Determine target market, intended users and primary competition.
2. **Understand users:** Make a commitment to understand and involve the user.
3. **Design the total experience:** Use a multidisciplinary team to design everything a user sees or touches.
4. **Evaluate Designs:** User feedback is gathered early and often and used to drive product design.
5. **Assess Competitiveness:** Superior design requires continuous awareness of the competition.
6. **Manage by continual user observation:** User feedback is integral to product plans, priorities and decision making.

1.3 Process Design

Websters dictionary defines a process as "a method of doing something, with all the steps involved; a systematic series of operations in the production of something." (Guralnik, 1983). In an engineering context, a process can be defined as a series of operations required to take a set of inputs to produce a set of outputs (Figure 1.7). For example, the production of steel can be considered an engineering process. The inputs (iron ore, carbon, etc.) are taken through a series of operations resulting in steel and waste materials as the outputs. Another example is a pharmaceutical production where active and inactive ingredients (inputs) are taken through a series of operations to produce a medicine (output). A third example is the production of petroleum products. Crude oil and additives are the inputs, with gasoline and other products as the outputs.

Process design can therefore be defined as the *design of the series of operations necessary to convert the inputs to outputs*. In engineering disciplines process design is mainly found in chemical engineering, industrial engineering, bio-engineering, and material science.

Successful process design requires the knowledge of the how the process will react to process parameter changes, as well as controllable and uncontrollable

Figure 1.7 – *A simple representation of a process as operations that transform a set of inputs to a set of output(s)*

disturbances. This is achieved through a 'process model' that describes how the process behaves in relation to variables that affect it. Note that process models not only include physical variables such as materials, temperature, pressure, etc., but also economic variables such as cost and time. Process models

1. Allow design teams to determine if a candidate process will produce the desired output(s) within the prescribed budget and time frame.
2. Dictate the operation of the processes once installed. They allow the determination of new settings to achieve different outcomes to meet new requirements.
3. Facilitate automation. The models point to areas that can be automated and to those that cannot. In addition, as automation requires control, process models are used to mechanize or computerize the taking and interpreting of measurements, and the use of the model to calculate adjustments that must be made to inputs to continue getting the desired output(s), despite external disturbances (Nevins et al., 1989).

Disturbances come both in the form of inputs you cannot control (e.g., fluctuations in environmental conditions, material properties, etc.) and outputs that you cannot measure well enough (e.g., hidden flaws, very small changes, etc.). An enhanced view of the process model including disturbances is illustrated in Figure 1.8.

Finally, the key element of process design is the client or customer. The aim of the process and its outcome is to meet the needs of a client. Therefore, when designing a new or redesigning an existing process, a designer should not lose sight of why the process is needed and the output(s) the client wants.

1.4 Overview of the Engineering Design Process

Over the years, engineering design has come to be viewed more as a 'science' and less as an 'art'. Although engineering design still requires creativity and inventiveness (the 'art'), engineering design is a process that can be taught, learned and successfully implemented to solve engineering problems (the 'science'). De-

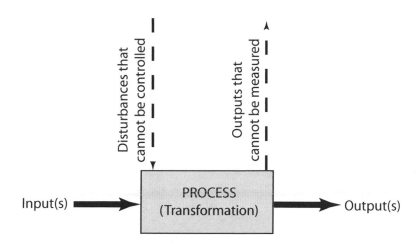

Figure 1.8 – *Enhanced process model with the addition of disturbances*

spite this general view of design as a science, there is quite a bit of variation in the steps and their labels. A survey of the literature detailing these differences are summarized in Figure 1.9.

Although no particular step categorization or step nomenclature is universally accepted, one can see a close correspondence among them. In the figure the shaded and unshaded bands group together steps that are similar. The primary difference within the bands is the degree of task segmentation. For example, whereas Asimow (1962) defines the initial step in the design process as *feasibility study*, Dym and Little (2003) divide this step into two, problem definition and conceptual design, while Dieter (1991) divides the step into three: definition of problem, information gathering and conceptualization.

Based on a comparative analysis of the literature, the engineering design process can be divided into five broad steps (Figure 1.10):

1. Determination of a need/problem definition
2. Conceptualization
3. Preliminary design and evaluation
4. Detailed design and testing
5. Production

It must be emphasized that engineering design is an *iterative process* requiring the repetition of most steps based on what is learned at a later stage. The primary iterations occur between the *conceptualization* and the *preliminary design and evaluation* steps. As a general rule of thumb, design changes due to iterations between steps become substantially more expensive to implement as

Asimow(1962)	Dym and Little (2004)	Dieter(1991)	Ulrich and Eppinger(2000)	Pahl and Beitz(1996)	Voland(2004)
FEASIBLITY STUDY Validate need, produce a number of possible solns, evaluate solutions	CLIENT STATEMENT The need	RECOGNITION OF NEED Typically arises from dissatisfaction with current state	PLANNING Articulate market opportunities; assess new technologies	PRODUCT PLANNING AND CLARIFICATION OF TASK Collecting customer information; generating initial product ideas; culminates in design specs or requirements	NEEDS ASSESSMENT Identify objective(s) and who will benefit from soln.
	PROBLEM DEFN. Clarify clients objectives and gather information on to develop statement of client wants	DEFN. OF PROBLEM Find the true problem; define broadly; culminate in formal problem statement			PROBLEM FORMULATION Identify the real problem to be solved; acquire and apply technical knowledge; identify design specifications and resources; prioritize design goals
PRELIMINARY DESIGN Determination of physical processes that govern main flows and conversions of materials, energy and information; building and testing of experimental models; determine optimal solns	CONCEPTUAL DESIGN Generate concepts or schemes of candidate designs	GATHERING INFORMATION Iteration with previous step results in revised problem statement	CONCEPT DEVELOPMENT Collect customer needs investigate feasibility of product concepts; build and test prototypes.	CONCEPTUAL DESIGN Find essential problems; establish function structures; searching for working principles and structures; selecting principal solution	ABSTRACTION AND SYNTHESIS Develop abstract (general) concepts or approaches to solve the problem; generate detailed design alternatives or solutions
	PRELIMINARY DESIGN Identify principal attributes of the design concepts or schemes	CONCEPTUALIZATION Synthesis; taking elements of concept and arranging them in the proper order, sized and dimensioned in the proper way	SYSTEM-LEVEL DESIGN Generate alternative product architectures; define major subsystems and interfaces; refine industrial design.	EMBODIMENT DESIGN Layout design (arrangement of components and relative motion) and form design (shapes and materials of individual components)	
DETAILED DESIGN Design brought to point of a complete engineering descrip. of a tested and producible product	DETAILED DESIGN Redefine and detail the final design	EVALUATION Analysis of the design through computer simulation or physical prototyping	DETAIL DESIGN Define part geometry; choose materials; assign tolerances.	DETAIL DESIGN Yields the specification of production: laying down of arrangement, forms, dimensions surface properties of all parts; material specification; all documents and other production documents produced	ANALYSIS Compare and evaluate alternative designs; construct prototypes of promising designs; test/evaluate/refine; select best alternative
	DESIGN COMMUNICATION Document the fabrication specifications and their justifications	COMMUNICATION Document the fabrication specifications and their justifications	TESTING AND REFINEMENT Reliability, life and performance testing; implement design changes.		IMPLEMENTATION Develop final solution and distribute to client/customer/user(s); after successful testing of prototype, proceed with full production

Figure 1.9 – A summary of literature survey on proposed steps for engineering design process. Similar steps are grouped together using the shaded and unshaded bands.

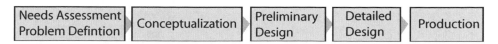

Figure 1.10 – *Division of the engineering design process into five broad steps.*

they move further along in the design process. Each of these general steps will be elaborated upon in the following sections.

Determination of the Need/Problem Definition

The need for a product typically arises from three distinct scenarios.

1. **The need to design a new product or process that will solve a particular problem or need where none other exists.** For example, Thomas A. Edison's[3] designed and developed the incandescent light bulb to provide light *(the need)* using electricity *(the new product where none existed)*.

2. **Redesign: To design a product or process that improves on an existing one.** Examples of improvements include lower cost, higher efficiency, lower pollution and better ergonomics. Of the three design scenarios, this is the most common one. A great example of this is found in the personal computer (PC) industry. The need for higher computational speed in PCs resulted in a tremendous rate of research, development and introduction into the market of new processors, starting with the Intel 8086 processor with speeds of 20MHz in the early 1980s to Intel Pentium IV processors clocking speeds over 3.2 GHz by the end of 2003. The higher speeds were achieved via a combination of improving the circuit architecture of existing processors and refining the manufacturing process. Competition between rival manufacturers combined with a demand from users for faster computers continues to drive innovation in processor design.

3. **Technology-push product or process: To design a new product or process and generate a need for it.** A company develops a new proprietary technology and then seeks an appropriate market to apply it. For example, Gore-Tex manufactured by W.L. Gore Associates, has found its way into numerous products including fabric for outer-wear and artificial veins for vascular surgery (Ulrich and Eppinger, 2003). A more recent example is the internet company eBay Inc. (www.ebay.com). They devel-

---

[3]Thomas A. Edison was born in Milan, Ohio in 1847. He is considered one of the greatest inventors of all time. Prior to his death in 1931, he patented 1,093 inventions. One of his most famous quotes is 'Genius is 1 percent inspiration and 99 percent perspiration'. Another of his quotes relevant to this step in the engineering design process is 'I find what the world needs; then I go ahead and try to invent it'.

oped the software technology to allow individuals and companies to carry out on-line auctions and then, through effective advertising and marketing, created a need for the service. Today online auctions at eBay and other commercial internet sites are one of the great successes in e-commerce.

Designing a product or process to fulfil a particular need requires a well defined problem. Problem definition, a crucial part of the design process, includes the following elements:

- A condensed formal problem statement clearly stating what the design is intended to accomplish.
- A listing of the technical and non-technical design constraints.
- A breakdown of the problem into smaller manageable sub-problems, with a clear understanding of how they all fit together. Each of the sub-problems is then translated into functions and sub-functions.
- A compilation and ranking of customer or client needs. What exactly does the customer/client expect in the final product or process? For consumer products, this information is typically obtained through interviewing and surveying potential users and current users of similar products (if they exist). For non-consumer items, for example buildings and chemical plants, similar information is obtained from the clients (those who have contracted the work) and potential users (in this case, of the building or plant).
- A definition of the criteria to be used to evaluate the design. The design team must know at the beginning of the project how prototypes developed in the *preliminary design and evaluation* step will be tested and what outcomes will define success or failure.

Poor or incomplete problem definition will result in the development of a product or process that does not adequately meet the need. It is important that the design team is solving the 'correct' problem. Sufficient effort and resources must therefore be committed to this stage of the design process. All design team members must be involved to ensure that the entire team fully understands the problem and the constraints placed on the solution. The culmination of this stage is typically a revised problem statement that reflects the customers' or clients' needs and the constraints placed on the design. A more detailed description can be found in Chapter 7.

Conceptualization

The conceptualization stage is typically divided into two phases: external and internal searches. The **internal search** is when the design team develops several concepts from which the best suited to the defined need is selected. This stage is the creative, inventive and some would argue the most difficult part of the engineering design process. *A concept is a very preliminary description of the form, required principles and required technology for the solution.* It is normally expressed as a two- or three-dimensional sketch with a brief accompanying description.

Before the design team generates any of their own ideas, time must be committed to discovering what has been done in the past. **The external search** includes performing literature searches, looking at previous patents, talking with experts, and benchmarking existing similar products. The conclusion of the conceptualization stage results in the generation and selection of a few promising concepts that warrant further development. A detailed discussion of the conceptualization stage is presented in Chapters 8 and 9.

Preliminary Design and Evaluation

Concepts shown to be feasible are further developed. Evaluations are carried out on all the concepts, leading to the selection of one concept. Selection is based on all design criteria specified during problem definition, as well as cost estimates. In addition, system and component design requirements are established by literature searches, computations and analyses, and discussions with vendors. The design requirements established here will dictate the detailed design specifications. Great care must be taken to ensure that the requirements are attainable.

During this stage, working prototypes (where appropriate) are also constructed and evaluated. Based on the test results, parts of the design or the entire design may need to be redone.

Detailed Design

This stage of the design processes develops the technical drawings and system specifications for the project. During this stage

1. All hitherto undefined system specifications and design requirements are defined. These include

   - Operating parameters
   - Test requirements
   - External dimensions and interfaces that must be controlled
   - Maintenance and testability provisions
   - Material requirements
   - Reliability requirements
   - External surface treatment
   - Design life
   - Packaging requirements (Ertas, 1996)

2. Detailed manufacturing drawings are produced.
3. Detailed assembly drawings are generated showing how the parts and sub-assemblies are interconnected.
4. Testing is performed to evaluate components, validate computer models and the design itself. Evaluation ensures that all the conclusions reached during the preliminary testing stage are accurate. If errors are found or if components do not meet anticipated design requirements, a redesign is initiated.

Production

Prior to production, production process planning is carried out. This typically involves

- Design drawings and specification interpretation
- Production processes and machines selection
- Stock material selection
- Determination of production sequence of operations
- Selection of jigs (equipment needed to hold and position the part), fixtures (equipment needed to guide the tool during the manufacturing operation) and tools (the equipment used to perform the actual metal removing and forming operation).
- Determination of tool cutting parameters (e.g., speed, depth, feed rate)
- Calculation of processing time (Zhang and Alting, 1993)

These activities are then followed by the production of the product.

1.5 Design Reviews

Due to increased global competition in bringing products to market, new tools are continuously being introduced to ensure product quality within cost and time constraints. One of these tools is the design review (DR). Design reviews can formally be defined as

> Judgment and improvement of an item at the design phase, reviewing the design terms of function, reliability, and other characteristics, with cost and delivery as constraints and with the participants of specialists in design, inspection and implementation. (Japanese Industrial Standard JIS Z 8115-1981).

An alternative definition states that

Design review is a system that involves gathering and evaluating objective knowledge about the product design quality and the concrete plans for making it a reality, suggesting improvements at each point, confirming that the process is ready to proceed to the next phase. *–JUSE Design Review Committee, 1976* (Ichida, 1996).

Design reviews can be conducted formally or informally. Formal design reviews (FDR) follow standard policies and procedures. The project schedule shows the designated days for the DRs and those responsible to carry them out. FDRs are considered to be essential for consistent product quality and ensuring that the design meets all requirements. Informal design reviews (IDR) are used as the need arises, and therefore their effectiveness can vary greatly. Whether completed formally or informally, design reviews should include the following steps

1. Collect and compile relevant information
2. Define quality target
3. Evaluate product and process designs and supporting operations
4. Propose improvements
5. Define subsequent actions
6. Confirm readiness for the next stage

It is important to conduct DRs at every transition between stages of the design process. Depending on the stage, the information compiled, quality targets, evaluation techniques will change. Table 1.2 lists inputs and outputs of the DRs for different stages of design.

Conducting effective design reviews requires abiding by three principles

1. At every stage until a product reaches the market, use DR to evaluate its design in terms of quality, cost, and delivery.
2. Make use of the best available knowledge and technology from both in-house and outside sources.
3. Do everything possible to resolve problems as they arise, to avoid passing them downstream.

In Takashi Ichida's words, a member of the first design review organization established in Japan in early 1970s, " ... DR is not about creating great designs. It is about getting good designs right, making them work, and bringing them to market quickly and profitably." (Ichida, 1996).

1.6 Summary

This chapter is a brief introduction to design and an overview of the design process. The focus of this text is on the first two steps of the design process:

Table 1.2 – *Design reviews (DRs) at various stages of the design process. Adapted from Ichida (1996).*

Design Process Stage	Output	Design Review	Purpose
Needs assessment, problem definition	Comprehensive set of customer needs and design criteria	Requirements Review	Review the completeness and correctness of compiled customer requirements
Conceptualization	Design concept	Design Concept Review	Review the overall system to estimate the project target feasibility and evaluate the development plan
Preliminary design	Preliminary design	System Requirements Review	Review the suitability of specific requirements and evaluate subsystem development plans
Detailed design	Detailed design	Detailed Design Review	Evaluate the revised portions of the previously completed manufacturing documentation
Prototype testing and production	Production design and production quality tests	Final Design Review	Evaluate design and production documentation after quality tests are complete

Determination of the need/problem definition and *conceptualization.* These areas will be greatly expanded upon in later chapters, in addition to brief introductions on engineering materials and commercial manufacturing methods. It is only by having an understanding of the latter two topics that designers can develop good, workable and *buildable* designs. In addition, we strongly believe that common computational tools, such as spreadsheets, should be incorporated as much as possible into the design process. These tools provide immense assistance in performing routine calculations, leaving more time for the creative aspects of the design process. Consequently, the use of **Microsoft Excel** is integrated into discussions on most design tools within the text[4].

---

[4] Although we use **Excel**, any other spreadsheet program can be used.

References

ABET, 'Criteria for Accrediting Programs in Engineering in the United States, 1995-1996 Accreditation Cycle' Accreditation Board for Engineering and Technology, Inc., 1995.

Bertoline, G.R., Wiebe, E.N. and Miller, C.L., *Fundamentals of Graphics Communication*, 3rd Edition, New York: McGraw-Hill Higher Education, 2002.

Blumrich, J.F., "Design", Science, Vol. 168, pp. 1551-1554, 1970.

Cagan, J. and Vogel, C. M., *Creating Breakthrough Products: Innovation from Product Planning to Program Approval*, Upper Saddle River,NJ: Prentice-Hall, 2002.

Dym, C.L. and Little, P., *Engineering Design: A Project-Based Introduction*, Second Edition, New York: John Wiley & Sons, 2003.

Ertas, A. and Jones, J., *The Engineering Design Process*, Second Edition, New York: Wiley & Sons, 1996.

Guralnik, D.B. (Ed.) Webster's New World Dictionary of the American Language, NY:Simon and Schuster, 1983.

IBM, 'User-Centered Design', http://www-306.ibm.com/ibm/easy/, viewed May 2004.

Ichida, T. (Ed.), *A Method for Error-Free Product Development: Product Design Review*, Portland, OR: Productivity Press, 1996.

Nevins, J.L. and Whitney, J. (eds.) *Concurrent Design of Products and Processes: A Strategy for the Next Generation in Manufacturing*, New York: McGraw-Hill, 1989.

Pahl, G. and Beitz, *Engineering Design. A Systematic Approach*, 2nd Edition, London: Springer-Verlag, 1996.

Ullman, D.G., *The Mechanical Design Process*, Third Edition, New York: McGraw-Hill Higher Education, 2003.

Ulrich, K. and Eppinger, S., *Product Design and Development*, 3rd Edition, New York: McGraw-Hill, 2003.

Zhang, H. and L. Alting, *Computerized Manufacturing Process Planning Systems*, London: Chapman-Hall, 1993.

Bibliography

Cagan, J. and Vogel, C. M., *Creating Breakthrough Products: Innovation from Product Planning to Program Approval*, Upper Saddle River,NJ: Prentice-Hall, 2002.

Dieter, G.E., *Engineering Design: A Materials and Processing Approach*, 2nd Edition, New York: McGraw-Hill, 1991.

Dym, C.L. and Little, P., *Engineering Design: A Project-Based Introduction*, 2nd Edition, New York: John Wiley & Sons, 2003.

Ertas, A. and Jones, J., *The Engineering Design Process*, Second Edition, New York: Wiley & Sons, 1996.

Kusiak, A., *Engineering Design: Products, Processes, and Systems*, San Diego: Academic Press, 1999.

Nevins, J.L. and Whitney, J. (eds.) *Concurrent Design of Products and Processes: A Strategy for the Next Generation in Manufacturing*, New York: McGraw-Hill, 1989.

Otto, K. and Wood, K., *Product Design: Techniques in Reverse Engineering and New Product Development*, Upper Saddle River: Prentice-Hall, 2001.

Pahl, G. and Beitz, *Engineering Design. A Systematic Approach*, 2nd Edition, London:Springer Verlag, 1996.

Ullman, D.G., *The Mechanical Design Process*, Third Edition, New York: McGraw-Hill Higher Education, 2003.

Ulrich, K. and Eppinger, S., *Product Design and Development*, New York: McGraw-Hill, 1995.

Voland, G. *Engineering by Design*, 2nd Edition, Upper Saddle River: Prentice-Hall, 2004.

Part I

Engineering Design Aids

Chapter 2

Management of the Design Process

2.1 Introduction to Project Management

A project by definition is a series of activities and tasks that

- Have a specific objective to be completed within certain specifications
- Have defined start and end dates
- Have funding limits
- Consume resources, such as money, employee time, equipment time and materials
- Involve multi-functional tasks, such as marketing, design, and modeling (Kerzner, 2001).

Engineering design projects are completed by multi-functional design teams through a set of parallel or sequential tasks that are iterative in nature. A successful design project requires the design team to have a good grasp of *both* the technical *and* the management aspects of the design process. The measurement of success in managing a design project involves designing and communicating a product that will satisfy customer requirements at the desired performance/technology level while utilizing assigned resources effectively and efficiently, within an allotted time and budget. Poor project management may result in a mediocre final outcome, no matter how technically proficient the team members are.

Ulrich and Eppinger (2003) categorize project management activities into three primary tasks:

1. **Planning and scheduling.**

    - Define objectives

- List all tasks and estimate their durations
- Determine interdependent tasks
- Schedule tasks and necessary resources to complete them

2. **Directing**. Implementing approved plans to achieve project objectives.

- Assign tasks
- Review criteria for task completion
- Track progress
- Compare actual outcome to predicted outcome (e.g., task duration estimate vs. actual time spent for the task)
- Analyze impact
- Resolve issues and make adjustments
- Review completed work
- Close project

3. **Administering**. Developing and implementing team member policies and operational procedures for project management.

Numerous commercial project management software packages are available, such as **Microsoft Project**. For projects that are not very complex, a spreadsheet program like **Microsoft Excel** can be used to perform the same management function. The latter approach will be explained throughout this chapter.

In educational environments, the wide adoption of commercial project management software has been hindered by several challenges. These include (1) design faculty may not know or feel inclined to learn the software, (2) funds may not be available to purchase the software, and (3) teaching students how to use one more software tool would divert time away from existing course material. Additionally, requiring students to learn yet another program may present an unwelcome burden because they are already required to learn and use a significant number of software tools like word processing, spreadsheets, programming, illustration, image manipulation, website design, and solid modeling (Ogot and Okudan, 2004). The use of **Excel** to create automated project management tools,therefore, presents a practical and cost efficient alternative as the software is readily available on college campuses, and both students and faculty typically have a basic level of proficiency in its use. Discussions in the rest of the chapter will therefore focus on how **Excel**-based *Project Management Workbooks (PMWs)*, which incorporate several of the standard project management tools, can be used for small projects. A step-by-step tutorial on creating the **Excel**-based PMWs can also be found in Appendix A. The following sections will discuss the first two project management tasks in more detail.

2.2 Planning and Scheduling

Various tools are available to effectively plan a project to ensure its completion within the allotted time. The three sequential stages for project planning, with the commonly used tools shown in brackets, are

1. Listing all tasks that need to be done and estimating the duration for each (*Work Breakdown Structure*).
2. Deciding which tasks can be done in parallel and those that must be done sequentially (*Design Structure Matrix/Activity Networks*).
3. Setting approximate start and end dates for all tasks (*Gantt/Milestone Charts*).

An explanation of these three stages and the tools employed follows.

Task Enumeration: Work Breakdown Structures

Before a design team can plan a project, it needs to have an idea of the project scope, what tasks are to be performed, and estimates on how long it will take to complete them. The Work Breakdown Structure (WBS) divides the entire project into a series of tasks, and the tasks into sub-tasks. The level of decomposition, for example to subsub-tasks or subsubsub-tasks, will depend on the complexity of the project. Although no order is implied in the WBS, arranging the tasks in approximate sequential order simplifies the process of ordering tasks later on. The WBS presents the tasks in an organized form that allows team members to readily see and understand how the tasks fit into the overall project.

Work Breakdown Structures can be presented in a graphical or list format. The Project Management Workbooks (PMWs) implement the list format. A generic design project WBS implemented in the **Excel**-based PMW is illustrated in Figure 2.1. The primary or top-level tasks in this example are

1. Determine customer needs
2. Generate concepts
3. Begin detailed design
4. Build prototype
5. Test prototype
6. Document and report the design process and its outcome

Each of the primary tasks are further decomposed into sub-tasks, as shown in the figure. To the right of each sub-task description is its estimated duration. This information is used to schedule the tasks appropriately and ensure that the project is completed on time. The sum of the durations of all the individual tasks *is not* the time required to complete the project. As certain tasks can be done in parallel, the overall project time will be significantly shorter.

	A	B	C	D	E
			DATES AND DURATION		
1			Duration (days)	Planned Dates	
2	Electrical Bicycle Project			Start	End
3					
4		0.1 Begin Project	1	12/01/03	12/02/03
5	1.0 Determine Customer Need	1.1 Interview users to establish requirements	4	12/01/03	12/05/03
6		1.2 Search the literature for any regulatory requirements	5	12/01/03	12/06/03
7		1.3 Find competitive products and research their reviews	5	12/01/03	12/06/03
8		1.4 Create a hierarchical list of customer needs	4	12/04/03	12/08/03
9		1.5 Revise Problem Statement	3	12/06/03	12/09/03
10	2.0 Generate Concepts	2.1 Functionally decompose the project	5	12/07/03	12/12/03
11		2.2 Research the literature on similar subtask solutions	7	12/09/03	12/16/03
12		2.3 Generate concepts	5	12/15/03	12/20/03
13		2.4 Select promising concept(s)	4	12/20/03	12/24/03
14	3.0 Begin Detailed Design	3.1 Perform detailed analyses of concepts	9	12/25/03	01/03/04
15		3.2 Peform simulations	9	12/25/03	01/03/04
16		3.3 Material selection/availability	9	12/25/03	01/03/04
17		3.4 Component selection/availability	9	12/25/03	01/03/04
18		3.5 CAD Drawings	17	01/04/04	01/21/04
19	4.0 Build Prototype	4.1 Purchase materials and off the shelf components	6	01/04/04	01/10/04
20		4.2 Machine/manufacture components	18	01/18/04	02/05/04
21		4.3 Assemble Prototype	8	02/04/04	02/12/04
22	5.0 Test Prototype	5.1 Develop testing protocol	6	12/07/03	12/13/03
23		5.2 Perform tests	7	02/12/04	02/19/04
24	6.0 Documentation and Reporting	6.1 Preparation of first progress report	4	12/09/03	12/13/03
25		6.2 Preparation of second progress report	4	12/21/03	12/25/03
26		6.3 Preparation of final report	5	02/16/04	02/21/04
27		6.4 Preparation of final presentation (Poster)	2	02/21/04	02/23/04
28		7.0 End Project	1	02/28/04	02/29/04
29					

Figure 2.1 – *Sample Work Breakdown Structure implemented in the* **project management workbook**

The WBS is a dynamic document and should be continually updated to reflect current events in the project. New primary or lower level tasks can be added by inserting new rows. In addition, existing tasks can be deleted by deleting the corresponding row.

Dependency: Design Structure Matrix

The *design structure matrix (DSM)* helps determine the task order for completing a project. The DSM identifies which tasks provide input to others and which tasks are independent and hence can be performed in parallel.

A DSM is constructed by listing all tasks as row and column headings in an n x n matrix, where n is the number of tasks. The tasks are listed as close as possible to the order they should be completed in. A generic design project DSM is shown in Figure 2.2. The tasks are also listed along the diagonals to make the matrix easier to read. Reading across rows, 'Xs' indicate column tasks that provide direct input to corresponding row tasks. For example, reading across row D (*Concept generation/selection*), tasks B (*Customer needs*) and C (*Market survey*) provide direct input to task D and are therefore marked with Xs. Likewise, reading down columns, 'Xs' indicate row tasks that directly receive input from a particular column task. For example, reading down column D (*Concept generation/selection*), tasks E (*Detailed Design*) and F (*Simulation*) are marked. This means that tasks E and F receive direct input from task D.

	A	B	C	D	E	F	G	H	I	J	K
Begin Project **A**	A										
Customer needs **B**	X	B									
Market survey **C**	X		C								
Concept generation/selection **D**		X	X	D							
Detailed design **E**				X	E	X					
Simulation **F**				X	X	F					
Purchase components **G**					X	X	G				
Purchase materials **H**					X	X		H			
Manufacture parts **I**								X	I		
Assembly and testing **J**							X		X	J	
Finish project **K**										X	K

Figure 2.2 – *Design Structure Matrix*

Tasks E and F deserve closer attention. Looking at row E shows that task E requires input from both tasks D and F. From row F, we see that task F requires input from both tasks D and E. The two tasks E and F appear to require input from and provide input to each other. Such tasks are said to be coupled or interdependent. Another way of looking at it is that neither task can be initiated nor completed without the other. One way of simplifying the handling of such tasks in DSMs could be to consider them as one task rather than two separate ones.

The best way to fill out the DSM is row by row. Start off with the task in the first row, and mark each task column (on the same row) with an X to indicate direct input. Move to the next task row and repeat the process. Continue until the last task row has been filled in. An ordering of tasks will begin to emerge. The order will be used to assign appropriate start dates for each task. A screen shot of the DSM for the Work Breakdown Structure example is shown in Figure 2.3.

Activity Network Analysis

After the completion of the design structure matrix, the information contained in the matrix is used to form an *activity network diagram* for the design project. Network diagrams show all the project tasks along with their precedence structure, which takes into account interdependencies among tasks.

Activity networks can be formed using one of the two principal notations (Lock, 2003):

1. **The activity-on-arrow notation.** It is also often referred to as arrow networks. Some of the techniques that use this notation are

   - The arrow diagram method (ADM), which is often referred to as the Critical Path Method (CPM) or the Critical PAth Analysis (CPA).

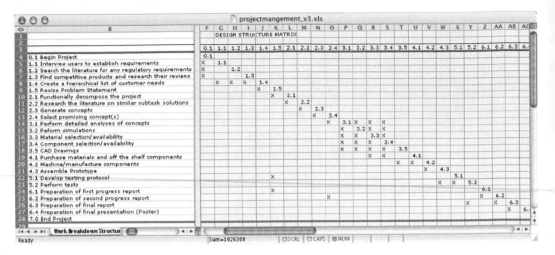

Figure 2.3 – *Screen shot of the* design structure matrix *for the WBS example incorporated into the* **project management workbook**

- Program Evaluation and Review Technique (PERT), which ADM notation but with probabilistic time estimates.

2. **The activity-on-node notation.** An example to this notation is the Precedence Diagram Method (PDM).

Precedence diagrams are the most commonly used of these techniques because they resemble engineering flow charts. They are widely supported by project management software (Lock, 2003), so they will be the notation used throughout this text.

Figure 2.4 displays a task in precedence notation. The flow of the work in precedence diagrams is always from left to right. Each task has a unique identification code, such as task A or task 1. All the tasks comprising a design project are linked by lines that indicate the task sequence using the information contained in the DSM.

Precedence diagrams may also contain task duration, earliest and latest start and finish times, as well as the float or the slack time information. An explanation of these quantities and how they are calculated by completing a time analysis in the precedence network is presented in the following section.

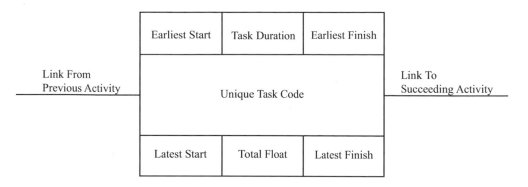

Figure 2.4 – *A task in precedence notation*

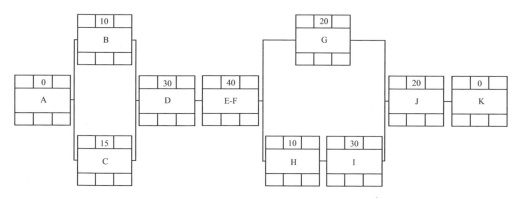

Figure 2.5 – *Activity network*

TIME ANALYSIS IN PRECEDENCE DIAGRAMS

Figure 2.5 shows the precedence diagram for the design project whose precedence structure is defined by the DSM in Figure 2.2. In the network, boxes[1] represent tasks and lines the flow. The letters in the boxes are the unique identification codes of tasks. In the middle cell above the identification code is the task duration. For example, the '10' in the box for task B indicates that it will take ten time units to complete task B before you can begin task D. Time units are typically days, weeks, months or quarters. Coupled tasks, for example tasks E and F, are placed in a single box. The number '40' in the box for tasks EF indicates the time required to complete *both* tasks.

With the *forward pass*, the overall project duration is calculated by adding activity duration estimates through the links, passing from left to right. While doing this, earliest start and finish times for each task are calculated and placed

---

[1]Circles can also be used to represent tasks in a network diagram.

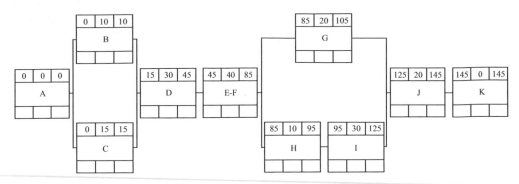

Figure 2.6 – *Forward pass calculations*

in the cells above the task code, on each side of the task duration as shown in Figure 2.4. The precedence diagram in Figure 2.6 illustrates this. In the figure, task A has a duration of 0 time units. If it is assumed that the design project starts at time 0, then the earliest start and finish times are going to be 0's as well. That is, 0+0=0, and the 0 sum indicates the earliest finish time for task A. Then, task A's finish time (0) is carried to the earliest start times for tasks B and C.

The earliest finish time for task B is calculated as 0+10=10 time units. In the same fashion, when all task nodes are visited, the earliest project completion time is found to be 145 time units. One point to remember is that when more than one task is feeding into a task, the later finish time should be carried to the succeeding task. This is to ensure that all preceding tasks are finished before starting the next in line. Such a situation exists for task D and J in Figure 2.6

The usage of a "dummy task" with 0 duration for the starting and finishing nodes in a network diagram is helpful for clearly showing one node as the starting point of the project and one for the end. In Figure 2.5, tasks A and K are dummy tasks.

There is more than one possible path through the network, and the path with the longest duration is known as the *critical path*. The critical path defines the *minimum* time required to complete the project. All other paths will be shorter than the critical path[2] and therefore end up with float times. The latter are defined as the time difference between a particular path and the critical path. The float times indicate the maximum delay that non-critical paths can experience before adversely affecting the overall timing of the project by interfering with the start time of succeeding activities.

In order to understand which tasks have floats on a path, a *backward pass* should be completed. Backward pass is accomplished through subtraction from

---

[2]If more than one path has the same longest time from the beginning to end of the project, they are all defined as critical paths.

Table 2.1 – *Activity paths and float times for the precedence diagram shown in Figure 2.5*

Path	Time	Float
ABDEFGJK	120	25
ACDEFGJK	125	20
ABDEFHIJK	140	5
ACDEFHIJK	145 (Critical path)	-

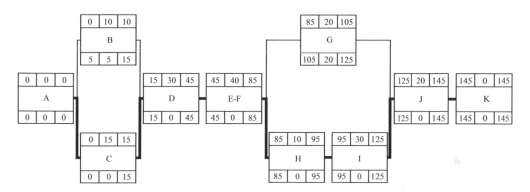

Figure 2.7 – *Backward pass calculations*

right to left. The subtractions should be repeated throughout the network, writing the permissable times and floats along the bottom row of all activity boxes. Where more than one path exists, the longest must be chosen so that the result after subtraction gives the smallest remainder (earliest time). The backward pass values are illustrated in Figure 2.7.

The path and float times for the precedence diagram shown in Figure 2.5 - 2.7 are listed in Table 2.1. The critical path (ACDEFHIJK) requires 145 time units to complete. Path ABDEFGJK contains some tasks that are done in parallel to the critical path and requires 120 time units. Consequently, completion of path ABDEFGJK can be delayed by up to 25 time units (the float time) without delaying the completion time of the overall project. Completion times and float times for the other paths are listed in Table 2.1.

On completion, the information from the precedence diagram is used to assign start and end dates to the tasks listed in the WBS. For this, calculated task floats can be used to effectively use all resources, such as employee time and equipment time. For example, in the precedence diagram in Figure 2.7, tasks B and C are parallel, and while C has no float time, B has 5 time units of float. Imagine that the employee who was assigned task B needs to help another project team, and thus can not start working on task B at the same time as the employee who is assigned task C. Because of the 5 time units of float, as long as the task B employee can start working on task B before 5 time units elapses

	Begin project	Customer needs	Market survey	Concept generation/selection	Detailed design	Simulation	Purchase components	Purchase materials	Manufacture parts	Assembly and testing	End Project		
TASK DURATION	0	10	15	30	40	0	20	10	30	20		**TOTAL**	**FLOAT**
TASK	A	B	C	D	E	F	G	H	I	J	K		
PATH 1	X	X		X	X	X	X			X	X		
	0	10	0	30	40	0	20	0	0	20		120	25
PATH 2	X		X	X	X	X		X	X	X	X		
	0	0	15	30	40	0	0	10	30	20		145	0
PATH 3	X	X		X	X	X		X	X	X	X		
	0	10	0	30	40	0	0	10	30	20		140	5
PATH 3	X		X	X	X	X	X			X	X		
	0	0	15	30	40	0	20	0	0	20		125	20

Figure 2.8 – *List format for an activity network*

from the time task C employee starts working, the project will not be delayed. Accordingly, when the task start and finish times are carried over to the WBS, float times should be taken into account.

Although precedence diagrams are graphical, their implementation in the PMW uses a text format. Adding the precedence diagrams to the PMW allows the design team to change the duration of time assigned to each task and immediately note its effect on the critical path time. Times can therefore be altered to ensure that the critical path time never exceeds the time allocated to the project. A list implementation of a precedence diagram corresponding to the graph in Figure 2.7 is shown in Figure 2.8. Each path is listed on a separate row, with the total and float times calculated at the end of the row. The tasks that form the path are indicated by an 'X' in the corresponding column, with the tasks performed sequentially along the path from left to right.

In the PMW's implementation of the precedence diagrams the tasks are listed in the rows (as was done for the WBS) and the paths are listed down the columns. A screen shot of the activity network corresponding to the WBS example is given in Figure 2.9.

An effective way to assign appropriate task durations and planned start dates is to split the work breakdown structure worksheet into two screens, as shown in Figure 2.10. On the left hand screen, display the tasks, task durations, planned start dates and planned end dates. On the right screen, display the activity network with the paths already defined. Go through and assign start dates and task durations, following one path at a time. Make sure that a task does not begin before its required input task(s) has been completed. In addition, as task durations are entered, the design team should keep tabs on the critical path to projectmangement_v3.xls

ACTIVITY NETWORK

	P1	P2	P3	P4	P5	P6	P7	P8	P9	P10
0.1 Begin Project	1	1	1	1	1	1	1	1	1	1
1.1 Interview users to establish requirements	4			4	4	4	4			
1.2 Search the literature for any regulatory requirements		5						5		5
1.3 Find competitive products and research their reviews			5						5	
1.4 Create a hierarchical list of customer needs	4	4	4	4	4	4	4	4	4	4
1.5 Revise Problem Statement	3	3	3	3	3	3	3	3	3	3
2.1 Functionally decompose the project				5	5		5	5	5	5
2.2 Research the literature on similar subtask solutions				7	7		7	7	7	7
2.3 Generate concepts				5	5		5	5	5	5
2.4 Select promising concept(s)				4	4		4	4	4	4
3.1 Perform detailed analyses of concepts				9	9		9	9	9	9
3.2 Peform simulations										
3.3 Material selection/availability										
3.4 Component selection/availability										
3.5 CAD Drawings				17				17	17	
4.1 Purchase materials and off the shelf components						6				6
4.2 Machine/manufacture components				18	18			18	18	18
4.3 Assemble Prototype				8	8			8	8	8
5.1 Develop testing protocol	6	6	6							
5.2 Perform tests	7	7	7	7	7			7	7	7
6.1 Preparation of first progress report						4				
6.2 Preparation of second progress report						4	4			
6.3 Preparation of final report	5	5	5	5	5		5	5	5	5
6.4 Preparation of final presentation (Poster)	2	2	2	2	2		2	2	2	2
7.0 End Project	1	1	1	1	1		1	1	1	1
TOTAL	33	34	34	100	89	28	45	101	101	98
FLOAT	68	67	67	1	12	73	56	0	0	11

Work Breakdown Structur

Figure 2.9 – *Screenshot of list format implementation of the* activity network *in the* **project management workbook**

projectmangement_v3.xls

	DATES AND DURATION			ACTIVITY NETWORK								
	Duration (days)	Planned Dates										
		Start	End	P1	P2	P3	P4	P5	P6	P7	P8	P9
0.1 Begin Project	1	12/01/03	12/02/03	1	1	1	1	1	1	1	1	1
1.1 Interview users to establish requirements	4	12/01/03	12/05/03	4			4	4	4	4		
1.2 Search the literature for any regulatory requirements	5	12/01/03	12/06/03		5						5	
1.3 Find competitive products and research their reviews	5	12/01/03	12/06/03			5						5
1.4 Create a hierarchical list of customer needs	4	12/04/03	12/08/03	4	4	4	4	4	4	4	4	4
1.5 Revise Problem Statement	3	12/06/03	12/09/03	3	3	3	3	3	3	3	3	3
2.1 Functionally decompose the project	5	12/07/03	12/12/03				5	5		5	5	5
2.2 Research the literature on similar subtask solutions	7	12/09/03	12/16/03				7	7		7	7	7
2.3 Generate concepts	5	12/15/03	12/20/03				5	5		5	5	5
2.4 Select promising concept(s)	4	12/20/03	12/24/03				4	4		4	4	4
3.1 Perform detailed analyses of concepts	9	12/25/03	01/03/04				9	9		9	9	9
3.2 Peform simulations	9	12/25/03	01/03/04									
3.3 Material selection/availability	9	12/25/03	01/03/04									
3.4 Component selection/availability	9	12/25/03	01/03/04									
3.5 CAD Drawings	17	01/04/04	01/21/04				17				17	17
4.1 Purchase materials and off the shelf components	6	01/04/04	01/10/04						6			
4.2 Machine/manufacture components	18	01/18/04	02/05/04				18	18			18	18
4.3 Assemble Prototype	8	02/04/04	02/12/04				8	8			8	8
5.1 Develop testing protocol	6	12/07/03	12/13/03	6	6	6						
5.2 Perform tests	7	12/09/03	02/19/04	7	7	7	7	7			7	7
6.1 Preparation of first progress report	4	12/09/03	12/13/03						4			
6.2 Preparation of second progress report	4	12/25/03	12/25/03						4	4		
6.3 Preparation of final report	5	02/16/04	02/21/04	5	5	5	5	5		5	5	5
6.4 Preparation of final presentation (Poster)	2	02/21/04	02/23/04	2	2	2	2	2		2	2	2
7.0 End Project	1	02/28/04	02/29/04	1	1	1	1	1		1	1	1
TOTAL	33			34	34	100	89	28	45	101		
FLOAT	68			67	67	1	12	73	56	0		

Work Breakdown Structure / Gantt Chart

Figure 2.10 – *Splitting the work breakdown structure worksheet window into two screens to facilitate assignment of task durations and start dates using the* precedence diagram *in the* **project management workbook**

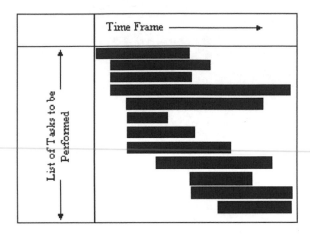

Figure 2.11 – *Gantt Chart*

ensure that it does not exceed the time available to the project.

The Gantt Chart/Milestone

The *Gantt chart* is the most widely used method in industry for project monitoring. Its advantages include

- Direct correlation of tasks with duration of time
- Straight forward integration of sub-tasks having separate scheduling charts
- Flexible time units ranging from daily to annual
- Visual representation for quick assessment of project progress

Figure 2.11 displays the general layout of a Gantt chart. The horizontal axis represents the time frame. Example time units include days, weeks, months or quarters. The vertical axis lists tasks in the order they are initiated. Tasks that begin earlier are placed near the top, while those started later are placed towards the bottom.

The *milestone chart* is similar to the Gantt chart in structure, except only symbols signifying the completion of a major task are included (Figure 2.12). There is no indication of task initiation or duration as in the Gantt chart. Milestones include report due dates, design reviews, and design phase completion. These two charts are often combined creating the Gantt-Milestone chart (Figure 2.13).

The PMW uses data entered in the work breakdown structure (task name, task duration, and planned start date) to automatically generate Gantt charts. Figure 2.14 displays a screen shot of a Gantt chart for a generic project in the

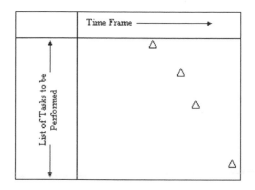

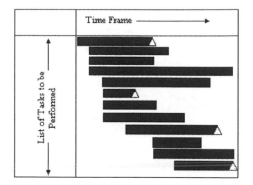

Figure 2.12 – *Milestone Chart*   Figure 2.13 – *Gantt/Milestone Chart*

PMW. As design teams enter or change planned start dates or task durations, the PMW automatically calculates the end dates and updates the Gantt chart. The duration of each task in the PMW is assumed to be days. Changes can me made within **Excel** to the planned end date calculations to allow design teams to work in other time units.

The number of days left for each task, based on the current date, is displayed at the end of each bar on the chart. In addition, the bars are automatically colored to reflect where the current date is on the chart. The portion of a task's bar to the left of the current date is shaded a different color from the portion to the right. With only a glance, therefore, the design team can readily see what proportion of time has expired since a task began and how much allotted time is left for completion of that task, allowing schedule adjustments be made if necessary.

Design Team Contact Page

The design team contact information is placed on a separate worksheet in the PMW. The worksheet lists all the team members with appropriate contact information (email, phone number and cell phone number), allowing team members to readily contact one another as the need arises during the course of the project. A sample contact page is shown in Figure 2.15.

Team Calendar

The team calendar should include all events that are relevant to the project, such as meetings, deadlines, and team member availability. An example calendar as implemented in the PMW is shown in Figure 2.16. Different text colors or cell shading should be used to highlight or categorize particular events.

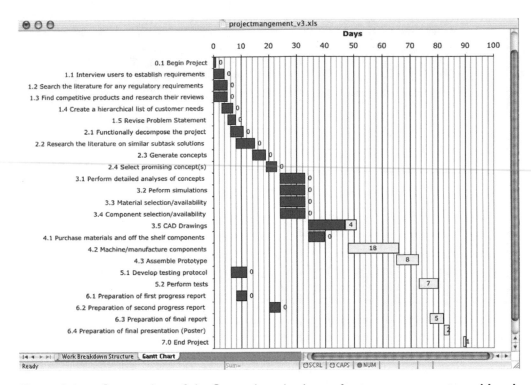

Figure 2.14 – *Screen shot of the* Gantt chart *in the* **project management workbook**

2.3 Directing

Directing a project keeps it on track by regularly comparing planned schedules with actual work completed and by making the necessary adjustments. Directing projects should include all of the following elements:

1. **Weekly coordination meeting**. The project team should meet *at least* once a week to discuss scheduling issues. This meeting could also be part of regular technical meetings. The outcome of the meeting should be

Naming worksheets in Excel
Double-clicking on the worksheet tab at the bottom of the **Excel** window allows you to type in a new title for the worksheet.

projectmanagement.xls

	A	B	C	D	E
1	**Team 25**	**The Go-Getters**			
2					
3	Name	Email	Phone	Cell Phone	
4	John Sanders	sanders@psu.edu	455-2343	234-1212	
5	Maranda Peters	peters@psu.edu	455-1290	243-2132	
6	Joshua Liberman	liberman@psu.edu	455-9034	243-8978	
7	Elizabeth Joans	joans@psu.edu	455-9090	-	
8					
9					
10					

Team Contact Information | Team Calendar | Work Breakdown Structure

Ready | Sum=0

Figure 2.15 – *An example contact page in the* **project management workbook**

projectmanagement.xls

	A	B	C	D
1	**DAY**	**DATE**	**CALENDER ENTRIES**	
2	Monday	10/20/03		
3	Tuesday	10/21/03	10 AM Administrative meeting	
4	Wednesday	10/22/03		
5	Thursday	10/23/03		
6	Friday	10/24/03		
7	Saturday	10/25/03	3 PM Meet in the engineering library to do background work	
8	Sunday	10/26/03		
9	Monday	10/27/03		
10	Tuesday	10/28/03	10 AM Administrative meeting	5 PM Revised Problem Statement D
11	Wednesday	10/29/03		
12	Thursday	10/30/03	John out of town to student conference	
13	Friday	10/31/03		
14	Saturday	11/1/03		
15	Sunday	11/2/03	12 noon John returns to campus	
16	Monday	11/3/03		
17	Tuesday	11/4/03	10 AM Administrative meeting	
18	Wednesday	11/5/03		
19	Thursday	11/6/03	2 PM Meeting w/ Prof. Johnson to go over aerodynamic aspects	
20	Friday	11/7/03		
21	Saturday	11/8/03		
22	Sunday	11/9/03		
23	Monday	11/10/03		
24	Tuesday	11/11/03	10 AM Administrative meeting	5 PM First Progress Report Due
25	Wednesday	11/12/03		
26	Thursday	11/13/03		
27	Friday	11/14/03		
28	Saturday	11/15/03		
29	Sunday	11/16/03		
30	Monday	11/17/03		

Team Contact Information | **Team Calendar** | Work Breakdown Structure

Ready | Sum=0

Figure 2.16 – *An example Team Calendar in the* **project management workbook**

a) Assessment of progress from the previous week.

b) Assignment of tasks for the coming week.

c) Re-evaluation of the WBS, adding or removing tasks or updating start dates and task durations.

At the first meeting the team should create their PMW with all the elements previously described: a team contact worksheet, a team calendar worksheet, a work breakdown structure, design structure matrix, activity network, and a Gantt chart[3]. The team should

a) Create their initial WBS. This is based on their best current estimate on what tasks are to be performed and their approximate duration.

b) Use the design structure matrix (DSM) to decide which tasks must be carried out sequentially and which ones can be done in parallel.

c) Utilize information from the DSM to create the activity network diagram. Based on the diagram, add a text-based version into the PMW and use it to calculate the critical path as well as float times for all other paths. If the critical path time (CPT), exceeds the available project time, the team must decide which tasks durations on the critical path to reduce in order for the CPT to be equal or less than the allotted project time.

d) Add the automated Gantt chart to the PMW.

2. **Selection of a coordination leader.** The role of the coordination leader is to lead team discussions during the meeting. She ensures that the meeting stays focused and concludes within a predefined time. This may be a rotating position from week to week. The leader may also be responsible for taking notes at the meeting, or he may delegate the responsibility to another team member.

3. **Assignment of tasks.** Design teams often have some team members not contributing their fair share, yet expecting to benefit from the success of the project. Assigning individual and mini-group tasks each week serves to ensure that (1) all work that should be done in the coming week is completed, (2) no team member carries an unnecessarily heavy work burden, and (3) *all* team members contribute fairly to the project.

4. **Assessment of tasks.** Each week, design teams must assess the extent of task completion from the previous week. Weekly assessment of task outcomes will determine if the current project direction is still appropriate, or if alternate paths have to be followed or solutions sought. It also ensures that the project does not fall behind schedule.

---

[3]Recall that a step-by-step tutorial to create all the elements in the PMW can be found in Appendix A.

5. **Communication.** The coordination leader should send an email shortly after the meeting to ALL team members and the project advisor(s), with the PMW attached. The email should include

   a) A list of present and absent team members.

   b) A summary of the current state of the project, indicating successes, challenges and failures.

   c) An indication in some detail of the tasks that were completed the week before and by whom. If a particular team member did not complete their task, that should be noted and an explanation given. Often a task may not be completed due to underestimation of task duration or inability to find a needed resource. Work not completed should be reassigned to the following week.

   d) Each team member's assigned tasks for the coming week. These tasks can be assigned to an individual or a subset of the team depending on the task type and scope.

The email memo in combination with the PMW will ensure that all team members and the project advisor(s) are aware of the current status of the project from week to week. Each team member will know exactly what tasks they are expected to complete that week, as well as what tasks their teammates are working on. The information in the memo and workbook will allow the project advisor(s) to provide immediate feedback on the project, addressing any design challenges that the group may be facing or taking care of trouble areas (for example, laggard team members or a project falling way behind schedule) before they become critical. A sample email memo from a group in week 13 of their project is shown in Figure 2.17.

2.4 Summary

The major tools used in project management have been presented in the context of **Excel**-based Project Management Workbooks. These tools help design teams to schedule major tasks necessary to complete the project on time. The same tools are used to monitor project progress, with adjustments being made as needed, and to regularly inform all team members of their responsibilities. Although the use of a spread-sheet based approach has been emphasized throughout this chapter, the same ideas would apply if other commercial project management software were used.

```
Week 13: 11/24-11/30

Administrative meeting held on 11/26
Present: Hank, Cedric, Zach, Evans, Jay   Absent: 0

We are on the last phase of the project.  Most of the effort is being spent
on documentation and preparing the animations and illustrations to be used
in the final report and final presentation.  No challenges or failures to
report.

Assessment of last weeks assignments:

Hank: Done (patent info, hose, engine running)
Cedric: Done (screen play, Ryobi.)
Zach : Done (2-stroke engine info, script on Otto cycle)
Evans: Done (illustrator-hub)
Jay: Done (script-flywheel)

Assignment for next week:

Zach: Fly-wheel animation, help hank on gas tank animation( find info on 4
      stroke)
Jay: work on putting together movie, script on blade hub assembly
Cedric: Background script
Evans: script on exhaust, drive train
Henry: modified gas tank animation, two stroke animation
```

Figure 2.17 – *Example coordination email sent to all group members and the project advisor after a group's weekly meeting*

References

Lock, D., *Project Management*, Burlington, VT: Gower Publishing Limited, 2003.

Kerzner, H., *Project Management: A Systems Approach to Planning, Scheduling, and Controlling*, New York: John Wiley & Sons, 2001.

Ogot, M. and Okudan, G., "Incorporating Project Management Methods in Engineering Design Projects: A Spreadsheet-based approach", *ASEE Annual Conference and Exhibition*, Salt Lake City, UT, 2004.

Ulrich, K. and Eppinger, S., *Product Design and Development*, 3rd Edition, New York: McGraw-Hill, 2003.

Bibliography

Boyle, G., *Design Project Management*, Burlington, VT: Ashgate Publishing Company, 2003.

Dominick, P.G., Demel, J.T., Lawbaugh, W.M., Freuler, R.J., Ginzel, G.L., and Fromm, E., *Tools and Tactics of Design*, New York: John Wiley & Sons, 2001.

Dym, C.L. and Little, P., *Engineering Design: A Project-Based Introduction*, Second Edition, New York: John Wiley & Sons, 2003.

Greasley, A., *Project Management for Product and Service Improvement*, Oxford: Butterworth-Heinemann, 1997.

Ogot, M. and Okudan, G., "Incorporating Project Management Methods in Engineering Design Projects: A Spreadsheet-based approach", *ASEE Annual Conference and Exhibition*, Salt Lake City, UT, 2004.

Ulrich, K. and Eppinger, S., *Product Design and Development*, 3rd Edition, New York: McGraw-Hill, 2003.

Exercises

1. A design team has been charged with the task of designing a portable basketball hoop. During the first meeting, the design team listed all the tasks that they foresaw will be required for completion of the project and an estimation of duration for each task (see Table 2.2). Your help is sought in planning the project schedule. From the tasks listed in Table 2.2

   a) Create a design structure matrix to determine task dependence and task order.
   b) Use the information from the DSM to create an activity network.
   c) From the activity network, determine the project completion time and the critical path.
   d) Calculate the float time for each activity.

Table 2.2 – *List of tasks for design of basketball hoop setup (not in sequential order)*

Task	Description	Duration (weeks)
A	Begin Project	
B	Testing of Final Design	1
C	Material Selection	1
D	Concept Generation	2
E	Final Concept Selection	0.5
F1	Detailed Design of Base Support	1.5
G	Background Research	1
H	Construction and Assembly	2
F2	Detailed Design of Backboard and Rim	2.5
F3	Detailed Design of Support Structures	0.5
I	Completion of Project	

2. Table 2.3 summarizes all the tasks necessary to complete a design project with a duration estimate for each task and the precedence structure.

   a) Draw the network diagram.
   b) Calculate the project completion time.

c) Indicate the critical activities.
d) Calculate the float time for each activity.
e) Explain how the project completion time changes if task F has a delay of 2 days due to material availability?

Table 2.3 – *List of design project tasks*

Task	Estimated Time (Days)	Preceded By
A	5	-
B	3	-
C	2	-
D	3	A
E	2	B
F	1	C
G	5	D,E
H	3	F
I	4	G,H
J	8	I

Chapter 3

Collaborative Design

3.1 Introduction

New product design efficiency and effectiveness are more important than ever in todays high-stakes business environment (Boujut and Laureillard, 2002; Flint, 2002). Gupta and Wilemon (1990) suggest that products that meet their development budget but come to market late generate substantially less profit than those that exceed their budget but come to market on time. Attesting to the continuing need to improve product design efficiency, Boujut and Laureillard (2002) state that "New organizations, based on concurrent engineering principles, after many years of experimentation within various companies and industrial domains, still suffer from a lack of efficiency." Although design efficiency as measured by time to market is critical to the success of new product development efforts, efficiency does not guarantee success. Indeed, Walsh (1990) state that 90% of new product development team efforts fail, and Flint (2002) state that " ... products continue to fail at alarmingly high rates." This indicates the continued importance of effectiveness in the product design process.

In general, product or process design and development is accomplished by cross-functional design teams. The makeup of these teams is seen as the strongest determinant of new design timeliness (Cooper, 1994) and product development success (Takeuchi and Nokana, 1986). Further, Sekine et al. (1994) suggest that design team activities control 70% of a companys product quality, cost and timeliness. Since the 1980s, cross-functional design teams have been widely used, although under different names such as concurrent engineering teams, simultaneous engineering teams, or integrated product teams (Clark et al., 1992). Product design teams can be considered a type of project team (e.g., Cohen and Bailey, 1997) in team typologies and are the most widely accepted means of bringing products from initial concept to the commercial stage - even for projects with a budget of $200 billion, spanning over 25 years, and requiring as many as 3000 engineers (McGraw, 2003). Experts from various disciplines and company departments such as design, manufacturing, quality testing and marketing work in these teams rather than individually to develop a quality product. The membership of the team depends on the type of product being developed, product characteristics, customer requirements, and other factors deemed important to the products development.

The mode of operation in the team should be collaborative. Collaboration requires each member to recognize and accept strengths and weaknesses of the other team members and share responsibility for group functioning and productivity (McGourty and De Meuse, 2001). In essence, these teams make complex decisions in the product-design stage so that downstream issues related to various attributes of the product such as manufacturability and serviceability are anticipated in the early stages of product development (Clark et al. 1992). Moreover, these teams facilitate the continuous communications related to the product as it evolves to satisfy customer and market requirements. In fact, recent work on engineering design indicates that design is a more social process than it was once thought to be (Leifer, 1997).

To meet the need for increased efficiency and effectiveness in the product design process, it is necessary for team members to have the right mental preparation to achieve and sustain a high performance level. This mental preparation involves an improved understanding of

1. The importance of becoming a team and the stages of team development
2. Challenges of working in a team
3. The most vital function in a team communication
4. Potential factors impacting team performance (diversity, personality, etc.)

The remainder of the chapter elaborates on these four items.

3.2 Conceptual Understanding of Teams and Team Development

A team is a small number of people with complementary skills who are committed to a common purpose, a shared set of performance goals, and a common approach for which they hold themselves accountable (Katzenbach and Smith, 1986). A group of people with any unifying relationship is not a team. The high degree of interdependence to achieve a goal and the dynamic exchange of information and resources among members are the main characteristics of a team. For product design teams, the main goal is to arrive at the best solution to a design challenge within cost and time constraints and to communicate the solution (in the necessary formats) to various stakeholders.

In general, teams (include product-design teams) have anywhere from 2 to 25 members, with normally 5 to 9 members being more practical. However, for complex product designs such as aircrafts, automobiles, a web of multiple design teams work together. It should be noted that as the number of team members increases, communication becomes more centralized as members no longer have an adequate opportunity to speak to each other or because the creation of adequate communication might yield inefficiencies.

It has been estimated that nearly all of the Fortune 500 companies use teams of some form (e.g., project teams, self-directed work teams, problem-solving teams, quality-circle groups) (Dumaine, 1994; Lawler and Cohen, 1992). Despite their wide use, however, they are not always successful. For example, according to a study, nine out of ten teams fail (McGourty and De Meuse, 2001). For product design teams in particular, Walsh (1990) states that 90% of new product development team efforts fail, and Flint (2002) states that "products continue to fail at alarmingly high rates." It is therefore important to study the characteristics of successful teams as part of preparing to be an effective team member.

A list of critical factors for becoming a successful team was compiled by Larson and La Fasto(1989) after they studied successful examples over a three-year period. These include

1. A clear, challenging goal that is understood and accepted by all members.
2. A result-driven structure with clear roles and responsibilities.
3. Competent, talented team members.
4. Unified commitment to team goals.
5. A positive team culture, which values honesty, openness, respect and consistency of performance.
6. Standards of excellence.
7. External support (for necessary resources) and encouragement.
8. Effective leadership.

Team members not only collaborate in all aspects of their tasks and goals, they share in management functions, such as planning, organizing, setting performance goals, assessing the team's performance, developing their own strategies to manage change, and securing their own resources.[1]

It is unrealistic, however, to expect people coming together to work on a design project to instantaneously become a team. A team identity develops over time. In fact, the Tuckman model (1965) theorizes that teams go through five observable stages. These stages – forming, storming, norming, performing, and adjourning – are elaborated in the following sections.

Forming

Forming is a stage of transition for team members where they go from individual to member status. Team members explore the boundaries of acceptable

---

[1]See *Chapter 2 Management of the Design Process* for project management tools and techniques.

group behavior while trying to understand (1) the nature of the task, (2) the resources required and available, (3) the way each member adds to the talent and experience pool, and (4) where, when and how to begin. Because there are so many unknowns in this stage, the team does not accomplish much, which is expected. Clark (2001) provides the following list for some observable behaviors and feelings:

- Excitement, anticipation, and optimism.
- Pride in being chosen for the project.
- A tentative attachment to the team.
- Suspicion and anxiety about the job.
- Defining the tasks and how they will be accomplished.
- Determining acceptable group behavior.
- Deciding what information needs to be gathered.
- Abstract discussions of the concepts and issues, and for some members, impatience with these discussions.

Storming

Storming is probably the most difficult stage for the team because there is no clear understanding for how to proceed. Members begin to realize that tasks are different and more difficult than they initially imagined. Impatience about the lack of progress may lead to arguments between members. Polarization among members might occur, resulting in departures from the team. These interpersonal pressures may leave little time and energy for achieving the team's goal(s). However, this is also the stage for all to begin understanding one another, and this understanding could take three to four meetings to achieve. Storming might include the following feelings and behaviors (Clark, 2001):

- Resistance to the tasks.
- Sharp fluctuations in attitude about the team and the project's chance of success.
- Arguments among members even when they agree on the real issues.
- Defensiveness, competitive, and side choosing behavior.
- Questioning of the wisdom of those who selected this project and appointed the other members of the team.
- Establishment of unrealistic goals.
- Increased tension.

Norming

At this stage, members accept the team, their roles in the team, and the individuality of other members. Conflict is reduced as previously competitive relationships become more cooperative. Ground rules for working together are established related to participation, decision-making and communication within the team. Examples for ground rules are

- Facts should be the main source for decision-making.
- Everyone should attend team meetings, be fully prepared and participative.
- Only constructive confrontation is allowed.
- Everyone has the same rank within the team, thus each persons opinion is equally important and worth considering.

As team members begin to work out their differences, they have more time and energy to spend on the project. Therefore, there is an observable performance increase. Some other observable feelings and behaviors might include (Clark, 2001):

- An ability to express criticism constructively.
- Friendliness, a confiding in each other, and a sharing of personal problems.
- A sense of team cohesion, spirit, and goals.
- Establishing and maintaining team ground rules and boundaries.

Performing

This stage finds the team settled in relationships and expectations. Team members have discovered and accepted each other's strengths and weakness and learned what their roles are. The team is now an effective, cohesive unit and thus performs well. Some noticeable feelings and behavior might be (Clark, 2001):

- Members have insights into personal and group processes, and better understanding of each other's strengths and weakness.
- Credit for success is shared.
- Members show close attachment to the team.

Adjourning

At this stage, members are proud of what they have accomplished. However, termination of the team might yield a sense of loss. Therefore, mixed emotions are observed. Many of the relationships that were formed continue long after the team dissolves.

A diagnostic check list for team health is given in Table 3.1. This check list takes into account Tuckman's (1965) stages of team development and Larson and La Fasto's (1989) critical success factors for teams. This diagnostic tool can be used at various stages of team development to understand the areas where improvement is needed for increased team success. When an area for improvement is identified, the whole team should put in the necessary time and effort to address it. The status of team health (using the diagnostic checklist) should, (1) be anonymously and individually completed by all team members, and then (2) a course of action to improve upon identified areas be charted at a follow up meeting, where all team members are present.

3.3 Challenges: Conflict Management, Performance and Motivation

While the "Diagnostic Check List for Team Health" is helpful in ensuring a team's march to success, it is important to know when and how frequently each item of the checklist is reviewed by the team. Various approaches can be adopted such as a periodic review at predetermined intervals, or as the need arises. If a diagnostic review for team health are not done periodically, however, team members should be aware of observable behaviors that are indicators of challenges to team health. These include:

- Absent or unprepared members at meetings.
- Domination by or non-participation of a few members.
- Majority of time is used for off-task issues.
- A few members want to do the entire project themselves because they do not trust others.
- Difficulty scheduling meetings.
- No clear focus or goals.
- Formation of subgroups excluding one or more members.
- Ineffective or inappropriate decisions or decision making processes.
- Suppression of conflict.
- Members not doing their fair share of the work.
- Lack of commitment to the teams mission by some members.

When faced with any of these challenges, a team has three issues to consider: (1) conflict management, (2) performance appraisal and management, and (3) motivation.

Conflict Management

If conflict never arises in design teams, it should be treated as a sign of dysfunction as teams can benefit immensely from seeing the problem or design at hand from different points of view. When faced with a conflict, however, it is important to know how to approach it with the ultimate goal of benefiting the team.

Table 3.1 – *Diagnostic Check List for Team Health*

Instructions: For your teams continued march to high performance, your candid opinion on the correctness of the following statements is crucial. Please indicate your opinion by checking the YES or NO response for each statement.

	YES	NO
I. Goals		
A clear mission statement is present		
Measurable objectives are set		
Objectives are prioritized		
Goals are set in all key task areas		
II. Roles		
Individual roles, relationships, and accountabilities are clear		
Style of leadership is appropriate for the team tasks		
Each individual has the skills or the knowledge to perform her key tasks		
III. Procedures		
Decision making procedure adopted is effective and efficient		
Information is effectively shared		
Key activities are effectively coordinated		
Individual and team outcomes, products, and services are of high quality		
Conflict is managed effectively within the team		
IV. Internal Relationships		
There are no areas of mistrust		
Feedback is constructive		
Relationships are not too competitive and not too supportive		
V. External Relationships		
Relationships with key external groups are effective		
Mechanisms are in place to integrate with each key group		
Time and effort is spent on building and monitoring key external relationships		

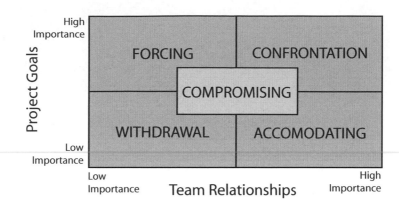

Figure 3.1 – *Blake and Mouton Conflict Model (1964)*

A widely accepted model of conflict management by Blake and Mouton (1964) identifies five potential courses of action a team might take when confronted with conflict given members' varying perceptions of the goal at hand and the relationship (Figure 3.1): withdrawal, forcing, accommodating, compromise, and confrontation. For design teams, the conflict management approach adopted by various members of the team might have direct implications on the product quality or team outcome. Selection of the best approach to any conflict situation is therefore very important. For most design teams confrontation improves the outcome.

When *withdrawal* is chosen as the approach to deal with the conflict, parties involved consider that neither the goal nor the relationship is important. They show no effort in communicating. This might be detrimental for a design team because the underlying reasons for conflict will have implications for the product that are going to stay unclear or somewhat unacceptable for some members of the team.

When the goal is seen as important but the relationship is not, one of the parties involved might adopt *forcing* as a strategy using various points of power such as rank, status, etc. This is an unhealthy sign for design teams as in many cases, teams cannot afford to lose members. One party forcing their will on another may create alienation, and eventually result in loss of members. Recruitment of new members requires additional time and money, slowing down the team and decreasing its efficiency.

If relationships between parties in conflict is seen as more important than the goal, *accommodating* can be adopted. When both the goal and the relationship are judged as important, and there is a clear pressure for meeting deadlines, a *compromise* can be worked out. However, the best approach for design teams should be *confrontation* based on facts. A confrontation is the direct expression of ones view of the conflict and an invitation to the other to do the same. In this

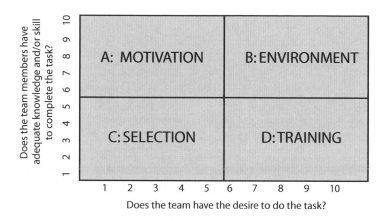

Figure 3.2 – *Performance Analysis Quadrant*

case, both the goal and the relationship are deemed important, and the conflict is formulated as a problem, for which the best solution is sought using factual information within the time and cost constraints.

Performance Appraisal and Management

In addition to conflict and how it is approached, there are four more potential causes of performance problems in design teams:

1. **Knowledge or skill level.** Team members are unable to perform their assigned tasks due to lack of appropriate skills or knowledge.

2. **Environment.** Performance problems may be caused by environmental factors, e.g. working conditions that may not be conducive to enhancing creativity.

3. **Resources.** Team members may not have access to the resources or technology needed to complete their tasks.

4. **Motivation.** Team members know how to get their task done but produce incorrect or incomplete work.

It is crucial to understand the situation correctly before taking any remedial steps. The Performance Analysis Quadrant (Figure 3.2) is a tool that can be used to identify performance problems (Clark, 2001). First, after asking two questions, (1) Does the team member have adequate job knowledge? and (2) Does the member have the desire to perform the job?, a numerical rating between 1 and 10 for each answer is assigned. Using these ratings, a team member can be placed into one of the four quadrants: A, B, C, or D.

- **Quadrant A (Motivation)**: If a team member has sufficient job knowledge but does not have the desire to complete the task, the performance problem may be categorized as a motivational problem. The consequences of the member's behavior, on both individual and team levels, should be brought to their attention. If the situation persists, a solution should be found that has the minimum impact on the team's performance.
- **Quadrant B (Resource/Environment)**: If a team member has both task knowledge and a favorable attitude, but his performance is still unsatisfactory, the performance problem may be due to lack of resources or time, or the work environment. Possible remedies include increasing the resources to minimize the possibility of a bottleneck that can limit others' progress.
- **Quadrant C (Selection)**: If a team member lacks both job knowledge and a favorable attitude, that person may have been improperly made a part of the design team. This may imply a problem with the member selection process, and suggests that a transfer or termination of membership be considered.
- **Quadrant D (Training)**: If a team member desires to perform, but lacks the requisite task knowledge or skills, additional training may be the answer.

It is important to note that regardless of the nature of the situation, prompt action should be taken to minimize team performance consequences. To monitor the performance level in a team, periodic (e.g., once a month) performance appraisals should be conducted. An example performance appraisal form is shown in Table 3.2. Based on the team's needs, the form should be modified and used for periodic performance appraisal.

Motivation

A person's motivation is an internal state or condition (sometimes described as a need, desire, or want) that sevres to activate or energize behavior (Kleinginna and Kleinginna, 1981). Influencing someone's motivation means getting them to want to do what you know must be done. A person's motivation depends upon two things (Clark, 2001):

1. The strength of certain needs. For example, if you are faced with being hungry and completing a solid model of a part you are designing for a design review: If you are starving you will eat. If you are only slightly hungry, you will finish the task first.
2. The perception that taking a certain action will help satisfy those needs. For example, you have two needs, the desire to go to lunch and the desire to complete the solid model you are working on. Your perception of those two needs will determine which one takes priority. If you believe that you

Table 3.2 – *Performance Appraisal Form*

Design Team Member Performance Evaluation					
Team/Company Name					
Name of Evaluator					
Name of Evaluatee					
For each item, circle the number that best reflects your evaluation of the participant's contribution to the team. 1 indicates lowest level of contribution, 5 indicates highest level of contribution.					
Committed to the goals of the team	**1**	**2**	**3**	**4**	**5**
Participated in the team deliberations					
Comments were clear / relevant / helpful					
Assisted with leadership functions					
Encouraged participation by other team members					
Helped keep discussions / worked on track					
When in conflict used factual confrontation					
Individual tasks done on time					
Individual tasks done accurately					
Individual tasks done completely					
Overall contribution in comparison to other team members					
General Comments					

could be fired for not completing the model, you will probably put off lunch and complete the task. If you believe that you will not get into trouble or perhaps you could still finish the task in time, then you will more than likely go to lunch.

3.4   Communication

Communication is the process of sharing information and feelings through an exchange of verbal and non-verbal messages. It involves a sender transmitting a message to a receiver. Effective communication occurs only if the receiver understands the exact information, idea or feeling –the message– that the sender intended to transmit. Many of the problems that occur in a design team are a result of people failing to communicate. Miscommunication leads to confusion and can cause a good plan to fail.

When a message is transmitted, both the content and the context are conveyed. *Content* is the actual words or symbols of the message, which is made up of spoken and written words. Meanings of words may be interpreted differently, therefore even simple messages can be misunderstood. Furthermore, many words have different meanings, especially in design teams where people with different backgrounds work together. Therefore, while a message is transmitted, the content may not be understood by everyone in the same way.

Context is the way the message is delivered and can be observed in the expression, body language, hand gestures, and/or state of emotion (anger, fear, uncertainty, confidence, etc.). Context can cause messages to be misunderstood.

It should be remembered that a message has not been successfully communicated unless the receiver understands it. To ensure effective communication feedback is necessary.

Barriers to Communication

Anything that prevents correct transmission of a message is considered to be a barrier to communication. Barriers can be physical or psychological. Some of them are:

- Culture, background, and bias
- Noise
- Perception
- Message
- Environmental distracters
- Stress

These barriers might result in modifications in the message, such as misunderstanding what one hears in a noisy environment. To limit and overcome the impact of communication filters, active listening and feedback should be practiced.

Hearing and listening are not the same. *Hearing* is involuntary and refers to the reception of aural stimuli. *Listening* is a selective activity that involves the reception and the interpretation of aural stimuli. Listening involves decoding the sound into meaning. Listening is divided into two main categories: passive and active.

1. **Passive listening** occurs when the receiver has little motivation to listen carefully. People speak at 100 to 175 words per minute (WPM), but they can listen intelligently at 600 to 800 WPM. Since only a part of our mind is paying attention, it is easy to think about other things while listening to someone.
2. **Active listening** is listening with a purpose. It requires that the listener attend to the words and feelings of the sender for understanding. Active listeners:

a) Rarely get preoccupied with their own thoughts when others are talking.
b) Plan responses after the other person has finished speaking, not while they are speaking.
c) Provides feedback, but without interrupting.

Feedback involves verifying that the senders message is understood and includes verbal and nonverbal responses to another person's message. Rogers and Farson (1969) list five main categories of feedback:

1. **Evaluative:** Making a judgment about the goodness of the other person's statement.
2. **Interpretive**: Paraphrasing, i.e. attempting to explain what the other person's statement means.
3. **Supportive**: Attempting to assist the other communicator.
4. **Probing**: Attempting to gain additional information, or to clarify a point.
5. **Understanding**: Attempting to discover completely what the other communicator means by her statements.

In a design team environment, communication has the utmost importance as it is the key function for

1. Understanding customer requirements
2. Communicating design concepts among team members
3. Coordinating design activities within the team

3.5   Potential Factors Impacting Team Performance

While it is evident that individual performance and communication problems affect a teams performance, the reasons are not always tangible (i.e., lack of resources) or solvable (i.e., long-term conflict among team members). For these situations, it is beneficial to train the team for potential factors that might impact communication and hence performance. Among these, two primary factors are diversity and personality.

Diversity

While having a diverse group typically enhances a design team's understanding of customer needs, it may also create potential barriers for communication. In the context of engineering design, Wulf (1998) emphasizes the importance of diversity as, "Every time an engineering problem is approached with a pale, male design team, it may be difficult to find the best solution, understand the design options, or know how to evaluate the constraints."

It should be noted that diversity in teams is not caused only by gender and race. Other sources can be:

- Background
- Accent
- Education (e.g., technical vs. non technical)
- Education level (e.g., high school diplomas vs. college degrees)
- Life style (e.g., urban vs. rural)
- Geographic origin
- Religions, beliefs, values

Diversity in a design team allows for a better understanding of unique customer groups, which in turn allows the team to improve the design of a product. Members should always be cognizant of the fact, however, that any of these diversity sources can also inhibit communication and hence performance.

Personality

Previous work on team member personality and performance has found that personality variables are generally predictive of team performance (Jackson, 1992; Moreland and Levine, 1992; and Morgan and Lassiter, 1992). However, it is not possible to directly adapt these insights to product or process design teams as various personality traits (e.g., openness, extraversion, etc.) might affect team performance in different ways depending on the nature of the team task. However, at the very least design team members should be aware of the impact of personality on the team processes (communication, delegation, and conflict management, etc.).

The most widely used personality type indicator is the five-factor model (FFM) (e.g., Tupes and Christal, 1961; Norman, 1963; Digman, 1990; Goldberg, 1990 and 1992), and its wide acceptance as a robust taxonomy of personality provides a common framework to examine how personality variables are related to both individual and team performance (Barrick and Mount, 1991; Costa and McCrae, 1988).

The FFM has five major trait domains of personality and six facets that define each domain. Together, the five domain scales and 30 facet scales of the FFM allow a comprehensive assessment of adult personality. The five domains are (Costa and McRae, 1992)

1. **Neuroticism.** Refers to the number and strength of stimuli that trigger negative emotions in a person. It is the extent to which a person is resilient (calm, composed, rarely discouraged, hard to embarrass, resistant to urges and stress) versus reactive (uneasy, quickly angered, easily discouraged, embarrassed or tempted, and susceptible to stress).

2. **Extraversion.** Refers to the number of relationships a person is comfortable with. It is the extent to which one is introvert (reserved, loner, stays in background, less active, low need for thrills and less enthusiastic) versus extravert (friendly, prefers company, assertive, active, craves excitement and cheerful).

3. **Openness.** Refers to the number of interests one has and the extent to which they are pursued. This factor categorizes people as preservers (focuses on here and now, has no interest in art, ignores feelings, prefers familiar, narrower intellectual focus, conservative) versus explorers (imaginative, appreciates art, values emotions, prefers variety, broad intellectual curiosity, open to examining values).

4. **Agreeableness.** Refers to the number of sources from which one takes norms. It is the extent to which one is a challenger (skeptical, guarded, reluctant to get involved, aggressive, hard headed and has feelings of superiority) versus an adapter (good natured, cooperative, forgiving and frank).

5. **Conscientiousness.** Refers to the number of goals one is focussed on. It is the extent to which one is flexible (unprepared, disorganized, casual, distracted and has low need for achievement) or focussed (thorough, achievement-oriented, reliable, organized, self-disciplined).

A few studies have investigated the effect of personality traits of members on design teams. Stevens et al. (1998) found significant differences between the number of decisions made and the degree of success for analysts, who evaluated ideas during the concept generation phase of a design activity, depending on their personality type. Though this study is not directly related to team performance, it does point to the impact of personality in design teams.

Two studies with more relevance (Kichuk and Wiesner, 1997 and Reilly et al., 2001) reviewed past research on team performance and hypothesized several correlations of personality traits with team performance. These hypotheses are summarized in Table 3.3. In the table, expected impact (positive, negative or no effect) is presented for both studies. Kichuk and Wiesner's (1997) study did not differentiate between product design teams for their innovative content, whereas Reilly et al. (2001) did and provided expected effects for incremental and radical product design teams. As seen in the table, while there is no agreement in what way various personality traits will affect design team performance, there is an agreement that they will. Therefore, it is prudent that team members get knowledgeable about personality traits, and constantly be aware of their negative and positive impacts on the team performance.

Table 3.3 – *Expected correlations of personality variables and design performance (developed from Kichuk and Wiesner (1997) and Reilly et al., (2001))*

Personality Variable	Design Teams	Incremental Design	Radical Design
Openness	Not provided	Zero	Positive
Neuroticism	Negative	Positive	Zero
Agreeableness	Not provided	Zero	Positive
Conscientiousness	Positive	Positive	Positive
Extraversion	Positive	Positive	Positive

3.6 Summary

While it has been practiced in team settings since 1980s, because engineering design is a complex activity there is still a need to increase design team efficiency and effectiveness. Therefore, it is necessary for design team members to have the right preparation to sustain a high performance level and achieve success. This preparation involves improving understanding of

1. The importance of becoming a team and the stages of team development
2. Challenges of working in a team
3. Communication in teams
4. Potential factors impacting team performance (diversity, personality, etc.)

This chapter has presented an overview of these issues. While being aware of these issues is very important, it should be noted that "becoming a team player" occurs over time through experience in a society of teams. More importantly, however, an acceptance of "$1 + 1 > 2$" is required. In other words, even in a team of two people, if performing well collaboratively, outcomes might surpass those of two individuals working separately.

References

Barrick, M.R. and Mount, M.K. "The Big Five Personality Dimensions and Job Performance: A Meta-Analysis." Personnel Psychology, vol. 44, pp. 1-26, 1991.

Blake, R.R. and Mouton, J.S. *The Managerial Grid*, Houston: Gulf Publishing Co., 1964.

Boujut, J.F. and Laureillard, P. "A Co-operation Framework for Product-Process Integration In Engineering Design." Design Studies, vol. 23, pp. 497-513, 2002.

Clark, K.B., Chew, W.B. and Fujimoto, T. *Product Development Process*, Technical Report, Harvard Business School, Boston, 1992.

Clark, D., Leadership, Version 1, http://www.nwlink.com/ donclark/documents/ leadershipshareware.html , 2001.

Cohen, S. G. and Bailey, D. E. "What makes teams work: Group effectiveness research from the shop floor to the executive suite. Journal of Management", vol. 23, pp. 239-290, 1997.

Cooper, R.G. and Kleinschmidt, E.J., "Determinants of Timeliness in New Product Development." Product Innovation and Management, vol. 11, no. 5, pp. 381-396, 1994.

Cooper, R.G., "Debunking the Myths of New Product Development. Research Technology Management, vol. 37, no. 4, pp. 40-50, 1994.

Costa, Jr., P. T. and McCrae, R. R., "Personality in Adulthood: A 6-Year Longitudinal Study of Self-Reports and Spouse Ratings on the NEO Personality Inventory." Journal of Personality and Social Psychology, vol. 54, pp. 853-863, 1988.

Costa, Jr., P.T. and McCrae, R.R., Neo-PI-R: Professional Manual, Odessa, FL: Psychological Assessment Resources, 1992.

Digman, J.M., "Personality Structure: Emergence of the Five-Factor Model", in Rosenzweig, M.R. and Porter, L.W. (Eds.), Annual Review of Psychology, Annual Reviews, vol. 41, pp. 417-440, 1990.

Dumaine, B., "The Trouble with Teams", Fortune, September, pp. 86-92, 1994.

Flint, D.J. "Compressing New Product Success-To-Success Cycle Time: Deep Customer Value Understanding and Idea Generation." Industrial Marketing Management, vol. 31, pp. 305-315, 2002.

Goldberg, L.R. "An Alternative Description of Personality: The Big Five Factor Structure." Journal of Personal Social Psychology, vol. 59, pp. 1216-1229, 1990.

Goldberg, L.R. "The Development of the Markers of the Big Five-Factor Structure." Psychological Assessment, vol. 4, pp. 26-42, 1992.

Gupta, A.K. and Wilemon, D.L. "Accelerating the Development of Technology-Based Products". California Management Review, vol. 32, no. 2, pp. 24-44, 1990.

Jackson, S.E., "Team Composition in Organizational Settings: Issues in Managing an Increasingly Diverse Workforce." In S. Worchel, W. Wood, and J.A. Simpson (Eds.), Group Process and Productivity, pp. 138-173. Newbury Park, CA: Sage, 1992.

Katzenbach, J. and Smith, D., The Wisdom of Teams, Harvard Business Review Press, 1986.

Kleinginna, P.Jr. and Kleinginna, A. "A Categorized List of Motivation Definitions, with Suggestions for a Concensual Defintion", Motivation and Emotion, vol.55, pp. 263-291.

Kichuk, S. L. and Wiesner W. H. "The Big Five personality factors and team performance: Implications for selecting successful product design teams". Journal of Engineering and Technology Management, vol. 14, pp. 195-221, 1997.

Larson, C.E. and La Fasto, F.M.J., Teamwork: What Must Go Right/ What Can Go Wrong, Newbury Park, CA: Sage, 1989.

Lawler, E.E. III and Cohen, S.G., "Designing Pay Systems for Teams", ACA Journal, Autumn, pp. 6-18, 1992.

Leifer, L., "A Collaborative Experience in Global Product-based Learning", National Technological University Faculty Forum, November 18, 1997.

McGourty, J. and De Meuse, K.P., *The Team Developer: An Assessment and Skill Building Program*, New York: John Wiley and Sons, Inc., 2001.

McGraw, D., "The Skys the Limit." ASEE Prism, March, pp. 36-39, 2003.

Moreland, R.L. Levine, J.M., "The Composition of Small Groups. Advances in Group Processes", vol. 9, pp. 237-280, 1992.

Morgan, B.B.,Jr. and Lassiter, D.L., "Team Composition and Staffing" in R.W. Swezey and E. Salas (Eds.), *Teams: Their Training and Performance*, Norwood, NJ:Ablex Publishing, pp. 76-100, 1992.

Norman, W.T., "Toward An Adequate Taxonomy of Personality Attributes: Replicated Factor Structure in Peer Nomination Personality Ratings." Journal of Abnormal Social Psychology, vol. 66, no. 6, pp. 574-583, 1963.

Reilly, R.R., Lynn, G.S., and Aronson, A.H., "The Role of Personality in New Product Development Team Performance", Journal of Engineering and Technology Management, vol. 19, pp. 39-58, 2001.

Rogers, C.E. and Farson, R.E., *Active Listening in Readings in Interpersonal and Organizational Communications*, Boston: Holbrook Press Inc., 1969.

Sekine, K., Arai, K., and Bodek, N., *Design Team Revolution,* Portland: Book News Inc., 1994.

Stevens, G., Burley, J. and Divine, R. "Profits and Personalities: Relationship Between Profits from New Product Development and Analysts Personality", Proceedings of the Annual Conference of the Product Development and Management Association, Atlanta, GA., 1998.

Takeuchi, H. and Nokana, I., "The New Product Development Game". Harvard Business Review, Jan/Feb, 1986, Boston.

Tuckman, B.W. "Developmental Sequence in Small Groups", Psychological Bulletin, vol. 63, pp. 384-399, 1965

Tupes, E.C., Christal, R.E., *Recurrent Personality Factors Based on Train Ratings.* Technical Report ASD-TR-61-97, Lackland Air Force Base, USAF Personnel Laboratory, TX, 1961.

Walsh, W., "Get the Whole Organization Behind The New Product Development", Research Technology Management, Nov/Dec 1990, vol. 33, no. 6, pp. 32-36, 1990.

Wulf, W.A., "Diversity in Engineering", Bridge, vol. 28, no. 4, pp. 8-13, 1998.

Bibiliography

Clark, D., Leadership, Version 1, http://www.nwlink.com/ donclark/documents/ leadershipshareware.html , 2001.

Exercises

1. Establish a set of ground rules for working together effectively. Your set of rules should include at least 5 rules. Use Table 3.4 as a template.

Table 3.4 –

Rule #	Rule Description
1	
2	
3	
4	
5	
Please sign your name and date in one of cells below to show your absolute commitment to the team ground rules.	

2. Complete the Diagnostic Check List for Team Health (Table 3.1). Decide, as a team, when and how often you will revisit this assessment.

3. Review the performance appraisal form in Table 3.2, and make changes in the appraisal items after considering each during a team meeting. Decide how often you will appraise each others performance.

4. In order to understand each other's personality, complete the Five Factor Model-based online questionnaire that can be accessed from: http://www.cede.psu.edu/~ogot/personality/.

Chapter 4

Engineering Communication: Reports and Oral Presentations

'When we interview college graduates, we do little to analyze their technical skills. We assume all graduates from the schools we recruit have those skills. Instead GE looks at five other qualifications: communication and interpersonal skills, analytical ability, self-confidence, personal initiative, and the willingness to adapt to change.' (Dahir, 1993)

Stephen Tucker
Program Manager for University Recruiting
General Electric

4.1   Introduction

The ability to communicate the design to others (for example, customers, production engineers, management, etc.) is an important aspect of the design process. Dieter (1991) notes that engineers spend about 60% of their time discussing designs and preparing written documentation of designs and only about 40% of their time actually designing. Communication can be in the form of graphs, schematics, engineering drawings, detailed technical reports, as well as oral presentations. This chapter focuses on technical reports and oral presentations. A discussion on graphic communication (sketching, technical illustrations, and solid modeling) is presented in Chapter 5.

4.2 The Formal Engineering Report

An important part of engineering communication is the ability to communicate one's ideas in writing, typically through a *formal engineering report*. The report forms a permanent document describing the work that was done, how the results were obtained and the conclusions that can be drawn or the recommendations that can be made.

One of the challenges engineers face when preparing reports is to ensure a logical content flow. The material within the report should follow a chronological sequence within each section, with periodic *signposts* provided to let the readers know where they have been, and where they are heading.

Formal engineering reports are normally written in the *third person* and *past tense*. Self-referential pronouns should be avoided, only included when absolutely necessary. A typical formal report can be divided into the following sections:

1. The Title Page
2. Abstract
3. Introduction
4. Technical Approach/Procedure (depending on the focus of the report)
5. Results and Discussion
6. Conclusions
7. References
8. Appendices

The following sections look at these categories in more detail.

The Title Page

The title page must contain the title of the report, authors's names and affiliations, and the date. The actual title should be concise, definitive and typically should not exceed 10 words.

The Abstract

Generally 200 words or less, the abstract gives a brief overview of the main ideas contained in the report. The abstract should contain brief sentences describing the objectives, methodology, significant results and conclusions. It should not attempt to condense the entire subject matter into a few words for quick reading. A sample abstract from a student design project report is presented in Figure 4.1.

ABSTRACT

(1) The objective of the design project was to create a mechanism to control the face air velocity on a chemical fume hood. Currently, the two basic methods for solving the problem are direct measurement of face air velocity via sensors, or determination of the sash position and using a controller to estimate the face air velocity. The team decided that the latter method would (2) be the easiest and cheapest to implement. The final design concept utilizes a mechanical system to convert the translational motion of raising and lowering the sash to rotational motion that can be used to adjust a potentiometer. This change in resistance is then sent to the controller, where the sash position is identified and the corresponding air velocity calculated. The design is easy (3) to install, as it is completely external to the fume hood. Finally, the design is inexpensive and adaptable to different types of fume hoods making it an ideal solution for the problem.

Figure 4.1 – *A sample abstract from a student project that illustrates the broad required parts: (1) objectives, (2) methodology and (3) significant results and conclusions.*

The Body of the Report

The introduction, technical approach/procedure, results and discussion, and conclusion sections form the body of the report. The composition of a well-written report, like design, takes good planning and attention to detail. The following is an outline of several steps that should be followed:

1. Before starting to write the report, stop a moment and consider the following:

   - **The Subject:** How well do you know it?
   - **The Purpose:** Why is the report being written?
   - **The Audience:** Who will read the report? What are their expectations? How much information do they already have?
   - **Sources of Information:** What will be your information sources? What method of documentation will you use?
   - **The Format:** What is the expected length and style of the report?
   - **The Deadline:** When is the report due? Do you have enough time to *complete* and *proofread* the report before turning it in?

2. Create an outline to map out the organization and flow of the report.

3. Compose a first draft. Begin with the **introduction** aimed at catching the reader's attention. Try to relate the topic to something the reader knows or describe a related occurrence. Using an analogy, a quotation or a reference might also help gain reader's attention.

The **main body** of the report follows. Each *paragraph* should focus on a single main idea and follow a logical sequence. Use transitional words and phrases to connect ideas within paragraphs and sentences. Examples of transitional words and phrases include the following:

- **To show addition:** And, also, besides, further, furthermore, in addition, moreover, next, too, first, second.
- **To give an example:** For example, for instance, to illustrate, in fact, specifically.
- **To compare:** Also, in the same manner, similarly, likewise.
- **To contrast:** But, however, on the other hand, in contrast, nevertheless, still, even though, on the contrary, yet, although.
- **To summarize or conclude:** In other words, in short, in summary, in conclusion, to sum up, that is, therefore, hence.
- **To show time:** After, as before, next, during, later, finally, meanwhile, then, when, while, immediately.
- **Logical relationships:** If, so, therefore, consequently, thus, as a result, for this reason, since.

Finally, the **conclusion** should 1) remind the reader of the purpose of the report (without repeating it verbatim) and 2) summarize the main points. It should not include any new ideas.

On completion of the report, read through for overall flow and completeness, and then make the appropriate revisions. In doing so keep the following in mind:

- **The purpose and audience.** Have you accomplished your goal? Is it appropriate for your readers?
- **The focus.** Do your paragraphs express complete ideas? Do you have a clear and relevant introduction and conclusion?
- **The organization.** Considering the overall structure of your report, are your ideas ordered correctly? Do your paragraphs make sense?
- **Content.** Are your ideas well supported? What can be deleted or added?
- **Consistency.** Ensure that your use of words and formatting is consistent throughout the report. A detailed discussion on consistency follows.

CONSISTENCY

Maintaining a consistent style and format throughout a report makes it easier for the reader to follow and understand. For example if 'Figure 1' is used to refer to the first figure, then the same format "for referring to figures" should be used throughout the report: 'Fig. 2' or 'figure 3', would be inconsistent. Common areas to pay attention for their consistency include

1. **References to figures**. Common forms are Figure 1, Fig. 1, figure 1 or fig. 1. Select one style and stick with it throughout the report.

2. **Figure and Table headings.** The formatting used for figure and table headings varies widely. Select a form, for example by looking at a book, magazine or journal article, and then use it throughout the report. Notice that the figure and table headings in this book all have the same format. If a figure or table caption is a single sentence, do not place a period at the end. For example,

   Figure 1 - Plot of acceleration versus position of the input crank

   On the other hand, if the caption is composed of two or more sentences, then a period is placed at the end. For example,

   Figure 1 - Plot of acceleration versus position of the input crank. Values are absolute and do not account for direction.

3. **Words with multiple spellings**. Examples include *traveling* and *travelling*, or *sizeable* and *sizable*. Select one spelling and use it throughout the text.

4. **Latin words whose plurals can take on multiple forms**. Examples include *formulas* and *formulae*.

5. **Capitalization of words**. An example of inconsistency would be the use of *the University* in one location and *the university* in another.

6. **Symbols and abbreviations**. Depending on the context, a symbol or an abbreviation may be used instead of the full word. In either case, once the choice is made, the symbol or abbreviation should be used throughout the report. For example, it would be inconsistent to state that, *the reading dropped by 50%* in one section of the report, and *the stress levels increased by 34 percent* elsewhere. Other common choices include *degrees* and *°*, *meters* and *m.*, *and so on* and *etc.*, *that is* and *i.e.*, and *the United States* and *the U.S.*

7. **Numbers.** There are several common recommendations for how numbers should be presented in your reports. Ross-Larson (1982) suggests using

   - Words for all single and double digit numbers (e.g, nine) and numerals for all others (e.g., 120).

- Numerals for all numbers that are followed by units or other qualifiers such as *percent.*
- Words for large approximate numbers such as *nine million.*
- A combination of numerals and words for large precise numbers such as *9.5 million.*

EQUATIONS

Equations should be numbered consecutively starting from (1). The equation number should be enclosed in parentheses and placed flush right on the same line as the equation. For example,

$$\alpha = \frac{\sqrt{(2absin\theta + 3a^2b^3cos\phi)}}{25} \tag{4.1}$$

This number is used when referring to an equation in the text. For example, *The angle of attack, α, can be calculated from Equation 4.1.* All symbols used in the equations and not previously defined should be explained right before the equation or right after. Finally, an extra blank line should be left above and below the displayed equation. If using **Microsoft Word**, equations can be typed in using the Equation Editor. To access the editor, use the menu commands, `Insert>Objects`. From the *Objects pop-up menu*, select **Microsoft Equation**.

FIGURES AND TABLES

Figures and tables must be numbered consecutively with a caption consisting of the figure or table number and a brief title. Captions must be placed *below figures* and *above tables.* Two blank lines should be left between the figures or tables, and the main text. Periods are not placed at the end of the figure or table captions, unless the caption consists of more than one sentence. This format is used throughout this document.

PROOFREADING

Proofread the report for grammatical, mechanical and punctuation errors. Common items to pay attention to include

1. **Sentence length.** Look for sentences that may be too long or awkward. Sentences should average twenty words or less. Shorter sentences tend to be clearer and easier to follow than longer ones.
2. **Subject-verb agreement.** For example in the sentence *All the team members **was** present at the meeting*, the subject and verb do not agree. Correct form is *All team members **were** present at the meeting.*

3. **Noun-pronoun agreement**. For example in the sentence *Each team member turned in **their** laboratory notebook*, the noun and pronoun do not agree. The correct form should be *Each team member turned in **his** or **her** own laboratory notebook*.

4. **Spelling.** The availability of spellcheckers in word processing software should help correct most spelling errors. Pay particular attention to

   - **Names.** They probably will not be in the spellchecker's database. Make sure that they are spelled correctly.
   - **Misused words or letter transpositions.** Examples of misused words include, cite/site, accept/except, affect/effect, allot/a lot, lead/led, than/then, and principal/principle. Examples of common letter transpositions include, from/form and on/no (Dominick, et al., 2001).

5. **Repetition of words.** Avoid using the same word over and over again, especially in the same paragraph. Use a thesaurus to look for alternative words with the same meaning or, even better, experiment with phrasing and, sentence structures.

6. **Shortened words.** Avoid the use of short forms for common phrases typically used in speech. Examples include, won't (*use* will not), don't (*use* do not), and shouldn't (*use* should not).

References

Professional associations, such as the Modern Language Association and the Institute for Electrical and Electronic Engineers (IEEE), have established guidelines for citing and referencing works by others. This text will focus on two referencing styles commonly found in engineering journals: parenthetical number system and parenthetical author-date. When using any referencing style, however, attention should be paid to formatting: indents, commas, italics, etc. Finally, a style manual should be consulted for exceptions and odd entries.

PARENTHETICAL REFERENCES: NUMBER SYSTEM

The *number system* uses square brackets to surround a number inserted in the text that corresponds to a reference in the listing at the end of the report. The references are listed in the order in which the citation appears in the text NOT in alphabetical order. Depending on the citation type, the following information should be included:

- **Book**
[#] Author, A.B., Author2, C.D. and Author3, E.F., *Title of Book*, City: Publisher, Year.

- **Journal article**
  [#] Author, A.B., Author2, C.D. and Author3, E.F., "Title of Article", Journal name, vol. #, no. #, pp. #-#, Month, Year.
- **Conference article**
  [#] Author, A.B., Author2, C.D. and Author3, E.F., "Title of Article", *Name of conference*, pp. #-#, Full date of conference, City, Country, Year.
- **Website**
  [#] Author, A.B., Author2, C.D. and Author3, E.F., "Title of Article", *Full website address directly to article*, Viewed on Date.
- **Patent**
  [#] Inventor, A.B., Inventor2, C.D. and Inventor3, E.F., "Title of Invention", Patent no. #, Month Day, Year.

An example reference list as it would appear at the end of a report is shown below.

References

1. Swanson Inc., "Online Users Manual for ANSYS 5.0", *http://www.ansys.com/manual*, viewed on March 1999.

2. Muriru, P.K. and Daewoo, R., "Prediction of the Heat Transfer Characteristics of a Multi-Flame Injector", Combustion and Flame, vol. 100, no. 2, pp. 123-135, 2002.

3. Zacharia, M. and Daudi, P.K., *The Effect of Multi-materials on Conventional Finite Element Formulations*, New York: Wiley and Sons, 2001.

4. Peters, L., Johnson, M., and Davidson, K., "A Novel Approach to Four-Bar Synthesis", 10th *ASME Design Automation Conference*, pp. 234-250, Pittsburgh, PA, 2001.

5. Wen-Cheng, C., "Electric Bicycle", US Patent no. 5,368,122, November 29, 1994.

Within the text, references are cited and numbered consecutively in the order they appear. The citation can take on several forms depending on the context. A single citation can take on one of two forms:

It was observed by Kariuki [12] that the average deviation from the mean was negligible.

or

It has been observed that the average deviation from the mean is negligible [12].

For two citations, both numbers, separated by a comma, are enclosed in a single pair of square brackets.

It has been observed that the average deviation from the mean is negligible [12,13].

For three or more consecutive citations, the first and the last number, separated by a dash, are enclosed between a single pair of square brackets.

It has been observed that the average deviation from the mean is negligible [12-15].

For multiple consecutive and non-consecutive citations, both the comma and dash should be used.

It has been observed that the average deviation from the mean is negligible [10,12-15].

PARENTHETICAL REFERENCES: AUTHOR-DATE SYSTEM

In the author-date system, references are cited by including the author's last name and the document's publication year in the text. The format of the citation varies depending on the usage. A single author citation can be in one of two forms:

It was observed by Kariuki (1999) that the average deviation from the mean was negligible.

or

It has been observed that the average deviation from the mean is negligible (Kariuki, 1999).

For citations with two authors, both last names are included, separated by an 'and'.

It was observed by Kariuki and Njenga (1999) that the average deviation from the mean was negligible.

or

It has been observed that the average deviation from the mean is negligible (Kariuki and Njenga, 1999).

For citations with three or more authors, only the last name of the first author is included immediately followed by 'et al.' denoting additional authors. Note that a period is placed after the 'al.'

> *It was observed by Kariuki et al. (1999) that the average deviation from the mean was negligible.*

or

> *It has been observed that the average deviation from the mean is negligible (Kariuki et al., 1999).*

For multiple citations from different authors, the citations are separated by semi-colons.

> *It has been observed that the average deviation from the mean is negligible (Kariuki et al., 1999; Nenga, 2000; Otieno and Okwaro, 2000).*

Lastly, for multiple citations of works by *the same author(s) with the same publication year*, citations are distinguished from one another in the text by appending a lowercase 'a' after the publication year of the first cited reference, 'b' after the second citation, and so on.

> *It has been observed that the average deviation from the mean is negligible (Kariuki et al., 1999a; Kariuki et al. 199b; Otieno and Okwaro, 2000).*

References to all citations are listed at the end of the report in *alphabetical order according to the last name of the first author*. The format for the different works cited is very similar to that for the number system. The primary differences are that (1) the reference list is not numbered, and (2) the reference used is in alphabetical order. The required elements for the different types of referenced works appear below:

- **Book**
  Author, A.B., Author2, C.D. and Author3, E.F., *Title of Book*, City: Publisher, Year.
- **Journal article**
  Author, A.B., Author2, C.D. and Author3, E.F., "Title of Article", Journal name, vol. #, no. #, pp. #-#, Month, Year.
- **Conference article**
  Author, A.B., Author2, C.D. and Author3, E.F., "Title of Article", *Name of conference*, pp. #-#, Full date of conference, City, Country, Year.
- **Website**
  Author, A.B., Author2, C.D. and Author3, E.F., "Title of Article", *Full website address directly to article*, Viewed on Date.
- **Patent**
  Inventor, A.B., Inventor2, C.D. and Inventor3, E.F., "Title of Invention", Patent no. #, Month Day, Year.

An example of in author-date reference list as it would appear at the end of the report is shown below.

References

Muriru, P.K. and Daewoo, R., "Prediction of the Heat Transfer Characteristics of a Multi-Flame Injector", Combustion and Flame, vol. 100, no. 2, pp. 123-135, 2002.

Peters, L., Johnson, M., and Davidson, K., "A Novel Approach to Four-Bar Synthesis", 10th *ASME Design Automation Conference*, pp. 234-250, Pittsburgh, PA, 2001.

Swanson Inc., "Online Users Manual for ANSYS 5.0", *http://www.ansys.com/manual*, viewed on March 1999.

Wen-Cheng, C., "Electric Bicycle", US Patent no. 5,368,122, November 29, 1994.

Zacharia, M. and Daudi, P.K., *The Effect of Multi-materials on Conventional Finite Element Formulations*, New York: Wiley and Sons, 2001.

4.3 Plagiarism

A significant portion of any technical report involves work, thoughts, results or figures created by others. It is *imperative to reference all work, including figures, that are not the design team's own original work*, whether it is presented in paraphrased form or as a direct quote. Note, however, that the US Copyright Act of 1976 considers the 'fair use' of others work via direct quotation to be limited to a few paragraphs or statements. Lengthy quotations should be avoided, unless permission has been obtained from the copyright holder.

4.4 Report Formats

While reports can follow various formats, adopting a format for similar reports can increase writing efficiency. Guidelines for a generic technical report format are given below.

1. **The Font.** Use *Times New Roman* with a font size of 12.
2. **Headings and Subheadings.** Must appear throughout the report, dividing the content into logical parts. These parts migh have

   - Major headings. Numbered consecutively, 1.0, 2.0,..., typed in bold face with a font size of 16.
   - Sub-headings. Numbered consecutively, 1.1, 1.2,..., typed in bold face with a font size of 14.

- Sub Sub-headings. Numbered consecutively, 1.1.1, 1.1.2,..., typed in bold face with a font size of 12. It is not recommended to go beyond three heading levels.

3. **Page numbers.** Center on the bottom, the top, or on the bottom right of each page. The font and font size should be the same as the body of the paper.

4.5   Oral Presentations

The following are some guidelines for a successful presentation adapted from ASME (1991):

1. **Do not read your report**. Your audience is there to hear you talk, not listen to you read. Never read more than a few words on a slide. The audience can read much faster than you can speak, and it may be annoying for them to have you read what they have already read. The slides are primarily there to remind you, and your audience where you are in the presentation and to provide useful/clariying visual content.

2. **Have uncluttered slides.** View your slides as reminders about what you want to say and as visual aids to enhance the attention of your audience. The focus should be on what you are saying.

3. **Make notes before your presentation**. If you need to, use index cards. Organize the cards and number them clearly in the order they will be used.

4. **Divide your presentation into main ideas.** State each idea in a short sentence on its own index card. On each card add a series of keywords or phrases to remind you of what you want to tell your audience about each idea. Arrange the ideas in a logical order to facilitate understanding.

5. **Secure the attention of your audience**. Clearly state the purpose of your presentation. Confine it to one simple declarative sentence, such as, "My presentation is on a new device we have developed to produce energy from ocean waves." State something compelling about your subject. Use a question, such as, "Can a mechanical device be devised to harness energy efficiently from ocean waves?" A question could be your opening statement, which could then be joined to your statement of purpose with a connecting sentence or phrase.

6. **Use connecting sentences and phrases**. Just as connecting sentence and phrases are helpful to readers of a report, they are even more necessary when an audience is listening to a presentation, where audience members don't have the ability to stop and reorient themselves. Therefore, you must remind the audience of what they have just heard and prepare them for what they will hear. For example, 'Now that you have a clear idea of

the major problems experienced by current devices used to harness ocean energy, it is time to look at the concepts we have developed to alleviate them.'

7. **Be sure to summarize.** State your subordinate conclusions, confining them to one sentence if possible, or to a brief series of very short sentences. For example, "As you can see, the low efficiency achieved by today's devices is due to 1..., 2..., 3..." State your main conclusion in one simple sentence.

8. **Time your talk.** Rehearse your presentation and learn to handle your index cards naturally. When using slides, allow no more than one minute per slide. During rehearsal, if you find that you have exceeded your allotted time, cut down on the explanations or eliminate unnecessary slides. Continue to condense your talk until you are within the time limit, do not speed up.

9. **Never talk facing the screen**. If talking while using a pointer, stand at an angle that allows you to point at the screen without turning your back to the audience. This allows you to simultaneously point at the screen and face your audience.

10. **Communicate with your audience.** Look at your audience to get feedback from them. Look at their eyes and read their expressions. Communication is two-way even if you are the only one talking.

11. **State your main results carefully and without a lot of detail.** Be clear about what your results are. Minimize detail in favor of clarity. Make sure they get the main idea. They can always read your report or paper for the details.

12. **Use connecting sentences to introduce and remove slides.** When introducing a slide, "Perhaps I can make the point clearer by showing you a section view." Or for removing a slide, "Now that that is clear, let us move on to the next slide."

13. **Speak clearly.** Do not speak too quickly, too slowly or too softly as you may lose the interest of your audience.

14. **Avoid numeric tabular data on the slides.** This especially holds true if the absolute values presented are not as important as a trend they represent. Plot the data instead to show the trend more clearly.

4.6 Poster Presentations

Poster presentations differ in several ways from formal presentations in that

1. Multiple mini-presentations are given over a fixed period of time, as opposed to a single presentation.
2. There is no captive audience. People continuously wonder to and away from the presenter.

3. Interruptions are more frequent and are generally unconnected.

This section briefly covers the main points for creating a good poster and for giving a good poster presentation.

The Actual Poster

Poster sizes and styles range considerably depending on the space available for the presentation, the equipment available to make the poster, and the creative talents of the design team. Despite these variations, a few points are worth noting. Each point relates to a specific section of the poster given in Figure 4.2.

1. Create a poster title that stands out across the top of the poster (Box 1). The font size should be significantly larger than that of the text within the poster. Under the title list the following items: (a) the team members and their affiliation, (b) the presentation event (such as a design review, a conference, etc.), and (c) the date. Note that the latter two items are sometimes placed at the bottom of the poster.
2. Divide the poster into several columns that are read from top to bottom, and from left to right (Box 2 in Figure 4.2).
3. Only include the main points in the poster.
4. Make sure that there is a 'Conclusion' section at the bottom right portion of the poster. The section does not necessarily have to be titled as such, but it should state the main conclusion(s) or finding(s) of the presentation (Box 3 in Figure 4.2).
5. Include sufficient visuals in the presentation.
6. Use different font sizes and colors to increase the visual appeal of the poster.

The Presentation

When making a poster presentation, be aware that there is no captive audience and individuals will continuously join and leave the presentation. Individuals who walk away have either lost interest (a bad thing) or have heard enough and are simply moving onto the next poster. The presenter should therefore continually adapt to the changing audience dynamics, accommodating those who just joined without being too repetitive to those already there.

With such a fluid audience it is essential for the presenter to have a concise 'five minute pitch' of the main project ideas. If the presentation is too long, bored listeners may simply walk away. Anyone who needs more information will ask. In addition, the presenter should remember to use the visuals on the poster to illustrate the points being made, without blocking the poster with his body (Figure 4.3).

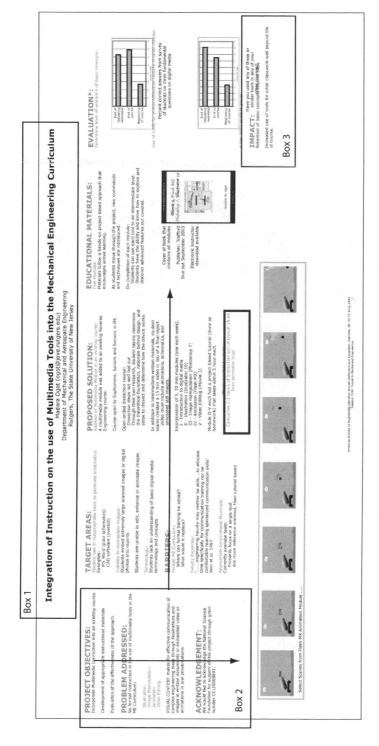

Figure 4.2 – Sample poster indicating key items

Figure 4.3 – *Using the visuals on a poster to make a point*

The presenter must engage individuals as they walk towards the poster. The following is a typical scenario with possible reactions to it. An individual walks up to the poster, says nothing to the presenter and begins to read the poster. The presenter should then give the person 10-20 seconds to read and then ask the person if they would like to listen to a presentation on the project. If the person agrees, the presenter can go ahead and give the 'five minute pitch'. Alternatively, if the person states that they are just browsing, then with a broad smile, the presenter should inform the person to feel free to ask questions at any time.

Finally, the audience will get interested in what the presenter has to say only if the presenter comes across as genuinely interested and excited about his project.

4.7 Summary

Engineers in design teams will be called upon several times throughout a project to participate in the documentation of the design and to make presentations. This chapter gives a brief overview of points to note while writing technical reports and when giving oral presentations. In both instances, effective technical communication can be achieved if all team members pay close attention to detail and provide the required input for the reports and oral presentations in a timely manner.

References

ASME Board of Communications, *Presenting your ASME paper In the best light*, ASME Manual MS-4A, New York, 1991.

Dahir, M., 'Educating Engineers for the Real World', Technology Review, vol. 96, Aug./Sep., 1993, pp. 14-16.

Dominick, P.G., Demel, J.T., Lawbaugh, W.M., Freuler, R.J., Ginzel, G.L., and Fromm, E., *Tools and Tactics of Design*, New York: John Wiley & Sons, New York, 2001.

Dieter, G.E., *Engineering Design: A Materials and Processing Approach*, 2nd Edition, New York: McGraw-Hill, 1991.

Ross-Larson, B., *Edit Yourself: A manual for everyone who works with words*, New York: W.W. Norton and Company, 1982.

Bibliography

ASME Board of Communications, *An ASME Paper*, ASME Manual MS-4, New York: ASME, 1991.

Dominick, P.G., Demel, J.T., Lawbaugh, W.M., Freuler, R.J., Ginzel, G.L., and Fromm, E., *Tools and Tactics of Design*, New York: John Wiley & Sons, New York, 2001.

Gibaldi, J. and Achtert, W.S., *MLA Handbook for Writers of Research Papers*, 3rd edition, New York: Modern Language Association of America, 1988.

Olsen, L.A. and Huckin, T.N. *Principles of Communication for Science and Technology*, New York: McGraw-Hill Book Company, New York.

Ross-Larson, B., *Edit Yourself: A manual for everyone who works with words*, New York: W.W. Norton and Company, 1982.

Turabian, K.L., *A Manual for Writers of Term Papers, Theses, and Dissertations*, 5th edition, Chicago: University of Chicago Press, 1987.

University of Chicago Press, *The Chicago Manual of Style*, 13th edition, Chicago: University of Chicago Press, 1982.

Chapter 5

Engineering Communication: Illustration and Solid Modeling

5.1 Introduction

Illustrations in engineering serve three main purposes, (1) visualization, (2) communication and (3) documentation (Bertoline et al., 2002). As engineers conceive different ideas for products and systems that can solve engineering problems, it becomes necessary to create graphics and illustrations to communicate those ideas. The illustrations must be easy to modify and enhance as ideas evolve and take form in the designers' minds. An illustration, usually in the form of a rough sketch, allows the designers to *visualize* the concept as it evolves. In addition to being a representation of the concept, these initial sketches typically contain notes to remind the designers about key aspects of the design or point out particular features.

As the design matures, more detailed drawings and models are generated to allow the *communication* of the design to other stakeholders, which may include other members of the design team, as well as manufacturing and production. These drawings are generally in the form of illustrations – two-dimensional (2D) and three-dimensional (3D) technical drawings, as well as 3D solid models. In addition, once the design is finalized, technical drawings and models are again used as a permanent record or *documentation* of the design.

This chapter will briefly discuss (1) the generation of hand-drawn sketches for visualization, (2) a general overview of the 3D solid modeling concepts, and (3) computer generated sketches for documentation. For the effective use of either illustration or solid modeling software, it is necessary to have a basic understanding of digital media and terminology. An introductory discussion follows.

5.2   Introduction to Digital Media

The use of digital media tools for illustration, animation, image manipulation and video editing is becoming the norm for non-professional as the costs of digital hardware and software tools continue to fall. It is now common for the average consumer to generate her own digital images and videos from digital cameras or to create digital illustrations and animations for reports, presentations and personal web sites. Technical presentations typically require the use of digital media tools, for example, to generate illustrations and manipulate images for reports, as well as to illustrate, animate, and edit video and images for presentations. To effectively use digital media tools it is necessary to have a basic understanding of digital media terminology and concepts. This section provides a brief introduction on how digital images are stored, viewed on the screen and printed. It also goes over several different image file formats, listing each one's strengths and weaknesses.

Pixels and Color Depth

Pixels, derived from the phrase 'picture elements,' are a series of tiny single color dots on the computer screen that together compose the image that you see. The higher the image resolution (pixels per inch), the smaller each pixel is and therefore the clearer (less grainy) the image appears. Figure 5.1 displays an image at four different resolutions: 300, 100, 50 and 25 pixels per inch. It's easy to see that as the resolution decreases, so does the clarity of the image.

The depth of color is an indication of the maximum number of colors that can make up your image. Note that each on-screen or printed image is composed of a collection of pixels or dots, respectively. The depth of color you select should depend on the capabilities of your computer system (or the system of your target audience) if the image is for on screen viewing or of your printer if the image is to be printed.

Image Dimensions

Digital illustrations and images typically have dimensions defined in inches, millimeters, pixels or points. One point (pt) is equivalent to 1/72 inches. You are probably familiar with using points during the selection of font sizes in word processing programs.

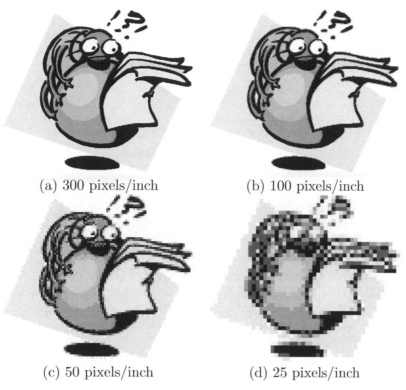

(a) 300 pixels/inch   (b) 100 pixels/inch

(c) 50 pixels/inch   (d) 25 pixels/inch

Figure 5.1 – *Image clarity decreases or quality degrades as the resolution (number of pixels per inch) is decreased*

1. **256 colors (8-bit)** represents the color of each pixel (or dot) by 8-bits. Note that a bit is the form in which a computer stores data and is either a 1 or a 0. Eight bits can therefore represent 2^8 or 256 different combinations or colors.

2. **High color (24-bit)** uses 8-bits to represent each of the three primary colors red, green and blue in an RGB image. You thus have a possibility of 256 x 256 x 256 or 16,777,216 combinations, i.e., millions of colors.

3. **True color (32-bit)** adds a fourth 8-bit stream of data, referred to as the alpha channel. This channel is used to assign 256 different levels of transparency to the image pixels.

Increasing the resolution (and therefore the clarity), however, comes at a price. Each pixel is a unique piece of information that defines the image and forms part of the image file. Increasing the resolution increases the number of pixels that defines the image, resulting in a larger file size. For example, the drawings in Figure 5.1 are approximately 2.0 inches by 1.7 inches in size with 256 colors, color depth. Each pixel is therefore represented by 8 bits. Drawing (d) in the figure is composed of (2 in. x 25 pixels/in.) x (1.7 in. x 25 pixels/in.) or 2,125 pixels, represented by 2,125 x 8 bits. But there are 8 bits in a *byte*, so the image file size will be 2,125 *bytes* or approximately 2 *KB*. Drawing (a), on the other hand, is made up of 306,000 pixels resulting in a file size of 299 *KB*!

Color Models

Two common color models are RGB (Red, Green, Blue) and CYMK (Cyan, Yellow, Magenta, Black). Deciding which format is more appropriate depends on the application of the final image.

1. **RGB** color model is based on the combination of three primary light colors (red, green, and blue) in different proportions and intensities to produce nearly all the visible spectrum colors. The RGB model is *additive* in that it creates color based on the light reflected back to your eye. For example, the addition of R, G and B produces white, indicating that all the light has been reflected back. The color your see on all computer monitors and televisions is based on this model. If your final image is to be viewed on a screen or monitor (i.e., not printed), then you should select this color format for your image.

2. **CYMK** color model is based on combining different percentages of cyan, yellow, and magenta. Cyan, yellow and magenta are obtained by subtracting red, green and blue, respectively, from white. The CYMK model is *subtractive* in that it produces color by absorbing some colors from the white incident light and reflecting the rest. It is the reflected color that you see. For example, the combination of C, Y and M will result in the absorption of all colors and appear as black. The CYMK model is primarily

Bits and Bytes

A *bit* is a single piece of information and is either a '0' or a '1'. A *byte* is composed of 8 bits.

- A Kilobyte (KB) is 2^{10} or 1024 bytes.
- A Megabyte (MB) is 2^{100} or 1024 KB or 1,048,576 bytes.
- A Gigabyte (GB) is 2^{1000} or 1024 MB or 1,073,741,824 bytes.

Figure 5.2 – *Inkjet printer cartridge*

used for printing. If for example, you look inside a color ink jet printer, you will notice two ink cartridge: a cyan, magenta, and yellow ink cartridge and a black cartridge (see Figure 5.2). To account for slight variations in the production of C, Y and M inks, black is added to the color model to give pure black. By mixing these colors in various proportions, nearly all colors in the visible spectrum can be obtained. If your final image is to be printed, you should use the CYMK color model for your illustration.

3. **Grayscale** uses shades of black to represent an object. The brightness or shade ranges from 0% (white) to 100% (black). All the images in this tutorial are grayscale.

In addition, on web pages, color is defined using Hypertext Markup Language (HTML) color codes. HTML is the code behind the web pages that is interpreted and displayed by web browsers like Netscape Navigator and Internet Explorer. In HTML code, colors are expressed as six-digit hexadecimal (i.e., 0,1,2,3,4,5,6,7,8,9,A,B,C,D,E,F) value strings. For example, the smallest value 000000 represents 'black,' while the largest value $FFFFFF$ represents 'white'.

Digital Image File Formats

Although there are a large number of digital file formats, only a small number are widely used. The most common are discussed here.

1. **GIF (Graphic Interchange Format)** is used primarily for artwork with flat, solid colors. The format stores a maximum of 256 colors using 8 bits of data per pixel (for on-screen viewing) or dot (for printed version). GIFs are supported by all web browsers.

2. **JPEG (Joint Photographic Experts Group)** format can store millions of colors giving superior quality for photographs or continuous-tone

images. JPEG files store color information as 24-bit data and tend to be larger than GIF or Portable Network Group (PNG) files. The file size can be reduced by reducing the quality of the image via compression. Note that the JPEG compression scheme discards data and that a compressed image's quality cannot be restored. It is therefore highly recommended to always keep original images and use copies for other tasks.

3. **TIFF (The Tag Image File Format)** format is extremely versatile, providing support for a wide variety of compression and data formats. It provides numerous options and is therefore a good choice for a variety of uses, including document storage, simple art, photographic and photo-realistic images.

4. **BMP (Bitmap - Windows OS)** is the default image format in Windows systems. Bitmap files typically store color information using 24-bit data and generally do not offer compression.

5. **PDF (Portable Document Format)** allows the viewing of documents without the fonts or software used to create it. Using a PDF reader, for example Acrobat Reader available from Adobe (http://www.adobe.com), files created on a MacOS, Windows, Linux or Unix platform can easily be viewed on any other. In addition, PDF is the default image format in MacOS.

6. **PNG (Portable Network Group)** format was designed as a replacement for GIFs. It provides superior quality than GIFs, with smaller file sizes (as compared to 8-bit PNG, PNG-8). Higher quality images can also be obtained using PNG-24 format that stores 24-bit data.

*All formats presented thus far store images in the form of pixels and are referred to as **raster images**.*

7. **SVG (Scaleable Vector Graphics)** format stores images based on lines and curves, not pixels. **Adobe Illustrator**, **Macromedia Flash** and **Solidworks** images, for example, are based on vector graphics. **Adobe Photoshop** images on the other hand, are raster images. Unlike raster images, SVG can be scaled up or down with no loss in quality.

5.3 Technical Sketching and Solid Modeling

One thing a company needs to succeed in today's global competition is an ability perform rapid product development, that is to identify customer needs and to quickly create products that meet those needs. Rapid product development has been especially important since the late 1980s. There have been vast improvements in the area, mostly focused on searching for ways to shorten the development process time. Among these, the advancement in design software is very significant. Accordingly, when preparing engineering students for similar responsibilities, integrating a solid modeler to teaching design is a must. Further, as solids models are typically created from design concepts that were first sketched, instruction on solid modeling in this text has been integrated with that of technical sketching.

Evolution of Solid Modeling

Solid modeling, developed in the 1970s, can be defined as a consistent set of principles for mathematical and computer modeling of 3D solids (Shapiro, 2002). It uses the mathematical abstraction of real artifacts that are transferred to computer representations (Requicha, 1980; Requicha, 1981) . Solid modeling was meant to be a universal technology for developing engineering languages that needed to maintain integrity for (1) validity of the object, (2) unambiguous representation, and (3) support of any and all geometric queries that may be asked of the corresponding physical object.

Early efforts in solid modeling focused on replacing engineering drawings with geometrically unambiguous computer models capable of supporting a variety of automated engineering tasks, including design and visualization of parts and assemblies, computation of their properties (mass, volume, surface), and simulations of mechanisms and numerically controlled machining processes (Requicha and Voelcker, 1982, 1983; Volecker and Requicha, 1993). Today, solid modeling is seen as an integral product development tool, as "... [it] allows everyone involved in the development of a new product – marketing/sales staff, shop-floor personnel, logistics and support staff, and customers – to add their input when changes can be made quickly and easily" (Schmitz, 2000).

Shapiro (2002) classifies product design related applications of solid modeling as

1. **Geometric design.** Involves visualizing, creating, modifying and annotating solid shapes.
2. **Analysis and simulation.** Applications include rendering and computing of mass properties.
3. **Dynamic analysis.** Includes the simulation of rigid body motion under externally applied forces and moments requiring the computation of mass properties (mass, moments of inertia, center of mass), solving for the instantaneous accelerations and velocities. Repetition of this computation at small discrete time intervals produces realistic motion simulation, such as 3D finite-element modeling in bulk forming processes (Li et al., 2001).
4. **Planning and generation.** Typically do not have a unique answer, but result in one or more acceptable solutions from a feasible solution space. Motion planning is an application example for planning.

Over the last several years there has been an increase in the use of solid modeling by industry for concept design (Tovey and Owen, 2000). Further, solid modeling tools today typically include collaborative tools allowing multi-location partners to work on the same design. Nam and Wright (2001) provides a good review on design collaboration using solid modelers.

Overall, solid modeling impacts a great variety of concurrent engineering activities, and its importance is increasing due to its wide acceptance. These activities include design sketches, space-allocation negotiations, detailed design, interactive visualization of assemblies, maintenance-process simulation studies, engineering changes, reusability of design components, analysis of tolerances, 3D mark-up and product data management, remote collaboration internet catalogs of parts, electronic interaction with suppliers, analysis (e.g., mechanism analysis or finite elements), process planning and cutter-path generation for machining, assembly and inspection planning, and product documentation and marketing (Requicha and Voelcker, 1983; Rossignac, 1999).

Ullman (2001) discusses the current stage of CAD systems as a design support system and indicates areas of opportunity for software developers to better support concurrent engineering design activities. These include

1. Ability to visualize function before the geometry is fully defined.
2. Extension of CAD systems to provide information about anticipated materials and manufacturing methods.
3. Generation of a running update of costs as parts and assemblies are changed in real time.
4. Integration of requirements and constraints into the development of parts and assemblies.

Despite these current inadequacies, a 2003 survey of design software users showed that only 30.4% of design practitioners use 2D CAD systems. The rest are either using 3D CAD systems (6.8%), implementing a hybrid usage of 2D/3D CAD systems (36.4%), or mainly using 2D CAD systems but evaluating 3D CAD systems (26.4%) (Green, 2003).

Technical Sketching - An Overview

Technical sketching is the process of producing a rough drawing representing the main features of a product. It is frequently used during the conceptual design phase. Tools for freehand sketching are paper, pencil and an eraser, while sketching, multiview or pictorial representations can also be used. Points to note while

- **Sketching straight lines** (Bertoline et al., 2002)

    1. Orient the paper to a comfortable position. Do not fix the paper to the surface.
    2. Mark the end points of the lines to be sketched.

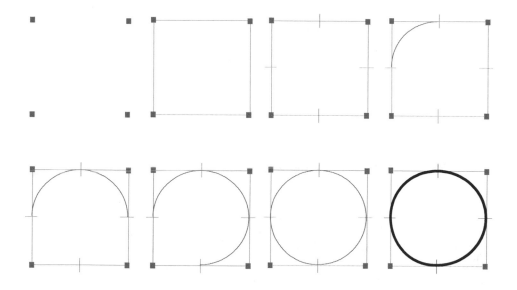

Figure 5.3 – *Steps in sketching a circle*

3. Determine the most comfortable method for sketching lines, such as drawing from left to right.

4. Use the edge of the paper as a reference point for making straight lines.

5. Draw long lines by sketching a series of connected short lines.

- **Sketching a circle or arc** (Bertoline et al., 2002). Figure 5.3

  1. Lightly mark the corners of a square with sides equal in length to the diameter of the circle or arc to be sketched.

  2. Lightly sketch the square.

  3. Mark the midpoints of the four sides of the square to give you four marks on the perimeter.

  4. Sketch the circle by creating four short arcs, each between the marks on the perimeter.

  5. Overdraw the arcs with a thicker line to complete the sketched circle.

The degree of precision needed in a given sketch depends on its use. Quick sketches to supplement verbal descriptions may be rough and incomplete. Alternatively, sketches that convey important information should be drawn as accurately as possible, such as the toothbrush design in Figure 5.4[1]. Although sketches are not drawn to an exact scale, approximate proportionality of features should be maintained.

Exercise 1

1. Critically observe products you use every day. Select one, and find five design flaws. Explain the flaws by conveying your ideas using sketches and short explanations.

2. For the flaws identified in the previous question, use sketches and brief explanations to propose design improvements.

Dimensioning

Dimensioning is the process of adding size information to a drawing. Dimensions can either indicate the size or the location of a feature. Dimensioning is governed by standards. The most widely used standard in the US, currently the ANSI Y14.5M-1982 standard, was developed by the American National Standards Institute (ANSI).

Accurate and unambiguous dimensioning is very important for the effective and efficient communication of the design. An example of a dimensioned object is shown in Figure 5.5. Note the different types of lines types used. These include

1. **Dimensions lines**. Thin, solid lines with arrowheads on both ends. Arrowheads indicate the direction and extent of a dimension. Dimension lines are usually broken in the middle to place the dimension value. Alternatively, the dimension value can be placed above or below an unbroken line.

2. **Extension lines**. Extend from an object line on a drawing to which a dimension refers, beyond the arrowhead of the dimension line. A gap of about 1.5 mm (1/16 in) should be left between the extension line and the object line. The extension line should extend about 3 mm (1/8 in) beyond the last arrowhead it touches.

3. **Centerlines**. Thin, alternating long and short dashes with about 1.5 mm (1/16) gaps in between. Centerlines are commonly used as extension lines for locating holes and other symmetrical features. When extended for dimensioning, centerlines cross over other lines without gaps.

4. **Leader lines**. Direct attention to a note or a dimension. Leader lines to circles or arcs should be radial so that if extended, they would pass through the center of the circular feature.

---

[1]Puzak, N., Sunkara, A., DeRoche, R., and Rotthoff, C., "Redesigning an Electric Toothbrush", final report for *Introduction to Freshman Design* course at Penn State, 2004.

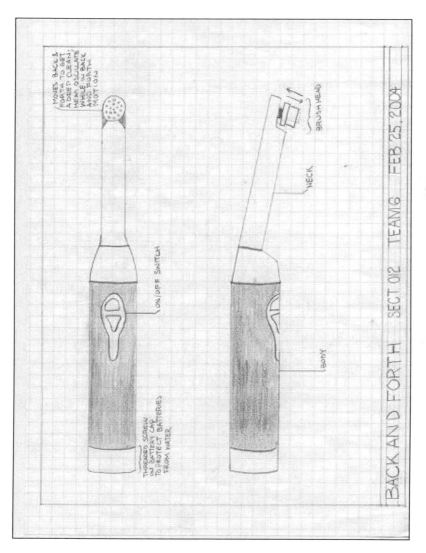

Figure 5.4 – A more detailed sketch that may be part of an informal written report or oral presentation

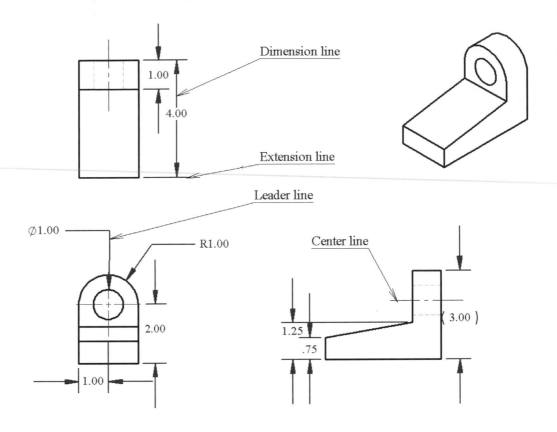

Figure 5.5 – *Dimensioning example showing different line types*

To ensure accurate and unambiguous dimensioning, the following guidelines should be observed:

1. Use decimal numbers for dimensions values with the appropriate number of decimal places permitted by tolerance limits. Fir example, one decimal place would be used for a tolerance of ±0.1.
2. Avoid redundant dimensioning of features. A minimal set of dimensions should be used.
3. Place dimensions in the view that most clearly describes the feature being dimensioned.
4. A visible gap should be placed between the ends of the extension lines and the feature to which they refer.
5. Avoid placing dimensions within object boundaries.
6. Unless specified otherwise, angles are assumed to be 90°.
7. Avoid dimensioning hidden lines.

8. Specify diameters, radii, counterbored or countersunk holes with the appropriate symbol preceding the numerical value – R for radius, ϕ for diameter, V for countersunk hole, and U for counterbored holes.

9. Leader lines for diameter and radii should be radial lines.

10. The size and style of leader line, text, and arrows should be consistent throughout the drawing.

11. Display only the number of decimal places required for manufacturing precision. For example, if the units in the drawing are inches, a dimension of '3.31' implies a desired manufacturing precision to one one hundredth of an inch.

Exercises 2

For each of the following problems in Figures 5.6 and 5.7, make a sketch of the three views to create a multiview drawing. Assuming that each square represents 0.5 inches, fully dimension the drawings.

Notes:

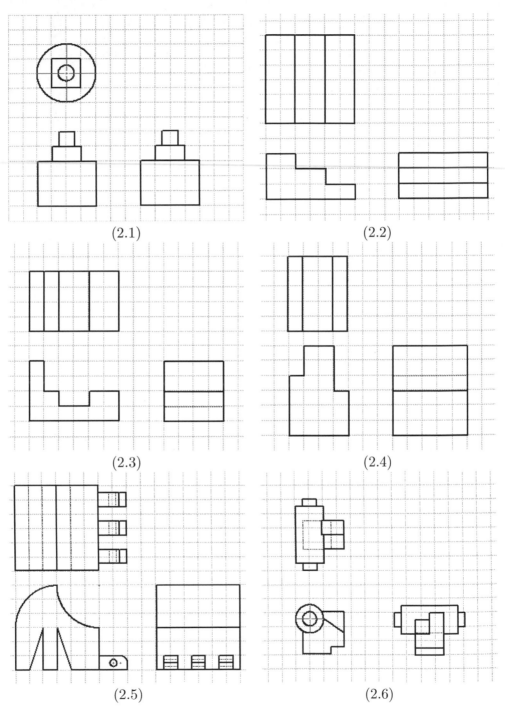

Figure 5.6 – *Multiview drawings for use in Exercises 2, 6 and 7*

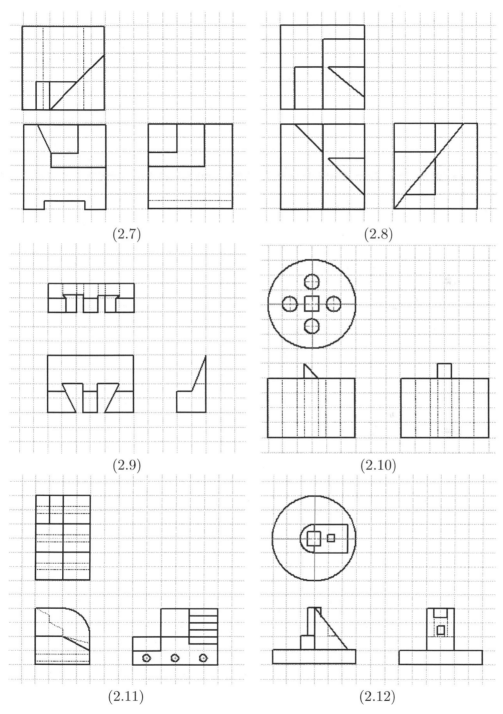

Figure 5.7 – *Multiview drawings for use in Exercise 2, 6 and 7*

Table 5.1 – *A comprehensive list of solid modeling functions. Adapted from Okudan and Rukowski (2004).*

Animation	Audit trail
Assembling parts	**Associativity (one way, two way)**
Complex blends	Calendar and task management
Creation of draft angles	Creation and retention of ribs
Cross sections	Decision making tools
Dimensioning	Document management
Extrusion	**Feature patterns (linear, circular)**
File conversion facilities	**Filleting, chamfering/ blending**
Isometric views	**Lofting**
Offset sections	Parametric relationships altering
Polls	Product data management systems
Project directory	**Rendering**
Requirements capture	**Revolve**
Section views	**Shelling/skinning**
Sweeping profiles along curves	Threaded discussion
Viewing and markup	Whiteboard

Introductory Concepts in Solid Modeling

Solid modeling supports the creation of parts, assemblies and automatic generation of the corresponding technical drawings. It was originally developed as a universal engineering language that uses various standards in representing parts and assemblies. These standards, as necessary, are discussed throughout the chapter. Solid modelers can provide a wide array of functions as listed in Table 5.1. The primary functions, indicated in bold in the table, are briefly discussed in the following sections.

The description of solid modeling functions is intended to enhance the reader's conceptual understanding of the functions rather than teach them how to use a specific software. The rational for this is that while the graphical user interface may differ from one solid modeler to another, the basic understanding behind feature creation does not. For example, extrusion of 3D models using 2D sketches are supported in virtually all solid modelers.

The assignments interspersed throughout the text also include references to the online tutorials found in the **SolidWorks** solid modeling package. The discussion is general, however, and applicable to any other package.

The creation of parts occurs using a sequence of solid modeling functions to define the parts' features. Features can be categorized as

1. **Sketched features.** Built from 2D profiles.
2. **Operation features.** Do not use sketches, but are directly applied to the part by selecting edges or faces.

To use any solid modeling software package effectively, one must be familiar with

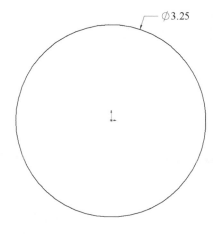

Figure 5.8 – *2D sketched profile*

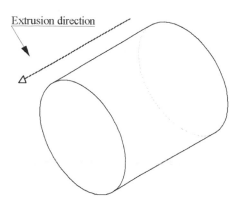

Figure 5.9 – *Extruded base feature (a cylinder) for coffee cup example*

the terminology. A summary of the common terms and their meaning follows.

1. **Axis**. An implied centerline that runs through every cylindrical feature.
2. **Plane**. A flat 2D surface on which 2D profile sketches are drawn.
3. **Origin**. The point where the three default reference planes intersect. The coordinates of the origin are $(x = 0, y = 0, z = 0)$.
4. **Face**. The surface or skin of a part. Faces can be flat or curved.
5. **Edge**. The boundary of a face. Edges can be straight or curved.
6. **Vertex**. The corner where edges meet.
7. **Conventions**. Commonly accepted methods or practices.

The followings sections will describe several of the common solid modeling functions with the use of a simple example, the creation of a coffee cup.

Extrusion

Extrusion is a sketched feature. It requires sketching a 2D profile on a plane and is followed by the extrusion of the sketch perpendicular to the plane. This feature is illustrated in Figures 5.8 and 5.9 for the first step in the solid modeling of a coffee cup: creation of a solid cylinder. First a 2D sketch of the cylinder profile is drawn (Figure 5.8) followed by extrusion of the profile to form a cylinder (Figure 5.9).

The first feature created is referred to as the base feature, to which all other features will be added. In this example, therefore, the extruded cylinder is referred to as the base feature. In general there are two types of extrusions:

1. **Extrude boss (build).** This feature adds material to the part and is what was used to create the base cylinder above.
2. **Extrude cut.** This feature removes material from the part. As such, it cannot be used to create the base feature.

Both boss and cut features require 2D sketched profiles that must be attached to an existing part, unless a boss feature is being used to create a base feature. The following points should be noted during extrusion

1. While sketching 2D profiles, pay attention to the status of the sketch. A Sketch's status can be

   - **Under-defined.** Additional dimensions or relations are required.
   - **Fully-defined.** No additional dimensions or relationships are required. A workable sketch must be fully-defined.
   - **Over-defined.** Contains conflicting dimensions and/or relations.

2. It is helpful to start a base feature with one of its vertices coinciding with the origin.

Shell

Shell, in an operation feature, that removes material from a selected face of a solid object resulting in a hollowed interior. The shell feature requires specifying the wall thickness that is maintained throughout the hollowing process. For the coffee cup example one of the circular faces of the cylinder is selected, and a wall thickness of 0.2 inches specified Figure 5.10. The shell feature then hollows out the cylinder maintaining the 0.2 inch thickness along all the cylinder walls and at the circular base to create the coffee cup body. Note that the face that is selected before the shell feature is executed is left open after the operation.

Fillet

Fillet is an operation feature used to round sharp edges and faces of a part either by removing or adding material (Figure 5.11).

1. **Convex fillet.** Applied to an outside edge and removes material.
2. **Concave fillet.** Applied to an inside edge and adds material.

Both fillet features requires specifying a radius as shown in Figure 5.11. For the coffee cup example the sharp edges at the cup lip are rounded with fillets of 0.1 inch radius (Figure 5.12).

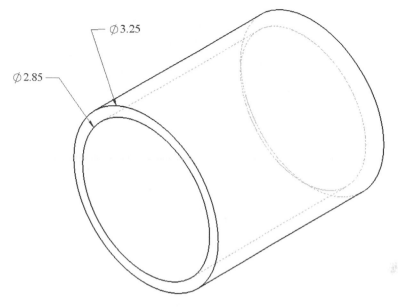

Figure 5.10 – *Shelled coffee cup with specified wall thickness of 0.2 inches*

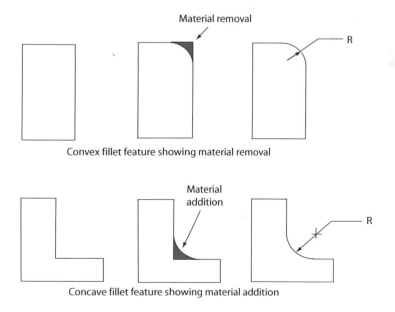

Figure 5.11 – *Illustration of convex and concave fillets*

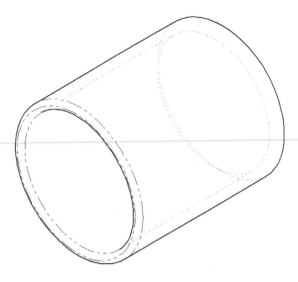

Figure 5.12 – *Coffee cup with filleted edges*

Exercise 3

1. Complete *Lesson 1: Parts*, from the **Solidworks** on-line tutorial.

2. Use the skills learned in the online tutorial to draw a solid model of a single light switch cover and a duplex outlet cover.

   a) Measure a single or double light switch cover.
   b) Using paper and pencil, manually sketch the light switch cover.
   c) Dimension the sketch.
   d) Review the switch and determine the solid modeling functions needed to create the solid model.
   e) Create a simple single or double light switch cover solid model. Save the file as *switchplate*.
   f) Create a simplified duplex outlet cover plate using a solid modeler. Save your file with the name *outletplate*.

Notes:

Assembling Parts

An assembly is created when two or more components are put together or *assembled* to form a complex object. Mates are relationships that align and fit the components together. Changes made to the components affect the assembly, and conversely changes made to the assembly affect the components.

The first component placed into an assembly in general is fixed. To move a fixed component, it must first be "unfixed." Subsequent components brought into the assembly are translated, rotated and the relations that limit their movement established.

The relations between components are created using the *mates function*, which aligns and fits together components in the assembly. Defining specific relations between faces, edges or vertices of the floating and fixed components creates relations that determine the relative location of components in an assembly and the relative motion between them. Relationships that can be defined include coincidence, concentric, parallel and vertical. Defined mates specify the movement in all six degrees of freedom (along and the X, Y and Z axes). An example of an assembly composed of six different parts is shown in Figure 5.13. Note that the four horizontal bars are the same, and therefore the same part is brought into the assembly four times.

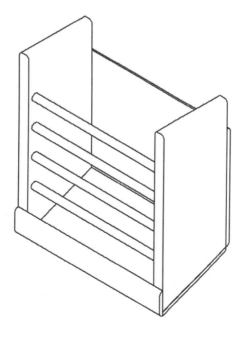

Figure 5.13 – *Example of a magazine holder assembly composed of six different parts*

Exercise 4

1. Complete *Lesson 2: Assemblies* from the **Solidworks** on-line tutorial.

2. Using the light switch cover created in Exercises 3, design and model two fasteners to complete the assembly. The fasteners' design should adhere to the following criteria:

   a) The fastener must be longer than the thickness of the switchplate
   b) The fastener must be 0.25 inches in diameter
   c) The head of the fastener must be larger than the hole in the light switch cover

3. Create the light switch cover/fastener completed. Save the assembly file as *switchplateassembly*.

Creating Engineering Drawings from Solid Models

Engineering drawings communicate size, shape and non-graphical information about the manufacturing processes such as drill, grind, heat treat, etc. Engineering drawings are based on projection theory that determines how 3D objects are represented on 2D medium. Projection theory is based on two variables: *line of sight* and *plane of projection*. A line of sight (LOS) is an imaginary ray of light between an observer's eye and an object. In parallel projection, all lines of sight are parallel. A plane of projection is an imaginary flat plane upon which the image created by the lines of sight is projected.

Orthographic projection is a parallel projection technique in which the plane of projection is positioned between the observer and the object and is perpendicular to the parallel lines of sight. Orthographic projection can produce

1. Pictorial drawings that show three dimensions of an object in one view (Figure 5.14).
2. Multiview drawings that show only two dimensions of an object in a each view (Figure 5.15).

Each will be described further in the following sections.

MULTIVIEW DRAWINGS

In multiview drawings, six principal views are considered: front, top, left, right, bottom, and rear (back). Each principal view represents the image that would be seen through a side of a glass cube totally enclosing the object. Figure 5.15 shows the standard arrangement of all six views of an object in accordance with the ANSI standard third angle projection, which is commonly practiced in the US. Note that the top, front and bottom views must be aligned vertically, and the back, left, front, and right views must be aligned horizontally.

The general characteristics of an object will determine which views are required to fully describe its shape. Typically, three standard views – front, top

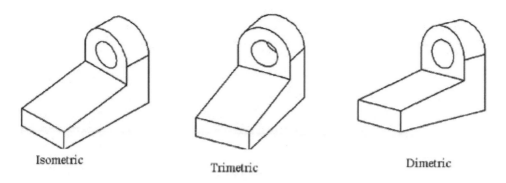

Isometric   Trimetric   Dimetric

Figure 5.14 – *Pictorial drawings*

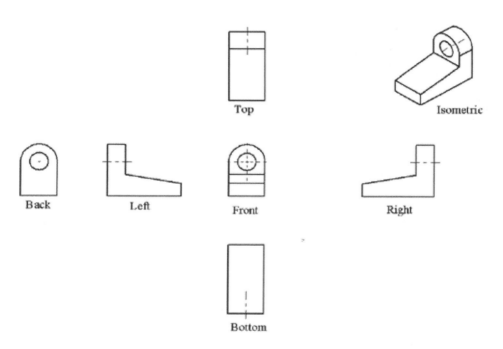

Top   Isometric

Back   Left   Front   Right

Bottom

Figure 5.15 – *Layout for multiview drawings*

103

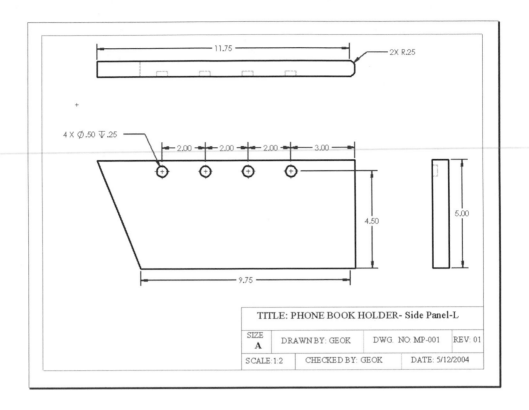

Figure 5.16 – *An engineering drawing showing the title block*

and right (or left) views – are used. For this case, the front and right (or left) view must be horizontally aligned, and the front and top views must be vertically aligned (Figure 5.16).

Multiview drawings also require a title block (Figure 5.16). Information placed in the title block may include company name, part number, part name, drawing number, revision number, sheet size, drawing scale, material and finish, and the name of the person who completed or checked the drawing.

The scale chosen to complete the drawing is indicated in the title block as a ratio. For example, for a full size drawing the scale will be indicated as 1:1, for a half size drawing the scale is 1:2. The ratio indicates the number of units on the drawing (first number) and the corresponding number of units on the object (second number). For reduction, scales of 1:2, 1:3, and 1:4 are common. For enlargement, scales of 2:1, 3:1, and 4:1 are typical.

Finally, several different line types are used on engineering drawings as illustrated in Figure 5.17. These include

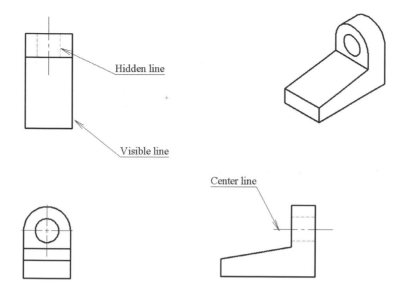

Figure 5.17 – *Engineering drawing illustrating the different line types*

1. **Visible or object lines.** Used to represent features that can be seen in the current view. Solid lines are used for visible lines.

2. **Hidden lines.** Used to represent features that cannot be seen in the current view. Dashed lines are used for hidden lines.

3. **Centerlines.** Used to represent symmetry and paths of motion, and to mark the centers of circles. Centerlines are represented by 1 inch and 1/8 inch alternating long and short lines, respectively, with a 1/16 inch gap between.

4. **Dimension and extension lines.** Used in combination to indicate feature sizes and locations on a drawing.

Exercise 5

1. Complete *Lesson 3: Drawings* from the **Solidworks** on-line tutorial.

2. The following exercise uses the drawing created in Question 1 of Exercises 3.
   a) Create a new A-size ANSI standard drawing template.
   b) Create a drawing for *Tutor2*. Use the drawing template you created in the previous task.
   c) Create Front and Top views. Add an Isometric view.
   d) Import the dimensions from the part.
   e) Create a note on the drawing to label the wall thickness (Wall Thickness=4 mm).
   f) Save your file as *tutor2drawing*.

3. The following exercise uses the drawing created in Question 2 of Exercises 3, which was saved as *switchplate*.

   a) Create an engineering drawing of the single or double light switch cover. The drawing should show the front, top and right views as well as an appropriately filled title block.

   b) Imagine that the fillet feature is too small to be clearly seen and dimensioned. Insert a detail view to show the dimension of the fillet feature.

   c) Save the file as *switchplatedetailview*.

4. For each of the drawings in Figures 5.18 and 5.19, draw the top, front and right views. Assume that each square is 0.5 inches. Unless otherwise stated, all holes go all the way through.

5. For each of the drawings in Figures 5.20 and 5.21, redraw given views and the missing view to create a standard multiview drawing with the top, front and right views. Assume that each square is 0.5 inches.

Notes:

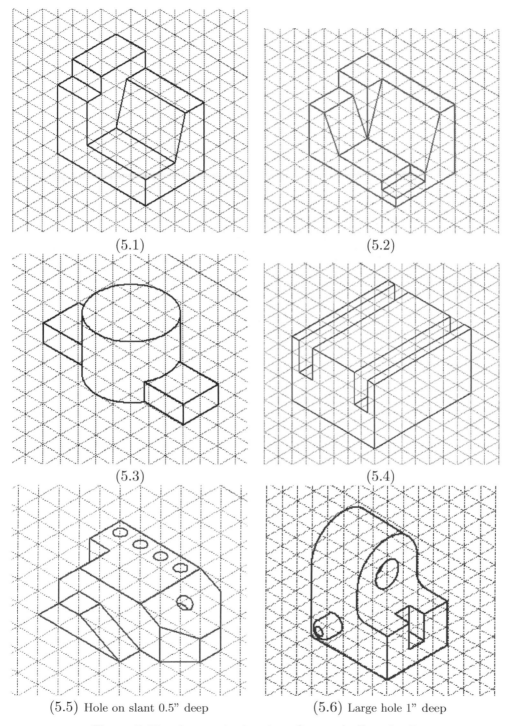

(5.1)

(5.2)

(5.3)

(5.4)

(5.5) Hole on slant 0.5" deep

(5.6) Large hole 1" deep

Figure 5.18 – *Isometric drawings for use in Exercise 5*

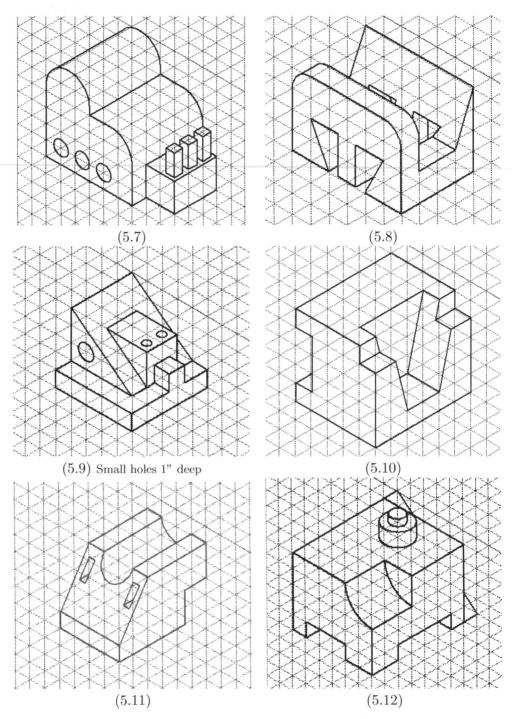

(5.7)

(5.8)

(5.9) Small holes 1" deep

(5.10)

(5.11)

(5.12)

Figure 5.19 – *Isometric drawings for use in Exercise 5*

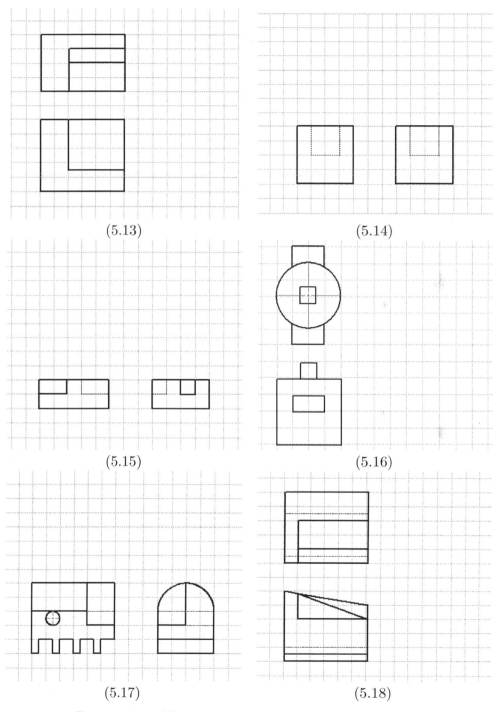

(5.13)

(5.14)

(5.15)

(5.16)

(5.17)

(5.18)

Figure 5.20 − *Missing view drawings for use in Exercise 5*

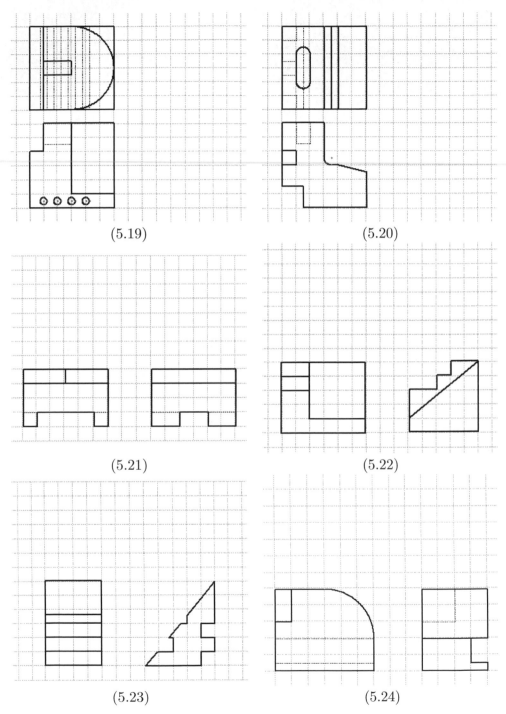

(5.19)   (5.20)

(5.21)   (5.22)

(5.23)   (5.24)

Figure 5.21 – *Missing view drawings for use in Exercise 5*

Pictorial Drawings: Isometric

Isometric drawings represent three views of an object in a single drawing, such as the object illustrated in Figure 5.22(d). In an isometric drawing, the three axes representing the object's width, height and depth are located as shown in Figure 5.22 (a), i.e. the height axis vertical and the width and depth axes are located 30° from the horizontal. To create an isometric drawing, the following guidelines should be followed:

1. All measurements parallel to any of the three major axes (depth, width and height) are true length.

2. For features that are not parallel to any of the major axes, points that define the features are located using parallel measurements. The non-parallel feature is then drawn by connecting the dots. Figure 5.22 shows several steps used to create an isometric view of an object with most features not parallel to any of the major axes.

3. Circles in isometric drawings will appear as ellipses. Using Figure 5.23 as a reference, the ellipses can be drawn using the following steps:

   a) Draw a parallelogram whose sides have the same dimension as the diameter of the desired circle. Note that the sides of the parallelogram are along the corresponding major axes.

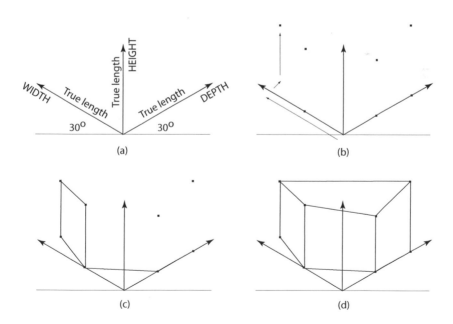

Figure 5.22 – *Steps to create an isometric drawing for features not parallel to the major axes*

111

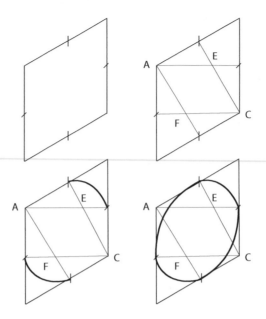

Figure 5.23 – *Steps to create a circle in an isometric drawing*

b) Find the midpoints of all four sides.

c) Draw faint lines between the closest opposing corners (marked A and C in the Figure) and both opposite side midpoints. This creates two intersecting points E and F.

d) Use a compass centered at E and F to draw two arcs as shown in the figure.

e) With the compass centered at A and C, draw two additional arcs to complete the ellipse.

f) Note that for a quick sketch the same can be achieved by connecting the midpoints of the parallelogram with the appropriate arcs.

Exercise 6

1. Manually draw an isometric cube with a 3 in dimension. Then place a tangent circle on each side of the cube that can be seen on the isometric view.

2. For each of the drawings in Figures 5.6 and 5.7, draw an isometric pictorial.

Pictorial Drawings: Obliques

Oblique pictorial views are commonly used for drawing objects that do not have a lot of detail on the top or right-side surfaces. The height and width axes are

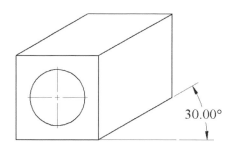

Figure 5.24 – *Cavalier oblique with full-size width dimension*

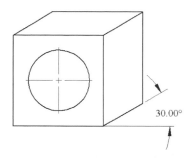

Figure 5.25 – *Cabinet oblique with half-size width dimension*

perpendicular to each other, and the depth axis can be at 30°, 45° or 60° to horizontal. In obliques, height and width dimensions are true length. Depth dimensions, on the other hand, can either be true length or half length creating *cavalier* (Figure 5.24) and *cabinet* (Figure 5.25) obliques, respectively.

For objects with large depth dimensions, the full-size depth in a cavalier oblique creates an unrealistic view of the object. For those objects, the cabinet oblique is preferred.

In order to minimize distortions of features in oblique pictorials, the object configuration guidelines below should be followed.

1. Place complex features such as arcs, holes or irregular surfaces parallel to the frontal plane. This allows these features to be drawn more easily and without distortion. For example, in Figure 5.25 by placing the face with the circle as the front view, the circle appears undistorted.
2. The longest overall dimension of the object should be parallel to the frontal plane.
3. If there is conflict between the previous two guidelines, the first one takes precedence.

Exercise 7

For each of the drawings in Figures 5.6 and 5.7, draw an oblique pictorial using the above guidelines.

Creating Part Configurations in Solid Modeling

Frequently, a number of part versions are needed with a variety of sizes. For such situations, it is not efficient to build each version individually. Most solid modelers offer a function to quickly build versions of a part such as application of *design tables* in **SolidWorks**.

To create multiple versions of a part for a selected number of part dimensions, alternatives that will define the part versions should be provided. Consider the example of the coffee cup. As for a part version, imagine that you would like to have a cup with a different base diameter. In this case, in addition to the original value of the diameter (3.26 in) seen in Figure 5.27, one other value can be selected, such as 5 in to build a multi-purpose cup that might be used to serve soup as well.

In addition to creating rapid part versions by assigning alternative values to various dimensions in a part, one can also create assembly versions. In these applications, rather than manipulating dimensions in a part, the number of parts in an assembly is manipulated. That is, certain parts in an assembly can be configured to be suppressed, or resolved.

Exercise 8

1. Complete the *Design Tables* **SolidWorks** on-line Tutorial.

2. Create a design table for *Tutor2* that corresponds to the four configurations of *Tutor1*. Save the drawing as *tutor2configurations*.

3. Create a cup in **Solidworks**. Create four configurations of the cup using a design table. Experiment with different dimensions.

Section Views

Section views use an imaginary cutting plane passing through the part to reveal interior features of an object that are not easily represented using hidden lines. Cutting plane lines, which show where the cutting plane passes through the object, represent the edge view of the cutting plane and are drawn in the view(s) adjacent to the section view. For example, the line A-A in the front view of the multiview drawing in Figure 5.26 indicates the location of the edge of the imaginary cutting plane used to generate section A-A.

A horizontal section view is one where the cutting plane edge is in the front view and the top view is drawn as a section. Alternatively, a profile section view is one where the cutting plane edge is in the front and top views and the side view is drawn as a section. Multiple sections can appear on a single drawing.

As shown in Figure 5.26, section lines or cross-hatch lines are added to a section view to indicate surfaces that are cut by the imaginary cutting plane. Different section line symbols can be used to represent various types of materials. However, the general-purpose section line symbol used in most section view drawings is that of cast iron. The cast iron section line is drawn at a 45° and spaced 1/16" (1.5mm) to 1/8" (3mm) or more depending on the size of the drawing, but can be changed when adjacent parts are in section. The spacing of section lines is equal or uniform on a section view. Section lines should be thinner than visible lines and should not run parallel or perpendicular to the visible outline.

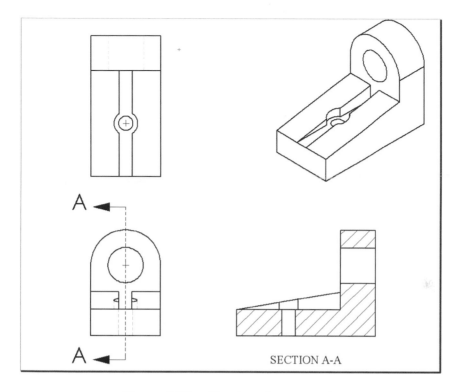

Figure 5.26 – *Section view example*

There are many different types of section views. Above, full sections were explained. However, there are times when it may be more appropriate to use a section view that is not a full section. For example, symmetrical objects can be sectioned using a half section. Or objects with only small areas that need to be clarified with the use of a section view can be represented with a broken-out section. Other types of section views are revolved sections, removed sections, offset sections and assembly sections.

Exercise 9

Complete the *Advanced Drawings* **Solidworks** on-line tutorial.

Revolve

The base feature created for the coffee cup can be created using another sketched feature: revolve. To complete a revolved base one needs to sketch a 2D profile and sketch a centerline or axis around which the 2D profile will be revolved.

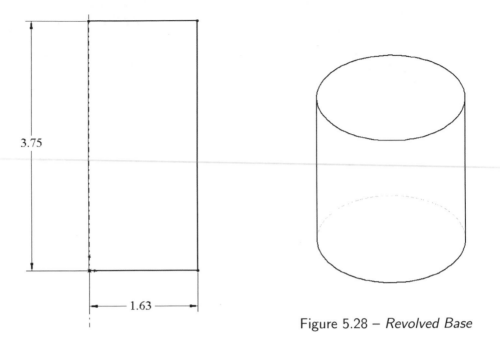

Figure 5.28 – *Revolved Base*

Figure 5.27 – *2D Profile of Revolve*

Angle of rotation should be specified. For the base of the coffee cup, the revolve feature is shown below.

Sweep

This feature is created by moving a 2D profile along a path. A sweep feature is used to create the handle of the coffee cup. The sweep feature requires two sketches: (1) sweep path and 2) sweep section (profile). Some important rules to remember when creating a swept feature are

1. The sweep path is a set of sketched curves contained in a sketch, a curve, or a set of model edges.
2. The sweep profile must be a closed contour.
3. The start point of the path must lie on the plane of the sweep section.
4. The section, path or the resulting solid cannot be self-intersecting.

Exercise 10

1. Complete *Revolves and Sweeps* **Solidworks** on-line tutorial.

2. Design a candle to fit the candlestick.

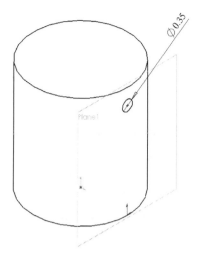

Figure 5.29 – *Sweep Profile*

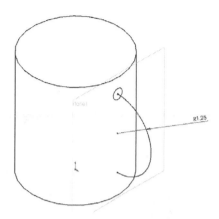

Figure 5.30 – *Sweep Path*

Figure 5.31 – *Completed Sweep*

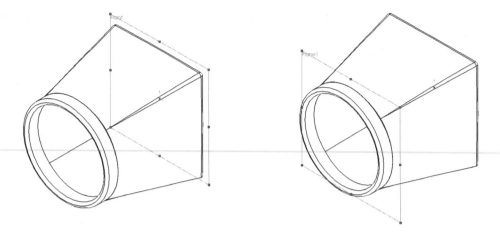

Figure 5.32 – *First plane*   Figure 5.33 – *Second plane*

a) Use a revolve feature as the base feature.
b) Taper the bottom of the candle to fit into the candlestick.
c) Use a sweep feature for the wick.
d) Create a candlestick assembly.
e) Use a design table to create various versions of the candle design.

Lofting

The loft feature allows the blending of multiple profiles. A loft feature can be a base, boss or a cut. To create the loft feature, first the planes required for the profile sketches should be created. Then, a sketch should be completed on each plane. As an example, a version of the coffee cup base is created in the following steps (see Figure 5.32-5.35).

1. Determine the position of the planes relative to each other.
2. Create the planes using the solid modeler.
3. Draw the necessary 2D profile on each plane.
4. Combine the sketched features using the loft function.

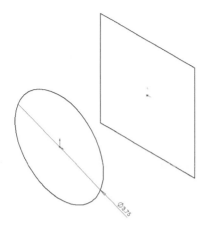

Figure 5.34 – *Sketches on plane 1 and plane 2*

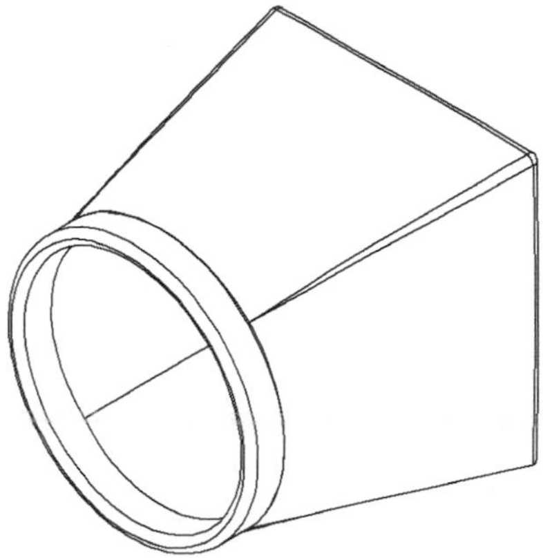

Figure 5.35 – *Lofted base*

Exercise 11

1. Complete *Lofts* **Solidworks** on-line tutorial.

2. Create the bottle shown in Figure 5.36. All dimensions are in inches.

Patterns

Pattern features are sketched features as well. To complete a pattern, a 2D sketch profile should be completed that will then be copied in a linear or circular fashion to the part that is being worked on. An example of a pattern is illustrated in Figure 5.37.

Chamfer

The chamfer feature is very similar to a fillet feature in that it is applied to external or internal edges, and it can remove or add material. Unlike fillets which round an edge, however, the chamfer feature bevels the edge as shown in Figure 5.38.

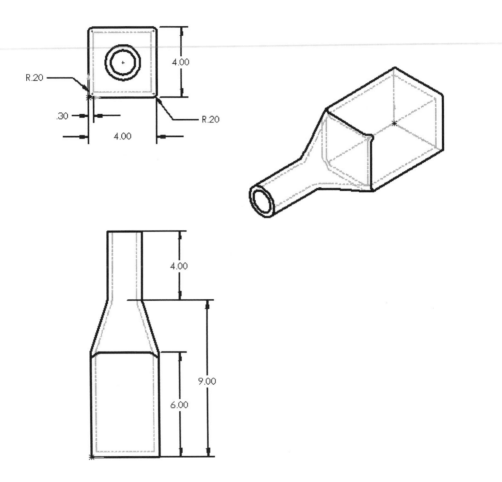

Figure 5.36 – *Bottle dimensions for Exercise 11*

Figure 5.37 – *Cup with linear and circular patterns*

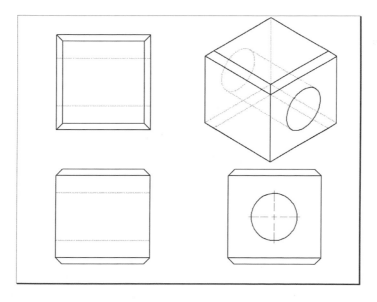

Figure 5.38 – *Object showing chamfered edges*

121

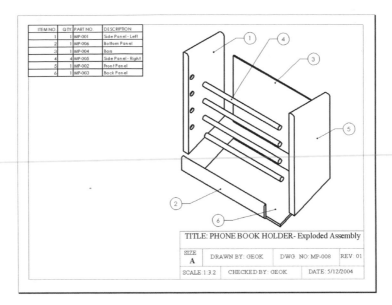

ITEM NO.	QTY	PART NO.	DESCRIPTION
1	1	MP-001	Side Panel - Left
2	1	MP-006	Bottom Panel
3	1	MP-004	Bars
4	4	MP-005	Side Panel - Right
5	1	MP-002	Front Panel
6	1	MP-003	Back Panel

TITLE: PHONE BOOK HOLDER- Exploded Assembly

SIZE A	DRAWN BY: GEOK	DWG. NO: MP-008	REV: 01
SCALE 1:3.2	CHECKED BY: GEOK	DATE: 5/12/2004	

Figure 5.39 – *Exploded assembly drawing of a phone book holder*

5.4 Working Drawings

Working drawings are the set of standardized drawings that are used to manufacture products. Through graphical and text information working drawings provide dimensions, form descriptions for all components, and specifications for the way the components are assembled. A complete set of working drawings for an assembly must include (1) detail drawings of each non-standard part[2] and (2) an assembly drawing that shows all the standard and nonstandard parts in relation to how they will all fit together to form the final assembly. The assembly drawing can be shown as an exploded view (Figure 5.39) or an unexploded view (Figure 5.40).

An assembly drawing should contain all of the following elements (Bertoline et al., 1998):

1. All the parts, drawn in their operating position.

2. The bill of materials (BOM). With reference to Figures 5.39 and 5.40, BOM is a table in the assembly drawing that contains detail (identification) number for each part, the quantity of the part needed for a single assembly, the name of the part, and if necessary, a description of the part. Detail

---

[2]Non-standard parts are those that are not commercially available and must be manufactured for the current application.

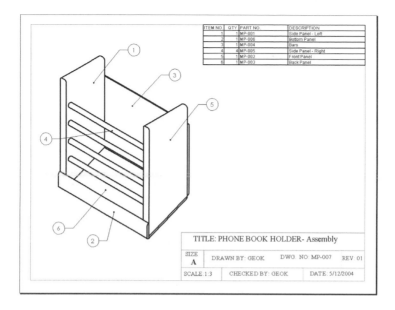

ITEM NO.	QTY	PART NO.	DESCRIPTION
1	1	MP-001	Side Panel - Left
2	1	MP-006	Bottom Panel
3	1	MP-004	Bars
4	4	MP-005	Side Panel - Right
5	1	MP-002	Front Panel
6	1	MP-003	Back Panel

TITLE: PHONE BOOK HOLDER- Assembly

| SIZE A | DRAWN BY: GEOK | DWG. NO. MP-007 | REV. 01 |
| SCALE:1:3 | CHECKED BY: GEOK | DATE: 5/12/2004 | |

Figure 5.40 – *Unexploded assembly drawing of a phone book holder*

numbers are typically composed of a string of numbers and letters. In Figure 5.39, the parts are numbered sequentially from MP-001 to MP-006.

3. Leader lines with balloons assigning a detail number to each part.

4. If required, leader lines with balloons pointing to areas or features of parts that require particular machining and assembly operations, or that indicate critical dimensions for these functions.

A detail drawing is a dimensioned, multiview drawing of a single part that describes the form, size and material in sufficient detail for the part to be manufactured (Bertoline et al., 1998). ANSI standards are followed when producing detail drawings. The detail drawings for the phone book holder assembly presented in Figure 5.39 are shown in Figures 5.41 to 5.46[3]. As previously noted, detail drawings of standard parts are not drawn, however, they are listed in the bill of materials for the assembly. A standard part in this context can be considered to be any part that is purchased from suppliers and used without any modifications in the assembly.

Working drawings are numbered using a standard numbering system established by company guidelines. These guidelines in general allow for drawing revisions to be tracked in databases. For example, in the title block of Figure

---

[3]Based on a phone book holder designed as a class project during the *Introduction to Engineering Design* course at Penn State by Amro Asad, Geoffrey Geise, Bryce Lambert, Garrett Risberg and Patrick Woolcock.

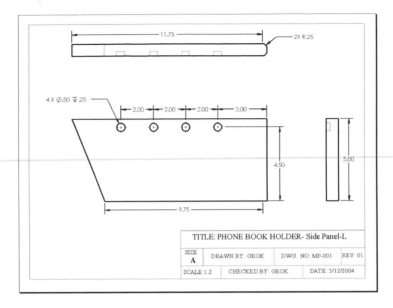

Figure 5.41 – *Detail drawing of left side panel*

5.39, the drawing number is MP-008 with a revision number of 01. In this case, the drawing number can be tracked as MP-008-01 or MP-008R01. If the drawing were revised, the second and third revisions would be numbered MP-008-02 and MP-008-03, respectively.

Exercise 12

1. Complete the *Assembly mates* **Solidworks** on-line Tutorial.

2. Complete the *Bill of Materials* **Solidworks** on-line Tutorial.

3. **Solidworks** provides numerous other features. These can be explored by completing the following on-line tutorials:
   a) Pattern features
   b) Animation
   c) Photoworks

Notes:

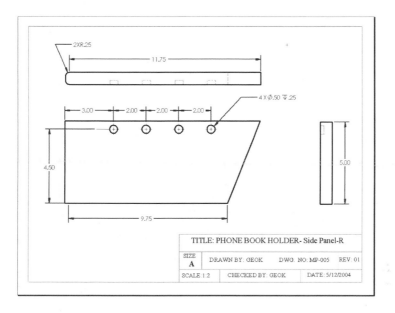

Figure 5.42 – *Detail drawing of right side panel*

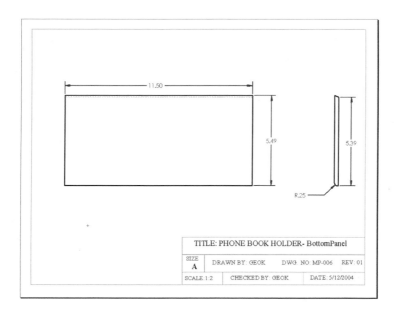

Figure 5.43 – *Detail drawing of bottom panel*

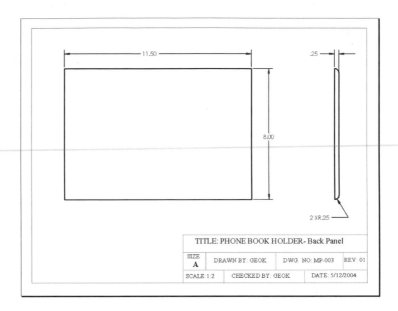

Figure 5.44 – *Detail drawing of back panel*

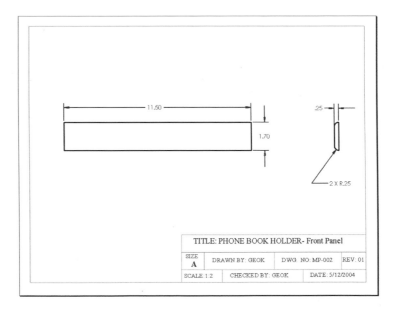

Figure 5.45 – *Detail drawing of front panel*

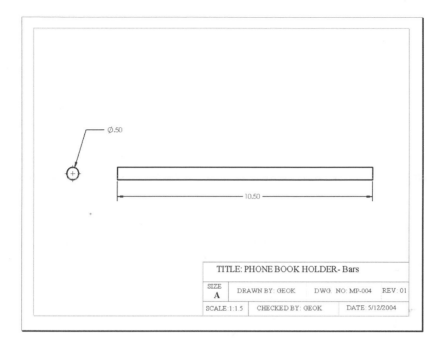

Figure 5.46 – *Detail drawing of bars*

5.5 Computer Generated Sketches for Documentation

In addition to hand-drawn sketches of the components and parts of a design, engineers frequently manually draw diagrams that layout the system, describe a concept or lay out the flow of a process. These hand-drawings provide a quick means of discussing and refining ideas with team members and other stake holders. Once the final conceptual or system diagrams have been finalized, computer software such as **Adobe Illustrator** or **Corel Draw** can be used to generate permanent versions of the diagrams appropriate for documentation and use in presentations. A project-based tutorial for **Adobe Illustrator** is presented in Appendix B. The entire tutorial takes about three hours to complete. It assumes no prior experience with computer illustration software.

5.6 Summary

This chapter summarizes introductory concepts in technical sketching and solid modeling as a means to communicate design concepts and provide a visual representation of technical information. Technical sketching allows the generation

127

of quick images that allow the designer to visualize concepts under consideration and to communicate those ideas to others. As ideas become more concrete, software tools such as **Adobe Illustrator** or **Solidworks** can be used to generate more permanent schematics and figures, or solid models, respectively.

Finally, technical sketching/drawing conventions and standards are interspersed as needed within the the solid modeling discussions. Although one can easily learn how to use a solid modeler, its effective use as a design tool requires unambiguous representation of parts and assemblies that can be achieved by having a sufficient level of understanding and the implementation of the conventions and standards.

References

Bertoline,G.R., Wiebe, E.N. and Miller, C.L., "Fundamentals of Graphics Communication", 3nd Edition, New York: McGraw-Hill, 2002.

Li, G., Jinn, J.T., Wu, W.T., and Oh, S.I., "Recent Development and Applications of Three-Dimensional Finite Element Modeling in Bulk Forming Processes", Journal of Materials Processing Technology, 113 (1-3), pp. 40-45, 2001.

Green, R., "CAD Manager Survey - Part 2". CADALYST, 20(12), pp. 26-27, 2003.

Nam, T.J. and Wright, D., "The Development and Evaluation of Syco3D: A Real-Time Collaborative 3D CAD System", Design Studies, vol. 22, pp. 557-582, 2001.

Requicha, A.A.G. "Mathematical Models of Rigid Solids: Theory, Methods and Systems", ACM Computing Surveys, vol. 12 no. 4, pp. 437-464, 1980.

Requicha, A.A.G. and Voelcker, H.B., "An Introduction to Geometric Modeling and Its Applications in Mechanical Design and Production", In Advances in Information Systems Science, Ed. J.T.Tou, Vol.8, Plenum Publishing, 1981.

Requicha, A.A.G. and Voelcker, H.B., "Solid Modeling: A Historical Summary and Contemporary Assessment", IEEE Computer Graphics and Applications, pp. 9-24, 1982.

Requicha, A.A.G. and Voelcker, H.B., "Solid Modeling: Current Status and Research Directions", IEEE Computer Graphics and Applications, 1983.

Rossignac, J. and Requicha, A.A.G., "Solid Modeling", In the *Encyclopedia of Electrical and Electronics Engineering*, Ed. J. Webster, New York:John Wiley and Sons, 1999.

Schmitz, B., Tools for innovation, Industry Week, May 15th, 2000.

Shapiro, V. "Solid Modeling", In *Handbook of Computer Aided Geometric Design*, Eds. G.Farin, J. Hosheck, MS. Kim, Elsevier Science Publishers, 2002.

Tovey, M. and Owen, J. "Sketching and Direct CAD Modeling in Automotive Design", Design Studies, vol. 21, pp. 569-588, 2000.

Ullman, D.G. "Toward the Ideal Mechanical Engineering Design Support System", Research in Engineering Design, 2001.

Voelcker, H.B. and Requicha, A.A.G., "Research in solid modeling at the University of Rochester: 1972-1987", in *Fundamental Developments of Computer-Aided Geometric Modeling*, L. Piegl (Ed.), London: Academic Press Ltd., pp. 203-254, 1993.

Chapter 6

Decision Making

6.1 Introduction

Engineering design is inherently a decision-making process where choices are constantly being made between alternatives, such as selection of concepts or components or the rating of client needs. Numerous methods have been developed to help design teams make the correct choices by using structured approaches. This chapter will focus on three widely used tools: pairwise comparison charts (PCCs), the analytic hierarchy process (AHP) and decision matrices,.

Throughout the chapter, emphasis will be placed on these methods, all of which can be used within a spreadsheet environment to (1) automate tedious calculations and (2) allow easy alteration of data or decision-making criteria.

6.2 Rank Order: Pairwise Comparison Charts

It is often necessary to numerically rank a set of objectives or evaluation criteria. Dym and Little (2003) propose using of *pairwise comparison charts (PCCs)*. PCCs are based on the premise that it is easier to differentiate between pairs of alternatives, such as A is better than B or A is similar to B, than it is to differentiate between a large set, such as A, B, C and D. PCCs, therefore, use a matrix structure to compare each alternative individually with every other (the pairwise comparison). The results from those comparisons are summed to obtain an overall rank-order.

PCCs can be generated using the following steps:

1. In Table 6.1, the n items to be compared are listed as row and column headings in an n x n matrix. An additional column is added to the end of the matrix to record the total score for each item.

131

Table 6.1 – *Structure of Pairwise Comparison Charts*

| Evaluated | Comparison Criteria | | | | | | Total |
	A	B	C	D	E	F	
A		-1	-1	-1	-1	-1	-5
B	1		-1	1	-1	0	0
C	1	1		1	1	1	5
D	1	-1	-1		0	-1	-2
E	1	1	-1	0		-1	0
F	1	0	-1	1	1		2

Key
A - Size
B - Weight
C - Strength
D - Cost
E - Availability
F - Manufacturability

2. The first row alternative is individually compared to all other column items. Scores of 1, 0, and −1, are assigned if the row item is better, similar or worse, respectively, than the column item.[1]

3. The row scores are totaled, yielding the overall score of the first alternative.

4. Steps 2 and 3 are repeated until overall scores have been calculated for all alternatives.

5. The rank order for the alternatives is compiled. The higher the overall row score, the higher the alternative's rank.

Consider the example presented in Table 6.1 comparing six attributes: size, weight, strength, cost, availability and manufacturability. Starting with the row A corresponding to 'size', a comparison is first made to weight (column B): *Is size more important (1), as important (0) or less important (-1) than weight?* In this case the size *is less important* than weight, and therefore a -1 is entered in the intersecting cell. The same question is asked along the entire row, and the corresponding numerical entries entered.

Note that the results in the lower triangular matrix must be the same magnitude and opposite sign to their corresponding cells in the upper triangular matrix. This is because entries in the lower triangle correspond to the same questions as the upper triangle, but asked in the opposite way. For example, for row B and column A, the comparison now becomes: *Is weight more important (1), as important (0) or less important (-1) than size?* As size was determined to be less important before (value=-1), that is the same as stating that weight is more important (value=1). Therefore, if the charts are created manually, care must be taken to ensure that this symmetry is maintained. This can be done by simply determining the entries for the top triangular matrix first, and then mirroring those results with opposite sign in the lower triangle. Table 6.2 provides

---

[1]Dym and Little (2003) proposed using numerical scores of 1,0.5 and 0 for comparisons that are better, similar or worse, respectively. We prefer the use of the 1,0,-1, scale as they are more intuitive and reflective of what they represent - positive, better; zero, similar; negative, worse. In addition, they result in a matrix that can be mirrored with opposite sign about the diagonal making matrix construction easier in **Excel**.

Table 6.2 – *Steps to create pairwise comparison charts in* **Excel**

1. Create a new file in **Excel**

2. Type in the row and column headings corresponding to the criteria you have, as shown in Figure 6.1.

3. After the last row, type in a key defining each letter.

4. Create an additional column heading 'Total' as shown in Figure 6.1.

5. First row sum: In the top cell under the *Total* heading, $H3$ in this example, type in the summation formula '$= sum(B3 : G3)$' (Figure 6.2). Note that your cell references may differ slightly depending on how many criteria you have defined.

6. Starting with cell $H3$, select all the cells in the 'Total' column, $H3$ to $H8$ in this example.

7. `Edit>Fill>Down`. The summation formula is copied into all the selected cells.

8. To add the gridlines around the chart, select the cells that correspond to the entire chart, $A1 : I8$ in this example.

9. `Format>Cells`. Click on the **Border tab**. This brings up the *Border pop-up menu* (Figure 6.3).

10. Click on the **Outline** and the **Inside** preset buttons. Click on the **OK** button.

11. The next few steps will shade in the cells where the same row and column criteria intersect. The shading provides a visual reminder that the bottom triangular matrix is a mirror image and opposite of the top triangular matrix. Select the intersecting cell corresponding to the criteria A row and column ($B3$ in this example.)

12. `Format>Cells`. Press the **Patterns tab** to bring up the *Patterns pop-up menu*.

13. Select and click on a shading color. Click on the **OK button**.

14. Repeat the previous two steps to shade all the diagonal cells in the chart (Figure 6.4).

15. Finally, using Figure 6.4 as a guide enter in all the formulas to have the lower triangular matrix mirror and multiply by -1 corresponding entries from the upper matrix. **Comparisons need only be made in the top triangular matrix; the lower triangle is updated automatically.**

16. Save the file.

Your generic PCC is complete. Using the **Save As** command to save a copy of the **Excel** spreadsheet allows the original to remain intact and be used as a template.

the steps to create these charts in **Excel** and therefore to automate entries for the lower triangluar matrix and calculation of totals.

Each criterion's overall score is calculated by summing the scores in each row. For this example, *size* would be the lowest ranked at -5 with *Strength* the highest ranked at 5. Note that the actual total values only provide rank-order. They determine which alternative is more or less important than another but provide no information on by how much. Table 6.2 provides step-by-step instructions for generating the PCCs in **Excel**. Substitute the criteria A, B, C, etc. with your own criteria.

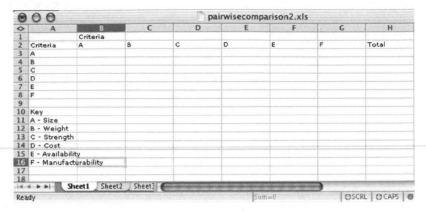

Figure 6.1 – *Initial layout for pair wise comparison charts in* **Excel**

Figure 6.2 – *Adding summation formula*

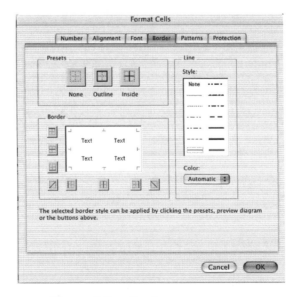

Figure 6.3 – Border pop-up menu

pairwisecomparison2.xls

	A	B	C	D	E	F	G	H
1		Comparison Criteria						
2	Evaluated	A	B	C	D	E	F	Total
3	A							0
4	B	=-C3						0
5	C	=-D3	=-D4					0
6	D	=-E3	=-E4	=-E5				0
7	E	=-F3	=-F4	=-F5	=-F6			0
8	F	=-G3	=-G4	=-G5	=-G6	=-G7		0
9								

Figure 6.4 – *Formulas to allow lower triangular matrix to mirror and be opposite to the top triangular matrix*

Table 6.3 – *Fundamental scale of relative importance. Adapted from Saaty (1986).*

Scale value	Explanation
1	Equal importance: Two alternatives are similar
3	Moderate importance of one over the other: Experience and judgment strongly favor one alternative over another
5	Essential or strong importance: Experience and judgment very strongly favor one alternative over another
7	Very strong importance: An alternative is strongly favored and its dominance demonstrated in practice (e.g., in similar products)
9	Extreme importance: The evidence favoring one alternative over another is of the highest possible order of affirmation
2,4,6,8	Intermediate values between the two adjacent judgments: When compromise is needed
1/n	Reciprocals: For inverse comparison

6.3 Relative Order: Analytic Hierarchy Process (AHP)

If a relative score is required for a set of qualitative alternatives, the analytic hierarchy process (AHP) can be used (Saaty, 1986; Saaty, 1994). The relative order determines by how much each alternative is better (or worse) than the others. In its simplest form, AHP is similar in structure to the pairwise comparison charts, except that the evaluations between two alternatives are based on a special rating system called the fundamental scale (Table 6.3). The *fundamental scale* captures individual preferences with respect to qualitative or quantitative attributes. The scale represents the relative weight given to one alternative over another. If two alternatives are similar, for example, they have a scale value of one.

Unlike the pairwise comparison charts, the AHP compares each alternative against itself, resulting in '1' entries along the diagonals. If an alternative has less importance than another, reciprocals of the fundamental scale values are used. For example if Alternative A is of *moderate less importance* than Alternative B, the comparison would be evaluated as 1/3. The latter statement is the opposite

Table 6.4 – *Structure of AHP pairwise comparison matrix*

Evaluated	Comparison Criteria						Total	Weight
	A	B	C	D	E	F		
A	1.00	0.33	0.20	0.33	0.14	0.33	2.34	0.04
B	3.00	1.00	0.33	4.00	0.20	1.00	9.53	0.16
C	5.00	3.00	1.00	3.00	6.00	4.00	22.00	0.36
D	3.00	0.25	0.33	1.00	1.00	0.25	5.83	0.10
E	7.00	-0.20	0.17	1.00	1.00	0.50	9.47	0.16
F	3.00	1.00	0.25	4.00	2.00	1.00	11.25	0.19

Key
A - Size
B - Weight
C - Strength
D - Cost
E - Availability
F - Manufacturability

of, *Alternative B is of moderate importance to Alternative A*, whose comparison value from the fundamental scale is 3.

Table 6.4 illustrates an example AHP matrix comparing six attributes: size, weight, strength, cost, availability and manufacturability. The last two columns of the matrix are the sum of each row, R_i, and the calculated weights, w_i, for each alternative, respectively. The weights represent the relative-order of the alternatives and are calculated from

$$w_i = \frac{R_i}{\sum_{j=1}^{n} R_j} \qquad (6.1)$$

n is the number of alternatives. Equation 6.1 simply divides the total of each row, by the sum of all the row totals to calculate the normalized row total, or weight, w_i. Note that all the weights must sum to one, i.e.

$$\sum_{i=1}^{n} w_i = 1 \qquad (6.2)$$

As reciprocals represent scores for alternatives that are of *less importance*, the lower triangular matrix of the AHP matrices are reciprocals of the upper triangular matrix. One then needs only fill in the top triangular matrix and use **Excel** to calculate the lower triangular matrix. *The weights in the final column of the matrix give an indication of how many times an alternative is considered more (or less) important than all the others.* In summary, pairwise comparison charts are used when only rank order is required. If relative weight order is needed, the AHP matrices are used.

Consistency Check

When the number of alternatives, such as design concepts, design criteria, etc., for pairwise comparisons increases, so does the likelihood of making inconsistent judgements. Therefore, utmost care should be given while making pairwise judgments. In addition, any change in the design requirements should be reflected in

Table 6.5 – *Average consistency index for randomly generated comparisons. Adapted from Saaty (1980).*

n	CI_{random}
2	0.00
3	0.52
4	0.90
5	1.12
6	1.24
7	1.32
8	1.41

the selection process by redoing the necessary pairwise comparisons. For example, during the early stages of a design project, environmental friendliness may not have been an important factor but could become so later on due to policy changes. In such a case the outcome of the design process will not satisfy the current needs unless the relative importance of the design criteria is evaluated again.

When there are many (more than eight) alternatives and the design project bears high-risk with a sizable budget, a consistency ratio (CR) can be calculated as a measure of the inconsistent judgements. CR is the ratio of a consistency index (CI) for a pairwise comparison matrix to the value of the same consistency index for a randomly generated pairwise comparison matrix

$$CR = \frac{CI}{CI_{random}} \qquad (6.3)$$

where CI is defined as

$$CI = \frac{\lambda_{max} - n}{n - 1} \qquad (6.4)$$

where λ_{max} is the largest positive eigenvalue of the pairwise comparison matrix. CI values calculated for randomly generated pairwise comparison matrices of order n are given in Table 6.5. The rule of thumb is that if the CR exceeds 0.10, pairwise comparisons should be redone to improve the consistency.

6.4 Relative Order: Decision Matrices

Decision matrices can vary in complexity from extremely simple to very complex. They allow the evaluation of alternatives relative to specified criteria and corresponding weighting factors. The basic structure of a decision matrix is illustrated in Figure 6.5. The design alternatives are listed down the rows and the evaluation criteria across the columns. To create a decision matrix

1. List all the design alternatives, each on a separate row.

137

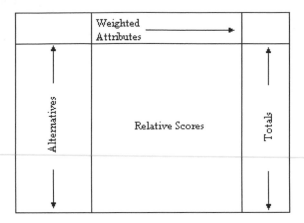

Figure 6.5 – *Structure of a basic decision matrix*

2. List the evaluation criteria, each on a separate row. Determine weights, w_i, for each criterion according to its importance. The AHP can be used for this purpose. Recall that the sum of all the weights must equal one.

3. Assign/calculate scores for each alternative based on each evaluation criterion. To allow comparison of scores that may be of different units, normalization must be performed. For example, if one of the evaluation criteria was cost and the other weight, it would be impossible to come up with a final score for each alternative, especially if the ranges for each criteria were very different. Normalization yields scaled parameter values within a single range (for example, 0-1, 0-10 or 0-100).

4. Sum each row of normalized scores to obtain the total score for each alternative. The sum provides a relative-order of the alternatives.

Decision matrices will contain criteria that need to be maximized (i.e., the larger the score the better) and that need to be minimized (i.e., the lower the value the better). Care must therefore be taken to ensure that the totals obtained by summing these different types of criteria make sense. Consequently, during normalization both types of criteria are treated differently.

For criteria that need to be maximized, such as material strength or loading capacity, normalization is performed by dividing each value in that criterion's column by the largest value

$$s_{max} = m\frac{ECV_i}{ECV_{max}} \qquad (6.5)$$

where s_{max}, ECV and ECV_{max} are the normalized maximization value, the evaluation criteria value and the largest evaluation criteria value, respectively. In addition, m can take on values of 1, 10 or 100 to achieve the desired normalized parameter range. For example if m equals 10, then all normalized scores will range from 0 - 10, with 10 being the best.

For criteria that need to be minimized, such as cost, weight or time to market, normalization is carried out by dividing the smallest value by each value in that criterion's column

$$s_{min} = m\frac{ECV_{min}}{ECV_i} \tag{6.6}$$

where s_{min} and ECV_{min} are the normalized minimization value and the smallest evaluation criteria, respectively. Since the smallest value is on the numerator, the smaller the value in the denominator, the higher the normalized score. As a result, after normalization the higher the score the better the alternative. Since both normalized scores (for criteria that need to be minimized and those that need to be maximized) are now based on the same range and both improve the higher the score, summation of each row will yield a total that makes sense.

The final score, T, for each alternative is the sum of the product of the normalized score for each evaluation criteria and its corresponding weight

$$T = \sum_{k=1}^{m} w_i s_i \tag{6.7}$$

where m is the number of evaluation criteria, w_i the weight assigned to each criterion, and s_i the normalized value for each criterion. Note that

$$\sum_{k=1}^{m} w_i = 1 \tag{6.8}$$

To illustrate the use of decision matrices, consider the case of a couple considering the purchase of a new vehicle. Table 6.6 lists the vehicles under consideration, the criteria to be used for evaluation and the relative importance (weights) they have attached to each one. Note that the criteria scores include both quantitative and qualitative values. For the qualitative criteria the couple agreed on four ratings: poor, OK, good and great.

A decision matrix was created in **Excel** and is presented in Table 6.7. In the first column, all the vehicles are listed. Beginning with the second column the criteria are listed in every other column (2,4,6, etc.). The skipped columns (columns 3, 5, 7, etc.) contain the normalized scores for each criteria. Note that for cost (the lower the better) is normalized using Equation 6.6 with $m = 10$ to obtain a range from 0-10 for the normalized score. The remaining parameters that need to be maximized are normalized using Equation 6.5 and the same value for m.

Table 6.6 – *Vehicles under consideration and evaluation criteria*

Criteria − >	Cost	Style	Fuel Economy	Warranty	Audio System
Importance − >	0.5	0.2	0.2	0.1	
Ford Outtahere	$23,000	Great	28 mpg	90K miles	OK
Dodge Speedster	$26,000	Great	32 mpg	100K miles	OK
Chevy Hellraiser	$19,000	Good	25 mpg	85K miles	Good
Datsun Seeyah	$22,500	OK	35 mpg	75K miles	Great

The couple assigned scores of 1, 2, 3 and 4, respectively to their four subjective ratings of poor, OK, good and great. The total scores for each vehicle are calculated by inserting the normalized scores and the corresponding importance values (weights) into Equation 6.7. Based on the couple's analysis, therefore, the Chevy Hellraiser would be the best vehicle for them.

Summary

During the design process, teams are often faced with the daunting task of making decisions based on a large number of alternatives and an equally large number of evaluation criteria. This chapter has presented an overview of three methods that can simplify decision making. A fourth approach commonly used for the evaluation and selection of design concepts, Pugh charts, will be presented in Chapter 9 together with a discussion of design-concept generation and selection. Finally, it must be emphasized that *the use of these tools cannot generate good designs unless they are fuelled with sound engineering judgment.*

References

Dym, C.L. and Little, P., *Engineering Design: A Project-Based Introduction*, 2nd Edition, New York: John Wiley and Sons, 2003.

Saaty, T.L., *The Analytical Hierarchy Process: Planning, Priority Setting, Resource Allocation*, New York: McGraw-Hill Book Co., 1980.

Bibliography

Dym, C.L. and Little, P., *Engineering Design: A Project-Based Introduction*, 2nd Edition, New York: John Wiley and Sons, 2003.

Saaty, T.L. "Axiomatic Foundation of the Analytical Hierarchy Process", Management Science, vol. 32, no.7, pp. 841-855, 1986.

Saaty, T.L. "Highlights and Critical Points in the Theory and Application of the Analytical Hierarchy Process", European Journal of Operational Research, vol. 74, pp. 426-447, 1994.

Table 6.7 – *Decision matrix example*

Weight ->	Cost		Style		Fuel		Warranty		Audio		Total
	0.45		0.20		0.20		0.10		0.05		
Ford	23000	0.83	4	1.00	28	0.80	90000	0.90	2	0.50	8.47
Dodge	26000	0.73	4	1.00	32	0.91	100000	1.00	2	0.50	8.37
Chevy	19000	1.00	3	0.75	25	0.71	85000	0.85	3	0.75	**8.65**
Datsun	22500	0.84	2	0.50	35	1.00	75000	0.75	4	1.00	8.05

Part II

The Engineering Design Process

Chapter 7

Problem Definition and Determination of Need

7.1   Introduction

This chapter focuses on the first stage of the design process (Figure 7.1), *problem definition* and *determination of the client's (or customers') needs.*

Before a team can begin working on a design problem, they must clearly define the problem to ensure that their final design solves the right problem. The form of the initial design task presented to a team can range from a short problem statement, such as the tire-cutter machine case study below, to a long detailed report. Irrespective of the length, the initial problem definition is only the starting point. The design team must expand on the problem definition to form a clearer understanding of the task before proceeding with the design process.

Once the problem has been adequately defined, the design team then determines the client's needs by drawing up a list of desired attributes for the final design. The attributes list at this stage should not imply a solution but simply present the desired qualities of the final design. This list should be drawn up after talking to current users of existing products (if available), potential users of the product (if the product does not exist), and from discussions amongst the team members. Each of these two areas, problem definition and determination of need, will be discussed in the remainder of the chapter. But first, the design of a tire-cutting device case study is presented, which will be used throughout

Figure 7.1 – *The first stage of the design process*

145

the chapter.

Case Study: Design of a Tire Cutter

The following case study is formed from excerpts from the final report of one of the capstone senior design projects at Rutgers University in 2003. It presents the initial problem definition. This case will be used throughout the text. The student group members were Nicholas Malinoski, Helen Moore, Mark Telesz, Jared Roszko and Robert Wotring.

Background and Problem Definition. Estimates of the number of "scrap" tires in stockpiles around the United States range from 500 million to 3 billion, with an additional 270 million tires scrapped each year. A scrap tire is one that is no longer used for its original purpose. Illegal or improper dumping and stockpiling of scrap tires pose serious health and environmental risks, as tire piles provide a breeding ground for rodents and mosquitoes, and are susceptible to fire from arson, lightning, and even spontaneous combustion. In addition, tire pile fires are extremely polluting and difficult to extinguish. As a result of these problems, nearly every state in the US has some form of scrap-tire management regulations, including charges for tire disposal. Many states also offer financial incentives for using scrap tires in products. Most states ban disposal of whole tires in landfills. While these regulations generally increase tire disposal costs, which has led to an increase in illegal dumping in some areas, they have also increased the overall reuse and recycling of tires. In 1990, only about 11% of scrap tires were recycled. By 2002 about 70% of scrap tires were recycled or exported. The remainder were stockpiled or shredded and buried in mono (single material) landfills or used as landfill cover.

Nearly 15 million scrap tires are chopped, ground, or powdered for use in a wide variety of products such as floor mats, adhesives, gaskets, shoe soles, and electrical insulators, or blended into asphalt for use in pavement binders and sealants or as an aggregate substitute. An additional 20 million whole or chopped scrap tires become fill and cover material in construction and landscaping, artificial reefs and breakwaters for beach erosion control, playground surfacing material and equipment, highway and race track crash and sound barriers, boat dock shock absorbers, and even house-building materials. Farmers and ranchers use about 2.5 million whole scrap tires for holding down covers on hay stacks, controlling erosion, protecting structures from livestock damage, or as rollers in corn husking equipment, as well as many other uses.[1] The design team was tasked with designing and constructing a portable machine that could be carried around on the back of a truck and used to cut off the tire side walls and to cut the tire treads of scrap tires at collection points prior to transportation. This initial processing at the point of collection reduces the volume occupied by each tire, allowing the transportation of more tires per trip and thereby reducing transportation costs. To be viable, the machine had to cost under $6,000.

---

[1]Energy Efficiency and Renewable Energy Network. Scrap Tire Recycling http://www.eere.energy.gov/, 2002.

7.2   Problem Definition

Problem definition ensures that design teams solve the 'correct' problem. Consequently, a significant amount of time should be spent on analyzing the problem before searching for a solution. This can be done in two steps, (1) problem analysis and (2) problem clarification.

Problem Analysis

The first step seeks to determine exactly what the design task is. It helps to determine if the true design problem is masked by the apparent problem presented in the problem statement. Domb (1997) recommends following a journalism approach to problem analysis by asking and answering the following questions:

1. Who has the problem?
2. What does the problem seem to be? What are the resources?
3. When does the problem occur? Under what circumstances?
4. Where does the problem occur?
5. Why does the problem occur? This question should be asked at least five times.
6. How does the problem occur?

With clear answers to each of the above questions, the design team will have a good grasp on the problem to be solved. Revisiting the tire-cutter case study, the questions could be answered as follows:

1. *Who has the problem?* Green Design, LLC.
2. *What does the problem seem to be?* Cost of transporting scrap tires from collection points to the recycling plants are prohibitive.
3. *What are the available resources?* Trucks for transportation and limited capital to pay for a method to reduce transportation costs.
4. *When does the problem occur? Under what circumstances?* Whenever scrap tires are picked up from the collection points and delivered to recycling plants.
5. *Where does the problem occur?* Not applicable.
6. *Why does the problem occur? This question should be asked at least five times.* Why is the transportation cost high? High cost of fuel, labor and wear and tear on transportation vehicles resulting in a high cost per tire. Why is there a high cost per tire? The number of tires that can be transported per trip is quite low. Why are the number of tires transported per trip low? Tires occupy a large volume, the bulk of which is air. Why do the tires occupy a large volume? They are currently transported as whole tires. Why are they transported as whole tires? Most collection

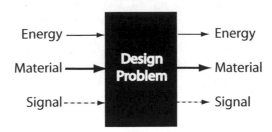

Figure 7.2 – *Energy, material and signal flows through a generic 'black box' design*

sites do not have machines to cut the tires, and existing machines are not portable enough to be carried in the collection truck.

7. *How does the problem occur?* Spending a large amount of capital for a relatively small number of transported tires.

Answering the questions quickly reveals what the problem is, why it occurs and how it occurs. The questions are general and are meant to apply to nearly any problem. Due to the generality, the answers may be repetitive or not applicable for specific problems.

Problem Clarification with Black-Box Modeling

The second step, problem clarification, uses black-box modeling to further analyze the problem. Analysis of engineering systems reveals that they essentially channel or convert energy, material or signals to achieve a desired outcome. *Energy* is manifested in various forms, including optical, nuclear, mechanical, electrical, etc. *Materials* represent matter. *Signals* represent the physical form in which information is channeled. For example, data stored on a hard drive (information) would be conveyed to the computer's processor via an electrical signal.

An engineering system can therefore be initially modeled as a black-box (Figure 7.2) with energy, material and signal inputs and outputs from the system. In black box modeling, energy is represented by a thin line, material flows by a thick line, and signals by dotted lines as shown. The engineering system, therefore, provides the functional relationship between the inputs and the outputs (Pahl and Beitz, 1996).

Problem clarification involves forming a clear understanding of the problem. The *overall problem* represented by the black-box can be broken down into smaller *sub-problems*. Breaking problems down allows solutions to complex engineering design problems to be found by considering simpler sub-problems. Design teams can then focus on the sub-problems critical to the success of the project first, deferring others. Sub-problems are then mapped to sub-functions for which a design is created. Combining of all the designs that achieve each of the sub-functions results in the desired system solution that achieves the overall desired function. Note that the functional decompositions and the resulting black-box diagrams are generic and do not commit the design team to any particular technological working principle.

Black-box modeling of existing systems that are to be redesigned, on the other hand, decomposes the existing system into sub-systems, as opposed to sub-functions. The sub-systems would then be translated to sub-functions from where the redesign proceeds. Black-box models can take on many forms depending on how and to what extent the design team decides to decompose the problem. As such it is recommended that design teams create several draft models gradually converging to a single diagram agreeable to all team members. Finally, at this early stage of the design process, it may not always be possible to decompose the problem and clearly define energy, material and signal flows (Ulrich and Eppinger, 2003). For those problems, simply listing the desired functions or customer needs may suffice.

Black-Box Modeling Examples

Several examples are presented to illustrate the use of black-box models for problem clarification.

TIRE-CUTTER

The tire-cutter device discussed in the case study could be represented by the black-box model shown in Figure 7.3. The desired device has been decomposed by function. Each of the sub-functions thus represents a mini-design problem. Note that the sub-functions are generic and do not imply a solution. For example, the sub-function *Store or accept external energy* refers to any external energy source making no mention of a specific type.

Alternatively, the problem could be decomposed by user actions following a similar approach to functional decomposition. In this case, the single 'black box' is decomposed into elements corresponding to the major user actions as illustrated in Figure 7.4.

AUTOMOBILE AIRBAG

The automobile airbag, when used in conjunction with a seat belt, provides protection to occupants during front end collisions. Airbag systems deploy when crash sensors located on the front of the vehicle detect high-rate deceleration. The sensors trigger the inflator module that through a rapid chemical reaction (a mini-explosion) rapidly releases nitrogen gas that fills the airbag. Typically the air-bag will be fully deployed within 1/20 th of a second after impact detection (Kowalick, 1997). An initial black-box model of the airbag system is

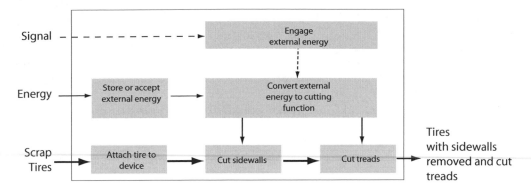

Figure 7.3 – *Tire-cutter example decomposed into functional sub-problems*

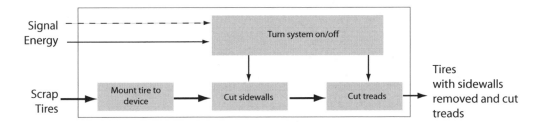

Figure 7.4 – *Tire-cutter example decomposed into sub-problems defined by user-actions*

presented in Figure 7.5(a). The entire airbag system is represented as a black-box with a single input, the mechanical energy from the automobile impact. Figure 7.5(b) illustrates the decomposition of the black-box into sub-systems, with the corresponding energy, material and signal flows.

The sequence of events, indicated by the arrows, starts with the detection of an impact that then signals the chemical reaction, rapidly releasing gas and mechanical energy into the airbag. The occupant then slams into the airbag, which in turn collides with the car interior. Note that the double arrows between the occupant and the airbag and between the airbag and the car interior are used to indicate that the mechanical forces are bi-directional.

COMPUTER HARD DRIVE

A computer hard drive is used to store and retrieve data (Figure 7.6). Within the hard drive, data is stored on a rotating magnetic disk, from which data is read using a read/write head. The head, situated at the end of a moveable actuator arm, can magnetize (write) or sense the magnetic field (read) on the disk. The

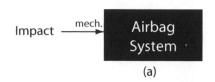

(a)

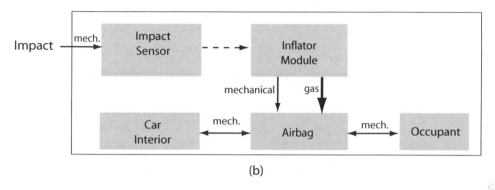

(b)

Figure 7.5 – *Black-box model of an airbag system*

Figure 7.6 – *Image of hard drive with cover removed to reveal magnetic disk, actuator arm and read/write head*

head floats on the airflow generated by the disk rotation, which maintains a very small gap between the two, preventing contact that may result in data loss. A black-box model of the hard drive in operation is shown in Figure 7.7.

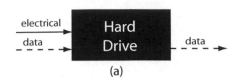

(a)

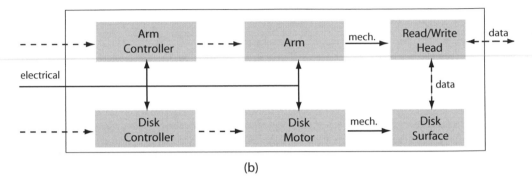

(b)

Figure 7.7 – *Black-box model of a computer hard drive*

7.3 Determination of Customer/Client Needs

To ensure success of a designed product or process, the design team must strive to meet the end users' expectations. For example, if a customer has a good experience with the product and if her expectations are met, she will buy it and recommend it to others. On the other hand, if the experience is poor, the customer will feel frustrated and have a negative view of the product (Cagan and Vogel, 2002). These feelings will be reflected in negative recommendations to others and poor sales.

The same principle are true for non-consumer products such as the design of a building or the design of an industrial process. If the clients are unhappy with the final design because their expectations not met, they probably will not proceed with construction. In addition, they are unlikely to provide the design team with additional work or recommend the team to others.

Determining customers' or clients' needs can be achieved through a three step process:

1. Gather and interpret data from customers, clients or end users.

2. Categorize and rank the needs.

3. Reflect on the results (Ulrich and Eppinger, 2003).

The first two steps will be elaborated upon in the following sections.

Table 7.1 – *Initial customer needs list for tire cutter example*

Portable
Easy to load on and off truck
Able to fit through standard doorway
Safe
Easy to load and unload tires
Flexible
Protection from cutting devices
Minimal debris
Adaptable to a wide range of tire sizes
Cut side walls and tread
Fast cutting operation
Small cycle time between each tire
Durable
Low noise
Debris contained
Retails under $6,000
Easy to operate
Light
Small footprint
Easy maintenance
Relatively maintenance free
Collapsible

Gathering and Interpreting Data

Data gathering for customer needs is usually done through interviews, focus groups, and if applicable, observation of the customer using an existing or competitor's product. *Interviews* and *focus groups* are similar because in a face-to-face interview the customers are directly asked about the product or process. At this stage the design team seeks to obtain a list of attributes that the customers or clients expect the product or process to have. The key difference between interviews and focus groups is that the former are conducted with a single customer, while the latter bring a group of people together with a moderator.

The third approach, observation of the customer using the product, may provide information to the design team on desired attributes that may not be obvious to the customer. For example, the design team could observe that too many steps are required to complete a task and that a reduction in the number of steps could be a desirable attribute in a redesigned product. Programming VCRs to record TV shows is a good example of this, where over the years through both customer observation and complaints, electronic companies have made the process much simpler.

In the tire cutter case study, the initial attributes list generated by the design team by interviewing potential customers is given in Table 7.1.

153

Table 7.2 – *Categorized customer needs list for tire-cutter example*

Objectives	Constraints
Portable	Retails under $6,000
Easy to load on and off truck	Small footprint
Safe	Able to fit through standard doorway
Easy to load and unload tires	
Flexible	**Functions**
Minimal debris	Cut side walls and tread
Adaptable to a wide range of tire sizes	Debris contained
Fast cutting operation collapsible	Protection from cutting devices
Small cycle time between each tire	
Durable	
Low noise	
Easy to operate	
Light	
Easy maintenance	
Relatively maintenance free	

Categorizing and Ranking Needs

A closer examination of the tire-cutter attributes list presented in Table 7.1 reveals that not all items are of the same type. In fact, they can be divided into three broad categories: *objectives* (or goals), *constraints* and *functions*. These categories are often incorrectly used interchangeably; however, they have different meanings and play different roles in the design process. Dym and Little (2003) provides the following definitions:

1. *Objectives or goals* define attributes that the design attempts to attain. For example, the tire cutter should be 'portable,' 'safe,' 'flexible,' 'durable.' All are desired attributes or objectives.

2. *Constraints* provide limitations or boundaries within which the final design specifications must lie. For the tire cutter, 'retails under $6,000' and 'able to fit through a standard doorway' are examples.

3. *Features or functionality* are the things that the design must do. In this case, tire cutter should be 'collapsible' and 'cut side walls and treads'.

The list in Table 7.1 can be re-written by grouping the attributes into each of the three categories (Table 7.2). Close examination of the objectives list reveals several similar attributes, for example, 'flexible' and 'adaptable to a wide range of tires.' Further, as the design process unfolds certain concepts may achieve some objectives to a greater extent than others. Based on the desired objectives, how should the design team decide which concepts to pursue and those to abandon? To address these issues the following steps should be taken once an initial objective list has been created:

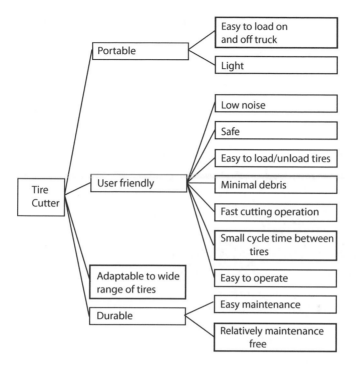

Figure 7.8 – *Objective tree for tire-cutter example*

1. Group similar objectives under major and minor headings to create a hierarchal objective list.
2. Add the constraints and functions as part of the tree or list for completeness. The constraints and functions should be clearly distinguishable from each other and from the objectives through the use of different fonts, colors, formatting, etc.
3. At similar hierarchal levels, rank or rate the objectives in order of importance.
4. Create a ranked objective tree or objective list.

These steps will be expanded upon in the following sections in the context of the tire cutter example.

HIERARCHAL OBJECTIVE LISTS AND OBJECTIVE TREES

Hierarchal objective lists and objective trees are functionally the same. The difference is that objective trees are graphical (Figure 7.8) while objective lists are text-based (Table 7.3). The number of levels varies from project to project but typically increases with the project complexity. As you move down the levels,

Table 7.3 – *Hierarchal objective list for tire-cutter example*

1. Portable
   1.1 Easy to load on and off truck
   1.2 Light

2. User friendly
   2.1 Low noise
   2.2 Safe
   2.3 Easy to load and unload tires
   2.4 Minimal debris
   2.5 Fast cutting operation
   2.6 Small cycle time between each tire
   2.7 Easy to operate

3. Adaptable to a wide range of tire sizes

4. Durable
   4.1 Easy maintenance
   4.2 Relatively maintenance free

the attributes become more specific. Design teams should therefore attempt to meet the higher-level objectives first.

The objective list[2] is often augmented by adding the constraints and the functions that were pruned from the original attributes list. Their addition gives a more complete picture of the design problem. Table 7.4 displays the hierarchal objective list for the tire cutter example, with the constraints and functions from Table 7.2 added. Note that the objectives, constraints and functions are formatted differently for easy identification.

RANKING THE OBJECTIVES

In its current form, the hierarchal objective lists assume that each attribute is of equal importance to the client. This is rarely the case. Determining a rank order is therefore important to ensure that higher ranked objectives are accounted for first during design decisions. Dym and Little (2003) proposed the use of *pairwise comparison charts* for this purpose[3] using following the guidelines:

1. Comparisons should only be made at the same level within the hierarchal structure. For example, comparing '3. Flexible' and '4.1 Easy maintenance' would be inappropriate.

---

[2]Due to the similarity of objective trees and objective lists, our discussion will focus on objective lists, but the same can be applied to objective trees

[3]*See*, Section 6.2 on Page 131 for a discussion of PCCs

Table 7.4 – *Hierarchal objective list for tire-cutter example*

1. Portable

    1.1 Easy to load on and off truck
    1.2 Light
    F.1 collapsible
    C.1 Small footprint
    C.2 Able to fit through standard doorway

2. User friendly

    2.1 Low noise
    2.2 Safe
        F.2 Protection from cutting devices
        F.3 Debris contained
    2.3 Easy to load and unload tires
    2.4 Minimal debris
    2.5 Fast cutting operation
    2.6 Small cycle time between each tire
    2.7 Easy to operate

3. Flexible

    3.1 Adaptable to a wide range of tire sizes
    F.4 Cut side walls and tread

4. Durable

    4.1 Easy maintenance
    4.2 Relatively maintenance free

    C.3 Retails under $6,000

---

2. Lower level objectives should only be compared with those under the same upper level objective. For example, comparisons could be made between 'Easy Maintenance' and 'Relatively maintenance free' but not between 'Minimal debris' and 'Light.'

3. These are *subjective rankings*, and it would be inappropriate to use the resulting numeric values directly in other calculations.

A pairwise comparison chart for the higher level objectives of the tire cutter example implemented in **Excel** is presented in Table 7.5. The resulting rankings (represented by no stars for the lowest, and in this case '**' for the highest) are then added to the hierarchal list to create a *ranked hierarchal objective list*. Similarly, the 'User friendly' sub-objectives were ranked (indicated by '+s') and the rankings added to the objective list (Table 7.6).

Table 7.5 – *Pairwise comparison chart for tire-cutter example*

Evaluated	Comparison Criteria				Total	Ranking
	A	B	C	D		
A		-1	1	0	0	*
B	1		1	1	3	**
C	-1	-1		-1	-3	
D	0	-1	1		0	*

Key
A - Portable
B - User Friendly
C - Flexible
D - Durable

Table 7.6 – *Ranked hierarchal objective list for tire-cutter example*

1. Portable (*)

   1.1 Easy to load on and off truck
   1.2 Light
   F.1 collapsible
   C.1 Small footprint
   C.2 Able to fit through standard doorway

2. User friendly (**)

   2.1 Low noise
   2.2 Safe (+++)
   F.2 Protection from cutting devices
   F.3 Debris contained
   2.3 Easy to load and unload tires (++)
   2.4 Minimal debris
   2.5 Fast cutting operation (+)
   2.6 Small cycle time between each tire (+)
   2.7 Easy to operate (++)

3. Flexible

   3.1 Adaptable to a wide range of tire sizes
   F.4 Cut side walls and tread

4. Durable (*)

   4.1 Easy maintenance
   4.2 Relatively maintenance free

C.3 Retails under $6,000

Table 7.7 – *AHP pairwise comparison matrix for tire-cutter example*

	Portable	User Fr.	Flexible	Durable	Total	Weighting
Portable	1.00	0.33	3.00	1.00	5.33	0.22
User Fl.	3.00	1.00	5.00	3.00	12.00	0.49
Flexible	0.33	0.20	1.00	0.33	1.87	0.08
Durable	1.00	0.33	3.00	1.00	5.33	0.22

WEIGHTING THE OBJECTIVES

If weights for each of the objectives are required, the AHP matrices can be used.[4] Similar to the rank-order formulation, comparisons should only be made at the same level within the hierarchal structure, and lower-level objectives should only be compared with those under the same upper-level objective. Table 7.7 illustrates the AHP pairwise comparison matrix for the four upper-level objectives and the resulting weights. Similar weight calculations were performed on the lower-level *user-friendly objectives*. The lower-level objectives of *portable* and *durable* previously found to be equal and are thus each assigned a weight of 0.5.

The weights for each objective are presented in the weighted hierarchal objective lists illustrated in Table 7.8. Next to each objective is a pair of numbers denoting (Absolute weight, Relative weight). At each hierarchy level, the first number, absolute weight, is the overall weight of that objective compared to all other objectives at that level. The second number is the relative weight (really applies only to lower-level objectives) of that objective compared to other lower-level objectives directly under the same upper-level objective. For example, *safe* has the weight pair (0.147,0.3). The absolute weight (0.147) is the product of the weight assigned to *safe's* upper level objective *user friendly* (0.49) and *safe's* relative weight (0.3). *Safe's* relative weight is obtained by comparing only the lower objectives under *user friendly*. Note that lower level objectives could be further broken down into subsub-objectives, subsubsub-objectives, and so on. The deeper the level of decomposition, the less importance should be attached to the absolute weight values.

Knowing whether to rank-order or weight-order the objectives will depend on the needs of the design team and the problem at hand. Always keep in mind that the numeric weights are based on *qualitative* decisions, and therefore small differences in weights between two alternatives may not imply a clear difference between them.

7.4 Revised Problem Statement

The initial problem statement formed the basis from which the *customer needs* were identified through the process discussed in this chapter. At the conclusion of that process, the design team should have a better understanding of the problem and have gathered and generated significantly more information than they

---

[4]*See* Section 6.3 on Page 135 for a discussion on AHP.

Table 7.8 – *Weighted hierarchal objective list for tire cutter example*

1. Portable (0.22,0.22)

    1.1 Easy to load on and off truck (0.11,0.5)
    1.2 Light (0.11,0.5)
    F.1 collapsible
    C.1 Small footprint
    C.2 Able to fit through standard doorway

2. User friendly (0.49,0.49)

    2.1 Low noise (0.0196,0.04)
    2.2 Safe (0.147,0.30)
        F.2 Protection from cutting devices
        F.3 Debris contained
    2.3 Easy to load and unload tires (0.098,0.20)
    2.4 Minimal debris (0.0196,0.04)
    2.5 Fast cutting operation (0.0539,0.11)
    2.6 Small cycle time between each tire (0.0539,0.11)
    2.7 Easy to operate (0.098,0.20)

3. Flexible (0.08,0.08)

    3.1 Adaptable to a wide range of tire sizes (0.08,0.08)
    F.4 Cut side walls and tread

4. Durable (0.22,0.22)

    4.1 Easy maintenance (0.11,0.5)
    4.2 Relatively maintenance free (0.11,0.5)

C.3 Retails under $6,000

---

had at the start. All this knowledge should then be condensed and incorporated into a *revised problem statement*.

7.5   Summary

The first stage of the design process, problem definition and determination of need, seeks to ensure that the design team is (a) solving the right problem and (b) developing a product or process that will meet the expectations of the client or customers. An overview of the main steps in the stage of the process have been elaborated upon using a tire-cutter example. On successful completion of this stage, the design team is well prepared to begin the next stage of the design process: conceptualization.

References

Domb, E., "Finding the Zones of Conflict: Tutorial", TRIZ Journal, no. 6, 1996.

Dym, C.L. and Little, P., *Engineering Design: A Project-Based Introduction*, 2nd Edition, New York: John Wiley and Sons, 2003.

Pahl, G. and Beitz, *Engineering Design. A Systematic Approach*, 2nd Edition, London: Springer-Verlag, 1996.

Ulrich, K. and Eppinger, S., *Product Design and Development*, 3rd Edition, New York: McGraw-Hill, 2003.

Bibliography

Dym, C.L. and Little, P., *Engineering Design: A Project-Based Introduction*, 2nd Edition, New York: John Wiley and Sons, 2003.

Ulrich, K. and Eppinger, S., *Product Design and Development*, 3rd Edition, New York: McGraw-Hill, 2003.

Chapter 8

Conceptualization I: External Search

8.1 Introduction

The next two chapters focus on the second stage of the design process, *conceptualization* (Figure 8.1). This stage can be divided into two broad sequential steps: external and internal search. The external search seeks existing solutions or partial solutions to the overall problem or to sub-problems of interest. The internal search, which is covered in the next chapter, uses the information obtained from the external search to generate solutions to the design problem. The internal search culminates in the selection of a few candidate designs that are then developed further.

Concept generation is the creative, inventive and some would argue the most difficult part of the engineering design process. *A concept is a very preliminary description of the form, required principles and required technology for the solution.* It is normally expressed as a two- or three-dimensional sketch with a brief accompanying description. Successful concept generation follows a structured approach to ensure that the entire design space has been adequately explored for different design alternatives. Common errors made by design teams during concept generation include (Ulrich and Eppinger, 2003)

- Considering only a few - typically one or two - alternatives, often suggested by the most vocal team members.

Figure 8.1 – *The second stage of the design process*

163

- Ignoring existing concepts found in similar and unrelated products.
- Lack of full-team participation during the conceptualization process, resulting in a lack of commitment by all members to the generated concepts.
- Poor integration of promising partial solutions to create a final concept.

Following a structured approach ensures that:

- Information is obtained from a wide variety of sources.
- Guidance is given in the exploration of alternatives.
- Mechanisms are in place for the integration of partial solutions.

The external search looks for existing solutions to the overall problem or any of the identified sub-problems. It reduces wasting resources by implementing existing solutions where possible, thereby allowing the design team to focus its energies on areas where no solutions exist. External searches are carried out by searching patents, consulting with experts, dissecting and benchmarking products, and searching the published literature, such as trade journals, journals, books, and articles on the Internet. In addition, consulting technical experts knowledgeable in one or more of the sub-problems can lead to significant insights. Technical experts include college faculty, industry professionals at companies that manufacture related products, and technical support personnel at companies that supply related products. At the conclusion of the external search the design team should have a good grasp of existing concepts that address the entire problem and each of the sub-problems.

8.2 Patents and Patent Searches

A patent is an official document provided by a government that provides specific rights to inventors and that exclude others from making, using or selling an invention. The patent allows the patent owner to prevent others from exploiting their patented invention. The actual patent consists of three essential components: a set of drawings (usually, but not always), an explanation of the invention, and the claims which define the extent of the intellectual property rights covered. Patents have a limited life, i.e., seventeen years in the US, after which anyone is free to exploit the information contained in it.

Patents are public documents and provide a valuable source for technical information, complete with technical drawings of ideas and concepts, that may or may not have been commercialized. These concepts may not appear anywhere else. Two online sources where free patent searches can be performed are

1. US Patent and Trademark Office: www.uspto.gov.
2. US patents, European patents A (applications) and B (issued), Japanese patent abstracts: www.patents.ibm.com.

Table 8.1 – *Portion of an art-function matrix for a child car seat. Adapted from Tasi and Tseng (2000).*

FUNCTION	ART				
	Ball, arc groove	Cam, spring teeth, lever, bar.	Slider, recess, groove	Recessed wall aperture, plate.	Bar, anchorage
Position indicator	US5058283				
Tightening strap		US5979982			
Safety strap guide			US5954397 US5964502	US5458398 US5286086	
Spring-back prevention					US5918934

According to the World Intellectual Property Organization (WIPO), patents cover 90%-95% of worldwide research results. Making good use of patents through thorough searches and analyses could reduce 60% of research time and 40% of research costs. Further, searching patents helps companies avoid getting into legal problems by inadvertently infringing on other's intellectual property rights (Idris, 2003).

As design teams search patents for solutions for each sub-problem, they should create an *art-function matrix* as shown in Table 8.1 for the example of child car seats. As relevant patents are found, the function performed by the patent is listed in the first column, and a new row is added for each function. The approach taken to meet the function (*the art*) for each patent is listed as subsequent column headings. For each function/art combination, the relevant patent is listed at the intersection. In this way, at a glance the design team can see the various approaches taken for each function related to the sub-problem. If the design team needs to look at a particular patent related to a function/art combination, the matrix quickly points them to the correct one(s).

8.3 Benchmarking

Benchmarking was developed in the late 1970s and early 1980s as a systematic process for quality improvement. Harrington and Harrington (1996) describe it as "a systematic way to identify, understand, and creatively evolve superior products, services, designs, equipment, processes, and practices to improve your organizations real performance". Three distinct methods of benchmarking are identified as

1. **Internal benchmarking** where all measurements and comparisons are done with an organizations own products or processes.

2. **Competitive benchmarking** which focuses on the products and processes that are in direct competition with those of the organization doing the benchmarking.

3. **Functional benchmarking** which focuses on functions rather than products or processes. It is more of a search for best practices in performing a particular function regardless of the product or process (Spendolini, 1992).

Major objectives of benchmarking pertaining to product design are:

1. Setting challenging but realistic product development, product performance goals
2. Defining how goals can be accomplished
3. Analyzing emerging technologies or practices
4. Searching for a breakthrough improvement

Case Study: Design of a Tire Cutter
The following case study is formed from excerpts from the final report of one of the capstone senior design projects at Rutgers University in 2003. It illustrates competitive benchmarking prior to concept generation. The student group members were Nicholas Malinoski, Helen Moore, Mark Telesz, Jared Roszko and Robert Wotring.

Benchmarking. As part of the external search, the design team looked at and tabulated products on the market that had similar requirements as the design problem and that cost less than $10,000. Three machines were investigated: a sidewall cutter and tire cutter from Tire Cutting and Recycling Equipment, Inc. (Figures 8.2 and 8.3, respectively) and a sidewall cutter from Glen Martin Engineering, Inc. (Figure 8.4). The design team was unable to find any machine within the specified price range that performed both tasks.

The capabilities of these machines were collected and tabulated as shown in Table 8.2. Information within the table was gathered from looking through the products' brochures, the companies' websites and talking to sales representatives. The information provided a baseline from which the design could be compared.

8.4   Product Dissection

Product dissection or teardown is a systematic process for taking apart and analyzing all components and subassemblies of a product (Otto and Wood, 2001). The type of analysis conducted during the dissection process can change based on the reasons for doing the dissection. For example, assembly analysis, material cost analysis or manufacturing processes and cost analysis may or may

Figure 8.2 – *Tire Cutting and Recycling Equipment, Inc. sidewall cutter*   Figure 8.3 – *Tire Cutting and Recycling Equipment, Inc. tread cutter*   Figure 8.4 – *Glen Martin Engineering, Inc. Sidewall cutter*

Table 8.2 – *Benchmarking for Tire Cutting Machines*

Feature	TC&RE Sidewall cutter	TC&RE Tread cutter	GME Sidewall cutter
Production rate (tires/hour)	150	150	240
Input Voltage (V)	120 AC	120 AC	120 AC
Weight (lbs)	300	290	310
Cuts side walls	Yes	No	Yes
Cuts treads	No	Yes	No
Retail Price ($)	4,495	4,495	3,999
Tire range	passenger car to light truck	passenger car to light truck	passenger car to light truck
Cutting mechanism	blade	disc	blade
Operation	hand	hand	foot pedal

not be necessary during the same dissection. The reasons for doing the dissection should direct the analyses included in the process. There are two primary reasons for product dissection: (1) benchmarking and (2) product redesign. In fact, product dissection is seen as a tool for product engineering and competitive analysis (Lamancusa and Gardener, 1999).

In the context of product redesign, dissection can be seen as a phase of product reengineering, where product reengineering is defined as "the examination, study, capture, and modification of the internal mechanisms or functionality of a product in order to reconstitute it in a new form and with new features, often to take advantage of newly emerged technologies, but without major change to the inherent functionality and purpose of the product" (Sage, 2002). Major objectives of the product dissection and redesign are

1. Improved maintainability
2. Improved reliability
3. Migration to a new technology
4. Functional integration, enhancement, or modification (Sage, 2002).

The relationship between product dissection, benchmarking and product redesign is shown in Figure 8.5. In the figure, benchmarking and product dissection are shown as ways to systematically uncover opportunities for product redesign.

Overall, the redesign process should start with internal benchmarking and the dissection of the product that is to be redesigned. This dissection will yield the current status of the product with its structure, properties, etc. Competitive benchmarking follows by dissecting competing products to uncover opportunities for product improvement. Finally, dissection of products that are not direct competitors but employ similar functions is performed and may yield additional product improvement opportunities.

Although all uncovered product improvement opportunities are analyzed in the redesign process, not all of them will appear in the final design as final features or be closely related to the companies business strategy, its resources and the needs of the population in the target market segment.

While what to dissect will change depending on the benchmarking activities (internal, competitive and functional), the major dissection steps remain the same. These include

1. Clarify and state the dissection mission
2. Get ready for dissection

   a) Record the manufacturer, model and the serial number
   b) List the design issues
   c) Shipping and handling examination
   d) Gather and prepare equipment to aid dissection

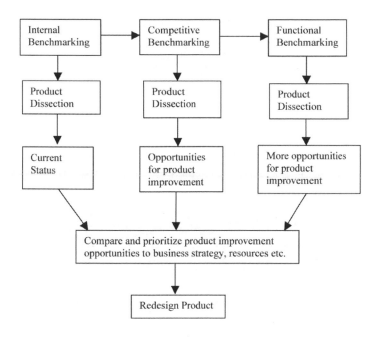

Figure 8.5 – *Product dissection and benchmarking for redesign*

3. Disassemble, measure, and analyze component functions
4. Form a bill of materials (BOM) and dissection report (Otto and Wood, 2001)

These steps are expanded on in the following sections, using the dissection of an electric toothbrush as an example, and drawing on ideas from Mickelson et al. (1995), Lamancusa et al. (1996), and Jorgensen et al. (1997).

Clarify and State the Dissection Mission

The most important step of product dissection is to clarify and state its mission. What are the objectives of this activity? What is expected to be uncovered? The answers to these questions will relate to product benchmarking and redesign, and might be similar to the objectives given in the benchmarking section. In addition, the dissection team might be given specific questions to answer. An example mission statement is shown in Table 8.3.

Getting Ready for Dissection

First, the manufacturer, model and serial number of the product to be dissected should be recorded. Then, design issues related to the product should be listed, i.e. list important performance criteria to be measured. In order to understand the design issues without disassembling the product, various answers to the questions related to the product and its operation should be recorded. Example questions include:

1. How many parts does the product have? Please list them.
2. What specific problems does this product solve?
3. What constraint limits have been placed on this model?
4. What safety features have been placed on this model?
5. How does this model work? Be specific.
6. What materials have been used to create this design? Why have they been chosen?
7. How might consumers use this design for a non-intended purpose? Would this damage the design? Could it increase salability?
8. What is the price range of the product?

While answering these questions, and if possible, the product should be operated if possible. Answers should be compiled as general product information. In addition, take measurements and sketch principal multiviews and a pictorial view of the product. Dimensioned multiview sketches for an electric toothbrush are given in Figure 8.6[1].

A customer needs analysis prior to dissection to provide specific information about the required product features and to rate the extent to which the dissected product meets them. For example, in Table 8.3 a preliminary list of product features is given for rating. These ratings, coupled with the general product information, will yield design issues. While examining product features, the means used to acquire and contain parts, as well as the shipping, distribution, and marketing of the product should also be recorded.

Finally, all tools required to complete the dissection should be identified and recorded in the dissection report. Tools might include test and measurement equipment, camera, multimeter, videotape, hardness tester, etc.

Disassemble, Measure, and Analyze Component Functions

In this step the assembly is taken apart. For each disassembled part, measurements are taken, and a solid model may be produced. Then, an exploded view of the assembly, using solid modeling or simply by taking pictures, is created. Several methods are available to help uncover subsystem functions and possible component redundancies or component combinations. Two common methods are (1) the subtract and operate procedure (SOP) and (2) force flow diagrams.

---

[1]Chaudhry, J., Moravek, M.K., Karch, A., and Gross, K., "Redesign of an Electric Toothbrush", 2004. Final report for class project in *Introduction to Engineering Design Course* at Penn State.

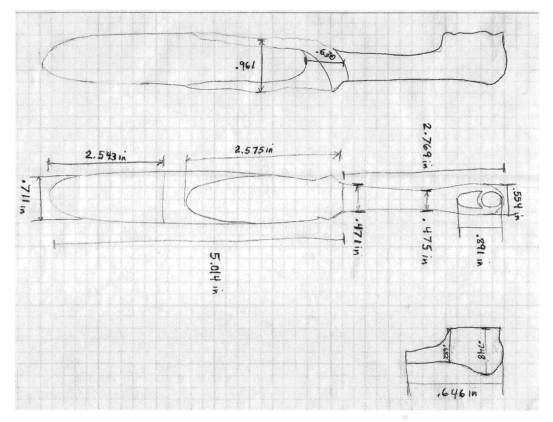

Figure 8.6 – *Multiview sketching while preparing for dissection*

The SOP seeks to find redundant components in an assembly through the identification of the true functionality of each component (Lefever, 1996). Steps of the procedure are

1. Disassemble one component of the assembly
2. Operate the system and verify the effect of the removal of the component
3. Analyze the effect. If the removal has no apparent effect, then leave it off and continue on to the next component
4. Deduce the subfunction of all the missing components
5. Replace each component, and repeat steps 1-4 n times, where n is the number of pieces in the assembly

Force flow diagrams represent the transfer of force through a product's components (Lefever, 1996). Steps for constructing force flow diagrams are:

Table 8.3 – *Product dissection report*

Product Dissection
Project Name: Electric Toothbrush Dissection
Product Engineers: Jane Learner, James Hardworker
Date:
Background and Reference Materials: http://www.hyperdictionary.com/dictionary/electric+toothbrush http://www.loc.gov/rr/scitech/mysteries/tooth.html http://www.asme.org/mechanicaladvantage/March2001/toothbrush.html http://www.howstuffworks.com/question292.htm http://www.howstuffworks.com/motor.htm http://www.travelproducts.com/store/electric.htm Example manufacturer: Philips Co., http://www.philips.com

1. Dissection Mission:

Dissection Purpose: To do a competitive benchmark to uncover the power generation and conversion system of the product dissected.

2. Getting Ready for Dissection:

Manufacturer/Model Number:

General Product Information: This model comes with two spare brush heads, a travel pack. It is powered by magnetic induction. It also can be powered by batteries.

Features:	
Aesthetics	
Cleaning	
Control location	
Cord location	
Cord storage	
Cost	
Ease of control	
Handle (Ergonomics)	
Noise	
Operating time	
Power	
Quality	
Safety	
Tip over stability	
Versatility, attachments	
Weight	

Required Tools for Dissection: Camera, multimeter, videotape, hardness tester etc

1. Identify the primary force flows (or energy) transmitted through a product
2. Map the force flow from the external source through each component of the product until the flow exists to ground
3. Document the result in a force flow diagram, where the nodes are components and the connections are forces
4. Analyze the diagram and label relative motion between components with an "R"
5. Decompose the diagram into groups separated by "R"s, and place these component groups in a box
6. Deduce the subfunctions and affected customer needs for each group
7. Develop creative, conceptual designs to combine the components in each group
8. Repeat for each force flow.

Form a Bill of Materials and Report the Dissection

During disassembly, the team should tabulate details of the product components. An sample format is given in Table 8.4. In addition, the assembly and component pictures taken and an exploded view of CAD drawings should be completed and added to the dissection report.

8.5 Biomimicry

Nature provides a wealth of examples that can be emulated to solve engineering problems. Biomimicry (derived from *bios*, meaning 'life,' and *mimesis*, meaning 'to imitate) is a new science that gives nature a closer look during the engineering design process. Biomimicry can be defined as "a science that studies nature's models and then imitates or takes inspiration from these designs and processes to solve human problems." (Benyus, 1997). Through billions of years of evolution, nature has devised ingenious ways to solve various problems. Biomimicry attempts to learn from these time-tested patterns and strategies and adapt them using current engineering materials to solve design problems.

Examples of products developed through biomimcry include

1. *Velcro.* Inspired by barbs on weed seeds.
2. *Swim suits.* Speedo developed very low water resistance body swim suits for the 2000 Olympic Games in Sydney, Australia by carrying out extensive studies on shark skins.
3. *Color without paint.* Paints used to enhance the visual appeal of objects, also causes harm to the environment. In nature, organisms use two methods to create color: internal pigments and the structural color. The latter is used, for example, by tropical butterflies, peacocks, and hummingbirds

Table 8.4 – *Product dissection report (cont..)*

Product Dissection Report- cont.									
Project Name: Electric Toothbrush Dissection									
Product Engineers: Jane Learner, James Hardworker									
Date:									
3. Bill of Materials				Disassembly method: Subtract and Operate Procedure (SOP): Yes, No. Force (Energy) Flow Diagram: Yes, No.					
Part#	Part Name	QTY	SOP Effect	Function	Mass (oz, g)	Material	Manuf. Process	Dimensions	Cost

4. Assembly (exploded view), component pictures and/or solid models:

Final Remarks:

in clever ways. A peacock, for example, creates an illusion of color by scattering light off melanin rods, and interference effects through thin layers of keratin[2]. A San Francisco company, Iridigm, is using structural color ideas from tropical butterflies to create a PDA screen that can be easily read in sunlight (Benyus, 1997).

8.6 Summary

The external explores the patents, the literature and the existing products for possible solution fragments for the current design problem. It also ensures that the design team does not reinvent concepts that are already on the market or protected through patents. At the conclusion of the external search, the design teams must have a good grasp of existing concepts that address the entire problem and each of the sub-problems. Armed with this information the team can begin generating its own concepts to solve the problem, the internal search.

---

[2]Keratin is also found in fingernails.

References

Benyus, J., *Biomimicry: Innovation Inspired by Nature*, New York, William Morrow and Co., Inc.,1997.

Harrington, H.J. and J.S. Harrington, *High Performance Benchmarking: 20 Steps to Success*, New York: McGraw-Hill, 1996.

Idris, K., *Intellectual Property - Powerful Tool for Economic Growth*, Geneva: WIPO, 2003.

Lamancusa, J. and Gardner, J.F., "Product Dissection in Academia Teaching Engineering the Way We Learned it", Proceedings of the International Conference on Engineering Educationm 1999.

Lamancusa, J., Torres, M., Kumar, V. and Jorgensen, J. "Learning Engineering by Product Dissection" ASEE Annual Conference Proceedings, (1996).

Jorgensen, J.E., Fridley, J.L. and Lamancusa, J.S., "Product Dissection A Tool for Benchmarking in the Process of Teaching Design", Proceedings of the Frontiers in Education Conference, pp. 1317-1321, 1996b.

Lefever, D. "Design for Assembly Technique in Reverse Engineering and Design", Proceedings ASME Design Theory and Methodology Conference, 1996.

Mickelson, S.K., Jenison, R.D. and Swanson, N. "Teaching Engineering Design Through Product Dissection", Proceedings of the ASEE Annual Conference, pp. 399-404, 1995.

Otto, K. and Wood, K., *Product Design: Techniques in Reverse Engineering and New Product Development*, Upper Saddle River: Prentice-Hall, 2001.

Sage, A.P., "Reengineering", in AccessScience@MsGraw-Hill, http://www.accessscience.com, 2002, viewed on 1/11/2004.

Sembara, K., http://www.asme.org/mechanicaladvantage/March2001/ toothbrush.html, 2001, viewed on 1/11/2004.

Spendolini, M. *The Benchmarking Book*, New York: Amacom, 1992.

Tsai, C-C and Tseng C-H, "Using TRIZ for an Engineering Design Methodology Course at NCTU in Taiwan", Triz Journal, vol. 3, 2000.

Ulrich, K. and Eppinger, S., *Product Design and Development*, 3rd Edition, New York: McGraw-Hill, 2003.

Chapter 9

Conceptualization II: Internal Search and Concept Selection

9.1 Introduction

Reflection should be carried out throughout the design process, especially before synthesizing solution ideas for sub-problems to generate design concepts. After the generation of design concepts, the best concepts for the design problem at hand will be chosen keeping in mind the design objectives, project timeline and the budget. During, reflection questions to ask include (Ulrich and Eppinger, 2003)

- Are there alternative ways to decompose the problem?
- Are there alternative function diagrams?
- Have all external sources been exhausted?
- Have all team members' ideas been accepted and integrated into the process (even if they are later rejected)?
- Has the solution space been fully explored?

Answering the last question affirmatively is not possible without answering the question, Have all team members' ideas been accepted and integrated into the process? Answering this question requires paying attention to the collaborative design issues, such as conflict management, communication etc., as discussed in Chapter 3. In addition, it should not be assumed that *if team members have ideas they will bring them to the team's attention.* Rather, there should be a deliberate effort to establish the right climate for the free flow of ideas and to budget time in the project schedule for the internal search before concept selection is undertaken. Further, designing a better product or process is not possible by only exhausting external search sources; it requires innovative thinking of the design team. This chapter discusses several methods that have been developed to stimulate creative thinking. The chapter closes with a discussion of concept selection.

9.2 Internal Search

The internal search involves the generation of new concepts by the design team to address each of the sub-problems identified in the problem clarification. New or original concept generation requires a creative process. Simon (1984) describes the creative process as a pattern that is not simply logical, linear, or additive. Further, he asserts that the process is often intuitive and inductive, involving creative thinking, or cross associational of two or more in-depth chunks of experience, know-how, etc.

Creativity is defined as "a modifiable, deliberate process that exists to some degree in each of us. It proceeds through an identifiable process and is verified through the uniqueness and utility of the product created." (Ford and Harris, 1992). Simon (1984) suggests that it takes 10 years or more for people to accumulate what he calls "50,000 chunks of experience," which enable them to be highly creative and recognize patterns, or familiar circumstances that can be translated from one place to another.

Creative thinking does not occur at a point in time; rather, it is a process. Wallas (1926) described creativity as a seven-step process:

1. **Encounter.** Identification of a problem to be addressed
2. **Preparation.** Gathering information about the problem
3. **Concentration.** Putting forth effort to solve the problem
4. **Incubation.** When a course of action is not clear
5. **Illumination.** An idea or solution becomes apparent
6. **Verification.** Proving that the solution is appropriate
7. **Persuasion.** Convincing someone else that the solution is viable for the problem

Recently Motamedi (1982) proposed a very similar creative process, also with seven steps: framing, probing, exploring, revelation, affirming, reframing, and realizing. Whatever the steps' names, from the discussion above it is clear both that creativity in a person develops over time and that creative idea/concept generation follows a process. There are several methods, which aim to enhance creativity, that allow design teams to facilitate a creative process, and hence generate ideas. These include brainstorming and its variations, Delphi method, checklists and morphological charts[1]. A discussion of each of these methods follows.

---

[1] A more structured approach to concept generation, *Algorithm for Inventive Problem Solving* will be introduced in Chapter 10.

Brainstorming

Developed by Osborn (1963), brainstorming requires team members to work both as a team and individually to create as many ideas as possible as potential solutions for a problem. To ensure success, the following guidelines should be followed.

- Establish the goal of generating a certain number of ideas, or a certain number of ideas in a certain amount of time.
- Pursue quantity first. The more ideas put forth the more likely the design space will be fully explored. Non-feasible ideas may trigger better feasible ideas.
- Suspend criticism, judgment or evaluation of the designs to allow the generation of a large number of concepts.
- Build on the ideas of others.
- Use both graphical and physical media to explain the design concepts. For example, the use of sketches adds significantly move clarity to concept descriptions than words alone.

Contrary to common practice, formal studies have suggested that a group of people working independently for a fixed amount of time will generate more and superior concepts than the same group working collectively for the same period of time (McGrath, 1984). *Design teams should therefore generate design concepts individually first, and then get together with their design teams to discuss and generate more ideas.*

Brainstorming Variations

Variation of brainstorming include Method 635, brainwriting, collaborative sketch, and gallery method. Developed by Rohrbach (1969), *Method 635* involves six people generating and writing down three potential solutions after studying a problem. After a set period of time, the solutions are handed to the participant's neighbor, who then either further develops the original three solution ideas or provides three new ones. The handing over of potential solutions to the next person continues until everyone in the group gets a chance to see each of the original three solutions and contribute to the document. Compared to brainstorming, Method 635 is more systematic, and while the originator of an idea is known it still allows everyone to take a look at the generated ideas and choose to develop or add to them. Finally, following Method 635 may generate less creative solutions because the mostly individual nature of the contributions does not take advantage of group discussions (Pahl and Beitz, 1996).

Brainwriting is similar to Method 635 except that there is no limitation on the number of people who are involved in the idea creation. As an improvement to both methods, Higgings (1994) suggests that individual columns be used to distinguish the ideas generated by each person in order to trace the development of concepts.

Collaborative sketch (C-sketch) method (Shah et al., 2001) is also similar to Method 635, except that its starting point is the initial presentation of a single design concept, as a sketch, by each of the team members.

Developed by Hellfritz (1978), the *gallery method* is conducted in the following steps:

1. Idea Generation: For 15 minutes the individual group members generate ideas and record them using sketches and text as they see fit.
2. Association: The results of idea generation step are hung on a wall as in an art gallery for everyone to see and discuss.
3. Idea Generation: The ideas and insights from step two are further developed individually by each of the group members.
4. Selection: All generated ideas are reviewed and classified before the most promising one is selected.

This method combines the individual work with positive group discussions, but in essence the organization and team building are similar to brainstorming (Pahl and Beitz, 1996). *Storyboarding* is a version of the Gallery Method, which uses 3x5 index cards to record the ideas and then hang them on the wall for the design team to visit and discuss each idea (Barr, 1988).

Delphi Method

Norman Dalkey and Olaf Helmer have been credited with developing the Delphi Method in 1953 in order to address military projects at the RAND Corporation (Lang, 1998). Since its creation, the method has been used frequently to make predictions, seek consensus and generate ideas. The technique recognizes human judgement as legitimate and useful input and allows experts to generate ideas systematically for a complex problem. An application of Delphi method involves the following steps (Fowles, 1978):

- Formation of a Delphi team: This team is responsible for choosing the experts, developing questions or problem statements, analyzing the responses and providing feedback, monitoring and reporting of the process.
- Selection of experts. Most studies use a panel of 15-35 people.
- Development of the first set of question or issues for idea generation.
- Transmission of the first set of questions to experts.
- Analysis of the first round of responses and feedback.
- Preparation and transmission of the second set of questions.
- Analysis of the second round of responses.
- Resolution.

- Reporting.

For product design applications, response probing can be like the following (adapted from Pahl and Beitz, 1996):

- First round of questions: What initial concepts do you suggest for solving the design problem presented? Please make as many suggestions as you can.

- Second round of questions: Here is a list of potential design concepts for solving the given design problem. Please go through this list and make further suggestions.

- Resolution: Here is the final evaluation of the first two rounds. Please go through the list and write down what suggestions you consider most practicable.

The group interaction in the Delphi method is anonymous, in the sense that comments, ideas, etc. are presented to the group without revealing their originators.

The internal search methods presented so far do not require starting the idea generation sessions with anything but the design problem at hand. The following techniques, however, exploit the use of analogies, metaphors, and checklists as triggers of idea generation.

Synectics

Developed by Gordon (1961), this method uses analogies and metaphors trigger idea generation (Voland, 2004). The method is based on the fact that the mind is more productive when dealing with a new or foreign environment. The analogous situation takes the individual away from the exact problem at hand and requires him/her to consider a related problem. This has a tendency to make the strange familiar, or the familiar strange. Synectics uses four different types of analogies:

1. Direct analogy: the current problem is directly related to a similar problem which has been solved.

2. Fantasy analogy: imagining that a solution already exists.

3. Symbolic analogy: use metaphors and similes.

4. Personal analogy: immersing self in the problem and viewing the problem from a different perspective.

Checklists and SCAMPER

The Checklist Method, developed by Osborn (1963), uses words and questions to trigger idea generation. While in the generic checklists there is no fixed set of questions, there is a version of checklists which is frequently used for product design problems, SCAMPER.

SCAMPER, developed by Michalko (1991), is a checklist that helps facilitate a creative process. The process follows the following steps:

- **Substitute (S).** What could you substitute?
- **Combine (C).** What could you combine?
- **Adapt (A)** What could be adjusted to suit a purpose or a condition?
- **Modify (M).** What would happen if you changed form or quality?
- **Put (P) to other uses.** How could you use it with a different purpose?
- **Eliminate (E).** What could you subtract or take away?
- **Reverse (R).** What would you have if you have reversed it?

A design team can use the process with the following questions to make changes to an existing product or to create a new one:

- Substitute (S) - energy, materials, signal, people
- Combine (C) - mix, combine with other assemblies or services, integrate
- Adapt (A) - alter, change function, use part of another element
- Modify (M) - increase or reduce in scale, change shape, modify attributes (e.g. color)
- Put (P) to another use
- Eliminate (E) - remove elements, simplify, reduce to core functionality
- Reverse (R) - turn inside out or upside down.

Morphological charts

One other tool for developing new concepts is the morphological chart. In this method, feature-based or functionally decomposed sub-problems, and possible solution ideas for each problem, are placed in a matrix. The chart's structure lists all the sub-problems or functions in the first row of the table and proposed solution concepts or means to achieve those functions in successive rows, as illustrated in Table 9.1[2].

Morphological charts can be populated entirely with text, graphics or a combination of both to depict the conceptual means to achieve each function. Each

---

[2]Whether the functions are listed in the first row and the means across columns or vice-versa is up to the design team. The morphological charts can be created either way.

overall design concept consists one entry from each column. Because of its structure, a morphological chart allows systematic exploration of the solutions' combinations. The use and structure of the morphological chart will be explained in the context of the following case study.

Case Study: Design of Light Tunnels for Passive Indoor Lighting

The following case study is formed from excerpts from the final report of one of the capstone senior design projects at Rutgers University in 2000. It illustrates use of morphological charts for systematic concept exploration after the selection of the top few concept solutions for each sub-problem. The student group members were James Celestino, Hank Daniecki, Mark Daniszewski, Eric Sermabekian, Andrew Thoms, Jey Won and Vernon Ghee.

Introduction. The Light Tunnel project involves designing and building a system for providing natural light into existing indoor structures. The use of natural sunlight as a supplement to or replacement for conventional lighting systems during daylight hours has a number of advantages, including lowering overall power consumption and aiding in plant growth. Current methods for introducing natural sunlight indoors include skylights and light pipes, both of which hold quite a few disadvantages. These methods involve cutting away large portions in roofs and thus require specialized, often expensive, installations. Both skylights and light pipes operate on the principle of creating a pathway for light to travel through. The Light Tunnel will be superior to these methods in that it will first focus the sunlight, then transport, and then diffuse the light at the target room. The problem was functionally decomposed into three sub-problems: collection of sunlight, transport of light, diffusion of light in room (Figure 9.1).

Morphological Charts. After idea generation, the most promising concepts for each of the three sub-problems were tabulated to create a morphological chart (Table 9.1). From the chart, different combinations of solutions could be investigated and evaluated to determine an overall solution concept. Five such combinations are illustrated in Figure 9.2. A few comments are warranted on some of the presented solution fragments. First, Fresnel lenses, shown in profile in the morphological chart, are used in lighthouses and overhead projectors and are well suited for energy directing systems[3]. Fiber optics, used in information technology, operate under the concept of light (as an electromagnetic wave) being 100% reflected at the interface of the glass or plastic tube and its surrounding cladding[4]. A fiber optic bundle can be much smaller than a reflective tube that transports the same amount of light. Finally, TIR lenses, shown in profile in the morphological chart, exploit the concept of total internal reflection to focus, diverge or collimate incident light, depending on the specific lens used. TIR lenses are capable of higher collection efficiencies and smaller sizes than Fresnel lenses[5].

---

[3]Talura The Fresnel Lens, http://www.talura.dk/optics/fresnel.htm, 2000.

[4]Hecht, E., Optics, 3rd edition, New York:Addison Wesley, 1998.

[5]Parkyn, W.A., and Pelka, D.G., "Compact Non-Imaging Lens With Totally Internally

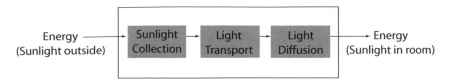

Figure 9.1 – *Functional decomposition diagram for light tunnel project*

Table 9.1 – *Morphological chart for light tunnel project*

Means	Function		
	Sunlight Collection	**Light Transport**	**Light Diffusion**
1	Fresnel lens	Fiber optic cables	Diffuser
2	Parabolic collector	Mirrors	
3	Prism	Reflective tubing	
4	Totally internally reflective (TIR) tubing		
5	Double convex lens		

In addition to being an idea generation tool, morphological charts can also be used to organize all ideas generated in a systematic fashion before the final concept selection. At the end of the concept generation, which is completed by exhausting both external and internal sources, several feasible solutions to each of the identified sub-problems can be organized into a morphological chart. However, it would be impractical to look at every possible solution combination generated from the developed concepts. For example, if three concepts were finalists for each of six sub-problems in a design, there would be a total of 729 (3x3x3x3x3x3) theoretically possible solutions! It would be impractical to

---

Reflecting Facets Nonimaging Optics: Maximum Efficiency Light Transfer, SPIE, San Diego, California, 1991.

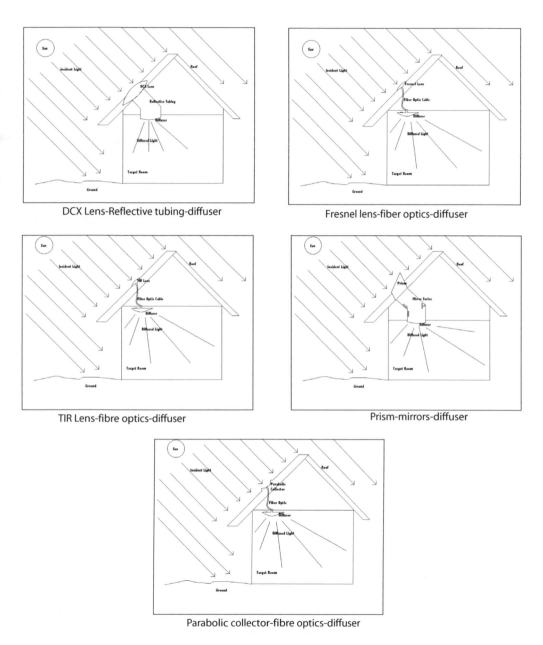

DCX Lens-Reflective tubing-diffuser

Fresnel lens-fiber optics-diffuser

TIR Lens-fibre optics-diffuser

Prism-mirrors-diffuser

Parabolic collector-fibre optics-diffuser

Figure 9.2 – *Five solution combinations from morphological chart*

evaluate the merits of each combination[6]. Therefore, a preliminary rating for each sub-problem solution might be appropriate.

9.3 Concept Selection: Pugh Charts

At end of the internal search, a small number of the most promising concepts should be selected for each of the sub-problems. Pugh charts, discussed in the next section, are particularly well suited for this task as the concepts are not well developed at this stage, making absolute evaluations very difficult.

Pugh charts are one of several tools that are typically used by design teams to evaluate and select concepts in an orderly fashion, reducing the likelihood of selecting wrong concepts or eliminating promising ones (Pugh, 1981). The technique is an iterative process and allows the relative comparison of concepts against a list of evaluation criteria. The relative evaluation allows differentiation between concepts during the early stages of design where absolute information may not be available. Use of the charts is best explained in the context of an example.

De-Orbit Stage for the Hubble Space Telescope Designed in the 1970s and launched by the space shuttle Discovery in 1990, the Hubble Space Telescope (HST) operates from a low earth orbit 600 kilometers (375 miles) above the earth's surface capturing images of distant stars and galaxies that would be impossible from earth due to distortions from the earth's atmosphere.

As the HST approaches the end of its useful life, conceptual studies are underway at NASA to determine ways to safely de-orbit and crash it into the Pacific Ocean. One such study was conducted in May-August 2003 at NASA's Marshall Space Flight Center to develop and evaluate several De-Orbit Stage (DOS) concepts that would be launched, attached to the HST, activated and used to safely de-orbit it.

One of the DOS subsystems is secondary power used to power systems that are always on, for example avionics, heaters, on-board computers systems, etc. Three concepts were considered:

1. Use of only non-rechargeable batteries.

2. Use of solar arrays and rechargeable batteries.

3. A hybrid of the first two concepts.

Detailed descriptions of all three concepts are presented in Table 9.2. The three concepts were evaluated on four criteria:

1. Size - the effect of each concept on the de-orbiting stage size (smaller = better).

2. Cost - the lower the cost the better.

---

[6]Not every theoretically possible combination would be feasible.

Table 9.2 – *Description of the three secondary power supply options for the Hubble Space Telescope De-Orbit Stage, Shuttle Option*

Concept	Description	Power (days)
1	All Batteries Bank of x non-rechargeable batteries. Other conditions dictate a stage height of approx. 6 feet (Figure 9.3).	m
2	Solar panel and rechargeable batteries y recharge-able batteries and z body mounted solar panel area. The area required for the body mounted solar panels forces an increase of the stage height from 6 feet to 9 feet (Figure 9.4).	Infinite
3	Hybrid Bank of x non-rechargeable batteries AND $y/2$ recharge-able batteries with $z/2$ solar panel surface area. The required surface area is $1/2$ that of Concept 2 and the stage height can be reduced to 6 feet, dictated by other conditions.	$2m+$

Notes: For the same voltage output characteristics, deliverable battery capacity increases as the discharge current decreases. As the current discharge of the non-rechargeable batteries for Concept 3 is cut in half (as compared to Concept 1), battery capacity and therefore duration of power supply will more than double.

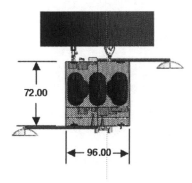

Figure 9.3 – *Illustration of the Hubble Space Telescope De-orbiting stage sized for use with only batteries. Visible are the internal fuel tanks and avionics locations, and the external communication antennas. The overall length of the stage is 72 inches dictated by factors other than the solar panel area (Drawing Sharon Fincher, NASA MSFC).*

3. Weight - contribution by each concept to the overall stage weight (lighter = better).

4. Power - duration of available secondary power (longer = better).

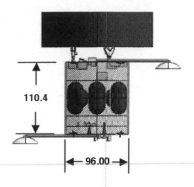

Figure 9.4 – *Illustration of the Hubble Space De-Orbiting Stage sized for use with body mounted solar panels and rechargeable batteries. The overall length of the stage is 110 inches, as dictated by the area required by the solar panels (Drawing Sharon Fincher, NASA MSFC).*

Pugh charts are used in an iterative manner. The charts are formed by listing each concept in a separate row in the first column with the evaluation criteria listed as headings of the successive columns (for example, Table 9.3). The basic five steps in using Pugh charts are outlined below in the context of the secondary power example.

1. Develop all concepts to the same level of detail and create a concept evaluation matrix.
2. Choose one of the concepts as the reference or baseline against which the others are compared. Evaluate each concept/criteria combination against the baseline, using '1 (superior), '0' (similar) or '-1' (inferior) ratings. For our example, Concept 3 was selected as the baseline for the first chart (Table 9.3).
3. Sum each row and evaluate the net score for each concept.
4. Change the baseline concept and repeat steps 2 - 3. A pattern of one or more strong concepts should begin to emerge. If not, your criteria are probably too ambiguous, or one or more concepts are subsets of each other. For our example, Concept 2 was selected as the second baseline. The generated Pugh chart is displayed in Table 9.4. Comparing Tables 9.3 and 9.4, one observes that after two iterations Concept 1 emerges as the strongest concept.
5. If one or more concepts remain strong, make them the baseline (if they have not been already) and redo the chart. If they still remain strong, they are likely to be your best candidates concepts. Concept 1 was strong after the first two iterations and was made the baseline. Repeating steps 2-3, a third Pugh chart was generated Table 9.5. Comparing Tables 9.3-

Table 9.3 – *Pugh chart for Hubble De-Orbit Stage concepts - Concept 3 baseline*

Concept	Size	Cost	Weight	Power	Sum
1	0	1	1	-1	1
2	-1[1]	-1[2]	-1[3]	1	-2
3	B	A	S	E	

Notes:

1. All solar requires double the surface area of concept 3. The craft height needs to be increased to account for this.

2. Solar panels are more expensive than batteries, and increased cost from larger craft size.

3. Large craft with more solar panels may be heavier than the weight of extra batteries on the battery only craft.

Table 9.4 – *Pugh chart for Hubble De-Orbit Stage concepts - Concept 2 baseline*

Concept	Size	Cost	Weight	Power	Sum
1	1	1	1	-1	2
2	B	A	S	E	
3	1	1	1	-1	2

Table 9.5 – *Pugh chart for Hubble De-Orbit Stage concepts - Concept 1 baseline*

Concept	Size	Cost	Weight	Power	Sum
1	B	A	S	E	
2	-1	-1	-1	1	-2
3	0	-1	-1	1	-1

9.5, Concept 1 still remains the strongest candidate after three iterations. Note that although Concept 1 fared slightly better than Concept 3, both should be further investigated due to the small difference between them. Based on all four stated criteria size, cost, weight and power Concept 2 clearly comes out as the weakest candidate in all three iterations.

Weighted Pugh Charts

Regular Pugh charts assume that all evaluation criteria are of equal importance and are therefore accorded equal weight throughout the analysis. For most situations, however, design teams require the flexibility to be able to rank or weight the importance of the evaluation criteria. Weighted Pugh charts assign weights to each of the evaluation criteria relative to the entire group. The weights, w_i, are between 0 and 1 with

Table 9.6 – *Use of AHP pairwise comparison matrix to generate weights for de-orbiting stage example*

	Size	Cost	Weight	Power	Total	Relative Order
Size	1.00	3.00	0.33	0.25	4.58	0.16
Cost	0.33	1.00	0.25	0.17	1.75	0.06
Weight	3.00	4.00	1.00	0.50	8.50	0.31
Power	4.00	6.00	2.00	1.00	13.00	0.47

$$\sum_{k=1}^{m} w_i = 1 \tag{9.1}$$

where m is the number of criteria.

The AHP pairwise comparison matrices provide a convenient way to determine the weights associated with each criteria (*see* Section 6.3 on Page 135 for a refresher on AHP). In the context of the hubble de-orbiting stage, the AHP pairwise comparison matrix and the weights associated with each of the design criteria are shown in Table 9.6.

The same process used to generate the Pugh charts as previously outlined is followed, except an additional row is added to show the weights associated with each evaluation criterion. Each concept's weighted total score, T_w is now obtained from

$$T_w = \sum_{k=1}^{m} w_i s_i \tag{9.2}$$

where s_i and m are the ith unweighted score and the total number of criteria, respectively. The three unweighted Pugh charts presented previously in Tables 9.3 to 9.5 were redone in **Excel** with the addition of the weights associated with each of the criteria. The new charts are presented in Tables 9.7 to 9.9.

With concept 3 as the baseline (Table 9.7), concepts 1 and 2 emerge inferior. With concept 2 as the baseline (Table 9.8), concepts 1 and 3 emerge equally superior. Finally, with concept 1 as the baseline (Table 9.9), concept 3 emerges as superior, with concept 2 as inferior. From the three iterations the concepts could be ranked as concept 3 (highest), concept 1, and concept 2 (lowest).

Whether to use weighted or unweighted Pugh charts will depend on the extent to which the different evaluation criteria can be differentiated during the early stages of the design process.

9.4 Summary

This chapter provided practical methods for design teams to complete an internal search as a part of the concept generation process and to select the best suited

Table 9.7 – *Weighted pugh chart for Hubble De-orbit stage - Concept 3 baseline*

Concept	Size	Cost	Weight	Power	Sum	Rank
Relative Order	0.16	0.06	0.31	0.47		
Concept 1	0.00	1.00	1.00	-1.00	-0.10	3
Concept 2	-1.00	-1.00	-1.00	1.00	-0.07	2
Concept 3	B	A	S	E		1

Table 9.8 – *Weighted pugh chart for Hubble De-orbit stage - Concept 2 baseline*

Concept	Size	Cost	Weight	Power	Sum	Rank
Relative Order	0.16	0.06	0.31	0.47		
Concept 1	1.00	1.00	1.00	-1.00	0.07	1
Concept 2	B	A	S	E		2
Concept 3	1.00	1.00	1.00	-1.00	0.07	1

Table 9.9 – *Weighted pugh chart for Hubble De-orbit stage - Concept 1 baseline*

Concept	Size	Cost	Weight	Power	Sum	Rank
Relative Order	0.16	0.06	0.31	0.47		
Concept 1	B	A	S	E		2
Concept 2	-1.00	-1.00	-1.00	1.00	-0.07	3
Concept 3	0.00	-1.00	-1.00	1.00	0.10	1

concept for the criteria at hand. Internal search methods stimulate members' creativity in order to help them arriving at superior ideas in quality and in quantity. The internal search is very important in that while an external search reveals what solutions are in already existence or what has been used in the past, truly original designs cannot be generated without creative thinking, and hence applying the internal search methods that stimulate it. One caution, however, is that a design team should not go on to the next step until after reflection the team can say "We have defined the solution space adequately."

Next was the concept selection stage. A number of practical methods are provided for this such as Pugh charts, weighted Pugh charts, pairwise comparisons, etc. Despite their numerical appearance, the design team should never overlook the fact that there are subjective ratings in concept selection. Therefore, sound engineering judgment should always be practiced. In addition, while making judgments, a design team should not seek consensus too quickly. There should be a free exchange of ideas and opinions between team members. The results of these exchanges, furthermore, should be subjected to careful, scientific scrutiny, even interrogation, as that is the best sign that the design team is functioning at its best.

References

Barr, V., "The Process of Innovation: Brainstorming and Storyboarding", *Mechanical Engineering*, November, 1988.

Ford, D.Y. and Harris, J.J.,III, "The Elusive Definition of Creativity", *Journal of Creative Behavior*, vol. 26 no. 3, pp. 186-198, 1992.

Fowles, J., *Handbook of Futures Research*. Westport: Greenwood Press, 1978.

Gordon, W.J.J., *Synectics: The Development of Creative Capacity*, New York: Harper and Row, 1961.

Hellfritz, H., *Innovation via Galerimethode*, Konigstein/Ts.: Eigenverlag 1978.

Higging, J.M., *101 Creative Problem Solving Techniques: The Handbook of New Ideas for Business*. Winter Park, FL: New Management Pub. Co., 1994.

Lang, T., "An overview of four futures methodologies", http://www.soc.hamaii.edu/~future/j7/LANG.html, 1998. Viewed June 2, 2004.

McGrath, J. E., *Groups: Interaction and Performance*, Englewood Cliffs, NJ: Prentice-Hall, 1984.

Michalko, M., *Thinkertoys*. Berkeley: Ten Speed, 1991.

Motamedi, K., "Extending the Concept of Creativity", Journal of Creative Behavior, vol. 26, no. 3, pp. 75-88.

Osborn, A.F., *Applied Imagination: Principles and Procedures for Applied Problem-Solving*", 3rd Edition, New York: Charles Schribner & Sons, 1963.

Pahl, G. and Beitz, W., *Engineering Design: A Systematic Approach*, 2nd Edition, London, UK: Springer-Verlag, 1996.

Pugh, S., "Concept Selection - A Method that Works", *International Conference on Engineering Design*, Rome, 1981.

Rohrbach, , B., "Kreativ nach Regeln - Methode 635, eine neue Technik zum Losen von Problemen", Absatzwirtschaft, vol. 12, pp. 73-75, 1969.

Shah, J., Vargas-Hernandez, N., Summers, J., Kulkarni, S., "Collaborative Sketching (C-Sketch): An Idea Generation Technique for Engineering Design", Journal of Creative Behavior, vol. 35, no. 3, pp. 168-198, 2001.

Simon, H. "What We Know About the Creative Process", Working Paper, Carnegie Mellon University, 1984.

Ulrich, K. and Eppinger, S., *Product Design and Development*, 3rd Edition, New York: McGraw-Hill, 2003.

Voland, G., *Engineering By Design*, 2nd edition, Upper Saddle River: Pearson-Prentice Hall, 2004.

Wallas, G., *The Art of Thought*, London: Watts, 1926.

Chapter 10

Systematic Innovation with TRIZ - Theory of Inventive Problem Solving

"[Altshuller] captured the essence of inventions from the past so there is a knowledge-based education people can readily tap into. TRIZ helps you to analyze the system.... Everything is a system and can be analyzed, and anyone can come up with ideas for how to improve them. I have seen engineers become addicted to it." (McGraw, 2004)

Zion Bar-el
CEO, Ideation International Inc.

10.1 Introduction

In the mid-1990s Ford Motor Company engineers were confounded by a design problem with a new model of the Ford Escort: incorporating an airbag into the steering wheel column resulted in unacceptable levels of steering wheel vibration during engine idle, which would certainly hurt sales and increase warranty costs. After unsuccessfully trying to come up with solutions through traditional brainstorming techniques amongst the design team members – for example the team tried adding a lead block to the steering column without success – management hired a TRIZ (pronounced 'trees) consultant to assist. Working with the consultant, engineers were able to relatively quickly come up with a solution. By attaching the airbag to the steering column with flexible connectors, they were able to use the airbag itself as a damper, significantly reducing the vibrations without increasing the weight (and cost) of the steering column (Raskin, 2003).

In 2000, the imaging and printing group at Hewlett-Packard had a very short time to introduce their next line of printers, including the Deskjet 990C. With very little testing time, they needed to predict how the printers output mechanism could fail during use and make appropriate design changes to mitigate against those failure modes. The design team turned to a TRIZ-based software tool, Invention Machines TechOptimizer, which resulted in several design changes being implemented in the final design. The Deskjet 990C printer was one of HPs besting selling printers in 2001 (Raskin, 2003).

So what exactly is TRIZ? TRIZ, the Russian acronym for Theory of Inventive Problem Solving, is a systematic approach to generating innovative designs solutions to seemingly intractable problems. It was first developed in Russia by Genrich Altshuller (*see* brief biography in Table 10.1) and is now used across the world. It is based on the analysis of thousands of Russian patents in the early sixties and seventies[1]. These original analyses distilled numerous solution patterns found across patents that can be successfully applied to solve new problems. In recent years numerous researchers have begun to analyze worldwide patents in all fields and to update the TRIZ tools.

TRIZ has been recognized as a concept generation process that can develop clever solutions to problems by using the condensed knowledge of thousands of past inventors. It provides steps that allow design teams to avoid the "psychological inertia" that tends to draw them to common, comfortable solutions when better, non-traditional ones may exist. With reference to Figure 10.1, a design team using TRIZ converts their specific design problem to a general TRIZ design problem. The latter is based on the analysis and classification of a very large number of problems in diverse engineering fields. The general TRIZ design problem points to corresponding general TRIZ design solutions from which the design team can derive solutions for their specific design problem. The power of TRIZ, therefore, is its inherent ability to bring solutions from diverse and seemingly unrelated fields to bear on a particular design problem, yielding breakthrough solutions.

The traditional approach to creativity using methods such as brainstorming, C-sketch, etc. (*see* Chapter 9) calls upon the designer to look inward for inspiration. This can be a daunting task. TRIZ, on other hand, invites the designer to use a ready pool of knowledge for inspiration (Ganguly, 2003). It does not discount the use of traditional approaches. On the contrary, TRIZ ensures that design teams use these traditional methods in a systematic and directed manner by carrying out intelligent idea generation in areas where other people have solved similar general design problems.

The solution patterns identified have been synthesized into numerous tools that form TRIZ. These include

1. Technical contradictions and the contradiction matrix
2. Physical contradictions and the separation principles
3. Standard solutions

---

[1]Altshuller's work focused primarily on mechanically oriented patents. In recent years numerous researchers have begun to analyze worldwide patents in all fields and to update the TRIZ tools.

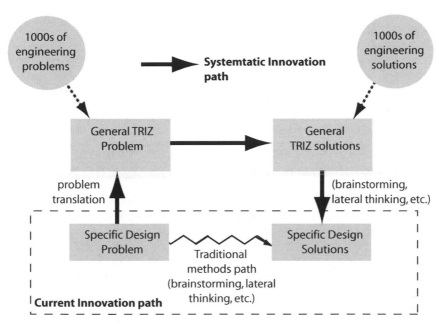

Figure 10.1 – *Generation of design solutions using TRIZ*

4. Laws of evolution
5. Physical effects

This introduction on TRIZ will focus on the first three tools and discuss how they can be used as an effective approach for structured concept generation.

10.2   Simplified Steps for Application of TRIZ Tools

To use the first three TRIZ tools effectively, the steps illustrated in Figure 10.2 should be followed. Despite the use of only three tools, the steps provide a powerful algorithm for structured concept generation. The steps in the algorithm are as follows:

1. Analyze the problem by **defining the contradiction zones** and creating an **EMS model**. This ensures that you understand the problem at hand and that you end up solving the right problem. In addition, at the end of this step you should have determined whether you have a physical or a technical contradiction(s). Define your **Ideal Final Result**.
2. If you believe you have a **technical contradiction(s)**, formulate it in terms of the **generalized engineering parameters**. Once complete,

Table 10.1 – Genrich S. Altshuller (1926-1998)

Genrich Altshuller, recognized as the father of TRIZ, was born on October 15, 1926 in Tashkent in the former Soviet Union. Altshuller was an avid inventor, receiving his first patent at the age of fourteen for an underwater diving device. His first major invention was in 1946 when he developed a way to escape from crippled submerged submarines without diving gear. The invention was immediately classified as a military secret. In the fifties and sixties, Altshuller analyzed over 200,000 patents and concluded that there were about 1,500 technical contradictions (similar to engineering trade-offs) that could be solved with a relatively small number of design principles. He published his first book in 1961, How to Learn to Invent, in which he criticized trial and error approaches to inventing and introduced readers to his first 20 inventive TRIZ principles. He published a second book in 1969, Algorithm of Inventing, in which he described 40 design principles and outlined the first algorithm to solve complex inventive problems. He published several more books in which he gave updated versions of the algorithms and discussed new inventive principles.

Altshuller wrote that, "Although people who had achieved a great deal in science and technology talked of the inscrutability of creativity, I was not convinced and disbelieved them immediately and without argument. Why should everything but creativity be open to scrutiny? What kind of process can this be which unlike all others is not subject to control? . . . . . . .What can be more alluring than the discovery of the nature of talented thought and converting this thinking from occasional and fleeting flashes into a powerful and controllable fire of knowledge." (Altshuller, 1988).

use the **contradiction matrices** to seek the most probable **design principles** to solve the problem. Recall that the contradiction matrices list the most probable solutions to solve your problem. If none of the recommended design principles work, go through each of the other principles in search of a solution. If you find a solution, you are done.

3. If no solution is found from the previous step or if the problem cannot be formulated as a technical contradiction, define the problem in terms of a **physical contradiction**. Redefine your **Ideal Final Result** in terms of the physical contradiction.

4. Apply the **Condensed Standards** to seek a solution. If a solution is found, you are done.

5. If not, use the **separation principles** to separate the physical contradictions. Apply the **Condensed Standards** to solve the new form of the problem. If a solution is found, you are done.

6. If not, revisit Step 1, and ensure the problem was defined correctly. Seek alternate forms of the contradictions and repeat all steps until a solution is found.

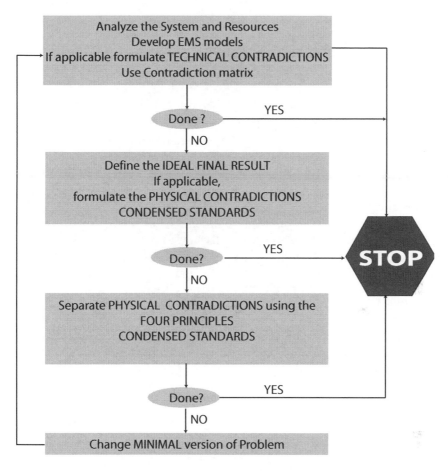

Figure 10.2 – *Flow diagram of a simplified algorithm for use of three common TRIZ tools*

Each of the terms referenced in the algorithm will be defined and discussed in the following sections.

10.3 Analyzing the System and Resources

Substance-Field Analysis (SFA) and Variants

A key concept in TRIZ is the modeling of all material objects (visible or invisible) as *substances* and all sources of energy (mechanical, chemical, nuclear, thermal, acoustic, etc.) as *fields*. A *function* (also known as substance-field) can therefore be defined as a substance, S_1, acted upon by a field, F_1, created by a second substance, S_2. The substance-field for a complete system can be represented with the notation

$$S_2 \frac{F_1}{description} > S_1 \qquad (10.1)$$

where the arrow shows S_2 having a positive or desired effect on S_1 through the field F_1. Note that in the TRIZ literature the graphical representations for the substance-fields varies greatly. Equation 10.1 merely presents a possible representation.

The parameters S_1 and S_2 are often referred to as *object* and *tool*, respectively, where the tool is acting on the object to create the desired effect. Models that do not have all three components (tool, object and field) are referred to as incomplete. By adding the missing element, a problem that may have been present in the system can be solved. Alternatively, if the tool has a harmful effect on the object, the straight field line would be wavy to indicate that harm is being done.

Royzen(1999) proposed the use of the Tool-Object-Product (TOP) analysis, a variant of SFA, as the next generation modeling approach. In TOP analysis a complete system has four elements: tool, object, field and product. The latter is defined either as a useful product (UP) or a harmful product (HP). The TOP analysis for a complete system can be represented with the notation

$$T \frac{F_1}{description} > O \quad => UP \qquad (10.2)$$

Equation 10.2 states that the tool creates the desired effect on the object via a field to produce a useful product.

Despite the appeal of both the SFA and the TOP models, they both require engineering designers to learn new modeling techniques, conventions and nomenclature and may therefore present a barrier to adoption. The following section will introduce energy-material-signals (EMS) models as a substitute modeling approach to SFA models. EMS models are based on black-box models for the process of problem decomposition introduced in Chapter 8.

The development of a good problem model, prior to concept generation, gives design teams a good understanding of the problem, including the relevant materials, systems, signals and sources of energy, and their interaction with each other. It is only with a clear understanding of the design problem that a good solution can be found.

Energy-Material-Signal Models in TRIZ

The EMS model extends the black-box model by incorporating symbols that indicate harmful and insufficient energy, material and signal flows within the system. In addition, symbols are also included to allow the modeling of multiple scenarios and discrete time-separated events. Table 10.2 lists the new symbols with a corresponding description. The use of the EMS model will be explained via several examples.

EXAMPLE 1: ICE DISPENSER

Developed from Tsuchikawa (2001). An ice maker/dispenser (Figure 10.3) consists of an ice-maker **(1)** with an ice-making cylinder **(1a)** located at the center and an ice storage compartment **(18)** on top of the ice maker. In the ice storage compartment, an agitator **(11)** is actuated periodically to prevent the ice from melting and sticking together to form large chunks that could prevent the smooth dispensing of the ice and that could result in the deterioration of the ice quality. The agitator rotates under three independent circumstances: (1) each time the ice is dispensed, (2) after a set period of time, and (3) each time the ice-making process is stopped (storage compartment is full). Situations arose where the agitator ran two or three times consecutively. The agitation could last twice as long, for example, if ice is dispensed right after the ice-making process stopped. On the other hand, it could last three times as long if, for example, the pre-timed agitation is initiated right after ice is dispensed, which occurs right after the ice making process is stopped. The extended run of the agitator results in some melting of ice due to the frictional heat generated between the ice cubes and the agitator fin **(11a)**, causing some fusion of the ice that the agitation was supposed to prevent. An EMS model of the problem is shown in Figure 10.4. In addition to showing the *harmful effect (mechanical energy)* between the agitator and the ice, the other relevant resources in the system are clearly indicated. Note that the EMS model contains materials (ice in storage compartment), subsystems (agitator, timer, ice-dispenser) and functions (accept external energy, make ice cubes).

EXAMPLE 2: FLUIDIZED BED COMBUSTION FURNACE EROSION

Developed from Lee et al. (2002). A fluidized bed combustion (FBC) furnace is illustrated in Figure 10.5. Coal is introduced into the furnace via a loop seal where it undergoes combustion. Coal that is not completely combusted is fed back into the heat cyclone from where it is returned to the furnace. High pressure air is pumped into the bottom of the furnace to facilitate the combustion of the coal. The furnace, however, experienced frequent shut-downs as a result of excessive erosion on the furnace walls from being bombarded by unburnt coal due to the high pressure air. An EMS model of the problem is illustrated in Figure 10.6. Note that the EMS model can also be used to model voids (the furnace interior).

EXAMPLE 3: AUTOMOBILE AIRBAG

Despite the success of airbags at saving lives, they have also resulted in numerous deaths to smaller occupants due to their deployment force. An EMS model

Table 10.2 – *EMS model symbols and description. System could mean an assembly, sub-assembly, function, user, and so on.*

Symbol	Description
S	The original system
S^c	A copy of the original system
S'	A modification of the original system
A	An additive can be material, energy, voids, systems, sub-systems or super-systems
E	The immediate system environment
- - - →	Signal flow
⟶	Material flow
→	Energy flow
[I]	When placed next to an energy, material or signal flow signifies that flow as being *Insufficient* to perform its desired task adequately.
⌒⌒	A wavy line representing any of the three flows (energy, material, signal) indicates that flow to be harmful to the system receiving it
↓	Placed above a flow indicates a decrease in the flow level from the original problem
↑	Placed above a flow indicates an increase the flow level from the original problem
⊣⊔⊢	Discrete sequential events
⊔	Each slot represents a different scenario and its effects on the EMS model

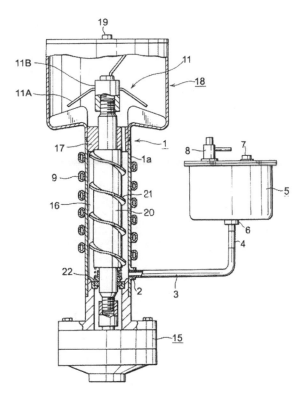

Figure 10.3 – *Schematic of Ice Dispenser Example (Tsuchikawa, 2001)*

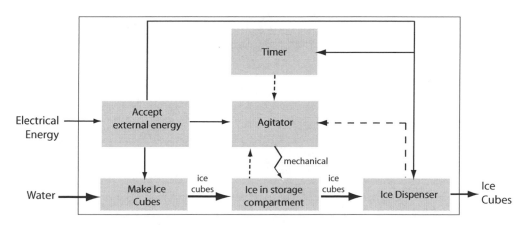

Figure 10.4 – *Energy-Material-Signal model of ice-maker showing* **harmful effect (mechanical energy)** *of agitator on the ice stored in the ice storage compartment*

201

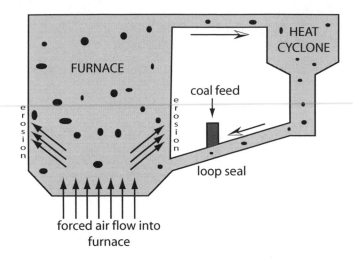

Figure 10.5 – *Schematic of fluidized bed combustion furnace*

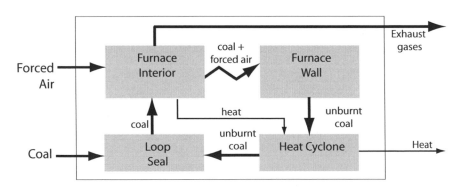

Figure 10.6 – *EMS model of fluidized bed combustion furnace showing the* **harmful effect (coal+forced air)** *on the furnace wall*

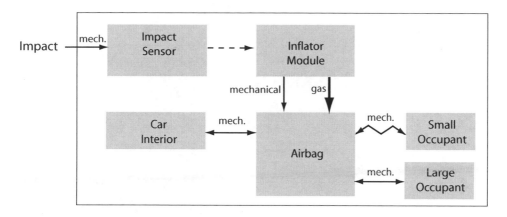

Figure 10.7 – *EMS model of a airbag system*

of the airbag system during a frontal impact is illustrated in Figure 10.7. A comparison of the black-box model and the EMS model shows that the main difference between the two is that in the EMS model, the generic occupant is now separated into a large and a small occupant, with the harmful effect of the mechanical energy on the small occupant shown. Within the context of the airbag system, one can clearly identify where the problem is that requires further attention. In addition, available system resources that could be used as part of the design solution are integrated into the problem clarification model. As such, a separate list of resources is not required, as is traditionally the case if SFA or TOP analysis methods are used.

To reduce airbag-caused fatalities, numerous automakers have installed de-powered airbags. The bags do not deploy as quickly as the original ones and therefore do not cause harm to small occupants. The problem with de-powered air bags, however, is that they are less effective during high speed crashes because the time it takes to achieve full deployment for the de-powered airbag is not fast enough to prevent the occupants (both large and small) from hitting the interior of the vehicle. The two scenarios, low-speed and high-speed collisions, are illustrated in the EMS model in Figure 10.8. Using the multiple scenario symbol, the low-speed impact scenario is represented by the top slots. In this scenario, the mechanical force from the airbag is sufficient to shield both types of occupants from colliding with the vehicle interior.

In the second scenario, high speed impact represented by the lower slots of the multiple scenario symbols, also results in airbag deployment. However, the mechanical force from the airbag is insufficient to prevent either occupant from colliding with the vehicle interior (harmful effect).

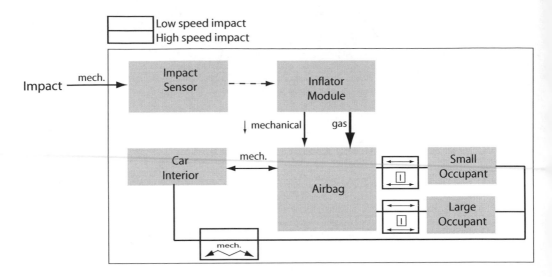

Figure 10.8 – *EMS model of proposed solution: De-powering of the airbag. This example illustrates the ability to model multiple scenarios in the same model.*

EXAMPLE 4: COMPUTER HARD DRIVE

Turning to the hard drive example, an area of concern arises when the computer is off and receives a hard external knock. Without the hard drive disk spinning, the head can be knocked off its rest position and data on the disk destroyed. In the rest position, the head is typically held in place by a magnetic latch. When the computer is powered on again, the airflow from the disk motion raises the head, and a permanent magnet/electro-magnet system situated at the arm axes of rotation (the pin) generates enough force to release the arm from the magnetic latch and to move the head to wherever data needs to be written or read (Royzen, 1999). An EMS model of this scenario is illustrated in Figure 10.9. In the model, one can track the sequence of events (the flow) from when the computer chassis receives a hard knock to the point where there is damage (harmful effect) to the disk surface by the read/write head. In the figure the magnetic field from the latch is insufficient, therefore indicating an area to be addressed by the design team.

A possible solution may be to use a stronger magnetic latch that would be more resistant to external shocks. This, however, may present its own problem by making it difficult for the arm to be released during start-up. The two scenarios are modeled in Figure 10.10, where the top slot in the multiple scenario symbols represents the hard knock scenario with the computer off, and the lower slot the computer start-up. Note that by increasing the magnetic strength of the latch, a desirable effect is achieved in response to reduction of damage from

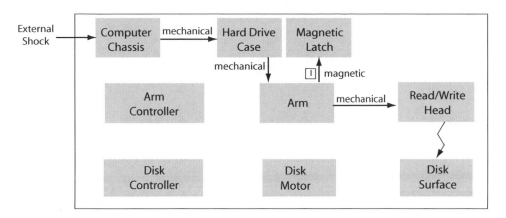

Figure 10.9 – *EMS model of hard drive when the computer is turned off. A hard knock on the computer dislodges the arm resulting in the head damaging the disk magnetic surface.*

external knocks, but it also produces an undesirable effect during system start up.

These examples illustrate how EMS models can be developed to show where undesirable effects occur in a system, focussing engineering designers' attention to those areas. In addition, the model presents the problem area in the context of the overall system allowing engineers to see what resources are available that may be used as part of a solution. Further, unlike traditional black-box modeling, the EMS models allow the inclusion of multiple scenarios within the same model.

Problem Clarification

TRIZ places great emphasis on making sure that design teams are solving the "correct" problem. Consequently, a significant portion of the approach focusses on analyzing the problem to get to the root cause of any contradictions (technical or physical) before embarking on a solution search. A technical contradiction is similar to the classic engineering trade off, i.e., as the design team improves one feature, another gets worse. A physical contradiction on the other hand is where a portion of the system is required to have two opposite characteristics or perform two opposite tasks. Each of these contradictions will be defined in greater detail later in the chapter.

There are two types of contradiction zones, (1) *operation zones* that define the space in which useful and harmful actions occur, and (2) *time zones* that define when the actions occur.

Domb (1997b) recommends following a journalism approach to defining these contradiction zones. This is similar to the approach presented in Chapter 8.

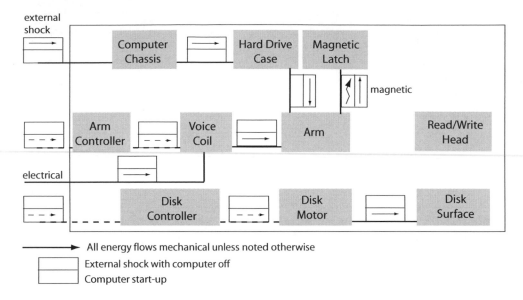

Figure 10.10 – *EMS model of hard drive with a stronger magnetic latch*

Revisiting the fluidized bed combustion furnace erosion example, the questions could be answered as follows

1. *Who has the problem?* LG Chemical Ltd, Korea.
2. *What does the problem seem to be?* Frequent shut down of the furnace due to degradation of the furnace wall. *What are the resources?* Presented in the EMS model in Figure 10.6.
3. *When does the problem occur? Under what circumstances?* Whenever the furnace has been in operation for an extended period of time with the forced air feed in operation.
4. *Where does the problem occur?* On the furnace wall adjacent to the air feed locations.
5. *Why does the problem occur? This question should be asked at least five times.* Why is there erosion on the walls? Because it is bombarded by fluidized coal particles. Why do the the coal particles collide with the wall? They are propelled by the air that is fed into the furnace. Why is air fed into the furnace? To facilitate the rapid combustion of the coal.
6. *How does the problem occur?* Coal particles are bombarded against the walls of the furnace by the compressed air that is fed into the system.

Answering the questions quickly reveals that the *contradiction operation zone* is at the inner walls of the furnace adjacent to the forced air feed, and that the *contradiction time zone* is whenever the furnace and the air feed are on. Further, from answering the questions one can conclude that this is a physical contradiction:

- The forced air speed should be high to facilitate rapid coal combustion
- The forced air speed should be low to prevent erosion on the walls

10.4   The Ideal Final Result

The Ideal Final Result (IFR) is a statement that defines the desired solution for the design problem, based solely on customer needs or functional requirements. It intentionally avoids the inclusion of current design features (if a design exists) to avoid biasing the concept generation. The IFR is stated in such a way that it

1. Removes the original system's deficiencies while preserving its strengths
2. Does not increase the original system's complexity
3. Does not introduce new deficiencies (Domb, 1997a)

Marconi (1998) suggests that the IFR can be stated in the following way:

> " **The resource** will eliminate the **negative effect** within the **operating zone** during the **operating time** without complicating the system while performing the **positive effect**."

where the bold items are substituted with a description of the current design problem.

The *resource* can be materials, energy, signals, systems, super-systems, and sub-systems already present in the current design or from the immediate environment, or it can be an addition to the system. In general, however, addition of new items into the design should be avoided. Different options are then substituted for the *resource* until reaching an IFR that adequately captures the problem to be solved. With reference to the Example 1: Ice dispenser, the IFR could be stated as

> "**A resource** will eliminate the **frictional heat generated by the agitator fins** within the **ice storage chamber** during the **time the agitator is in operation** without complicating the system while performing the **task of preventing ice-cubes from fusing together**."

In the context of Example 2: the fluidized bed combustion furnace erosion, the IFR can be stated as

> " **A resource** will eliminate the **erosion** within the **inner walls of the furnace adjacent to the forced air feed** during the **time the**

Table 10.3 – *General parameters used to describe engineering system metrics*

1	Weight of moving object	21	Power
2	Weight of stationary object	22	Energy loss
3	Length of moving object	23	Substance loss
4	Length of stationary object	24	Information loss
5	Area of moving object	25	Waste of time
6	Area of stationary object	26	Quantity of a substance
7	Volume of moving object	27	Reliability
8	Volume of stationary object	28	Accuracy of measurement
9	Velocity	29	Manufacturing precision
10	Force	30	Harmful action affecting the design
11	Stress or pressure	31	Harmful actions generated by the design project
12	Shape	32	Manufacturability
13	Stability of object's composition	33	User friendliness
14	Strength	34	Repairability
15	Duration of action generalized by moving object	35	Flexibility
16	Duration of action generalized by stationary object	36	Complexity of design object
17	Temperature	37	Difficulty
18	Brightness	38	Level of automation
19	Energy consumed by moving object	39	Productivity
20	Energy consumed by stationary object		

furnace is in operation with the forced air feed on without complicating the system while performing the **rapid combustion of the coal.**"

10.5   The 40 Design Principles

Characteristics of engineering systems can be described by a number of parameters that quantify or measure certain aspects of the design (metrics). Based on his analysis of patents, Althsuller developed a list of 39 such parameters (Table 10.3). He then reformulated the problems described by the patents in terms of these general parameters and noted their solutions. What began to emerge was that nearly all the solutions could be condensed into 40 general design principles.

As the next step, Altshuller postulated that if most design solutions to thousands of design problems could be condensed to 40 design principles, the reverse should be true: for a current design problem, the 40 principles can be used to find a solution. The principles are summarized in Table 10.2, and each is described in the following sections.

1 Segmentation

Similar to sub-division. Divide the object into as many sub-divisions as is practical.

Table 10.4 – *TRIZ 40 design principles*

1	Segmentation	21	Rushing through
2	Removal	22	Turning harm into good
3	Local quality	23	Feedback
4	Asymmetry	24	Go between
5	Joining	25	Self-service
6	Universality	26	Copying principle
7	Nesting	27	Inexpensive short life
8	Counterweight	28	Replacement of a mechanical pattern
9	Preliminary counteraction	29	Hydraulic or pneumatic solution
10	Preliminary action	30	Flexible or fine membranes
11	Protection in advance	31	Use of porous materials
12	Equipotentiality	32	Use color
13	Opposite solution	33	Homogeneity
14	Spheroidality	34	Discarding and regenerating parts
15	Dynamism	35	Altering an objects aggregate state
16	Partial or excessive action	36	Use of phase changes
17	Moving to a new dimension	37	Application of thermal expansion
18	Use of mechanical vibrations	38	Using strong oxidation agents
19	Periodic actions	39	Using an inert atmosphere
20	Uninterrupted useful action	40	Using composite materials

2 Removal

Remove the disturbing or the necessary part from the object. *See* Example 13 on page 223.

3 Local quality

Change the object's environment structure from homogeneous to non-homogeneous, or let different parts of the object perform different functions.

Example 5: Optical fiber sheathing tube. Developed from Stockman and Benton (1995). *Design Problem:* If an optical fiber or optical fiber cable is bent tightly, i.e. to a radius smaller than a critical value, there is a significant increase in the transmission loss. The fiber no longer acts efficiently as a waveguide for data transmission. These situations arise where the fiber may have to follow a tight bend, for example adjacent to a connector board. *Design Solution:* With reference to Figure 10.11, a sheathing tube (**12**) was designed to allow bending only to a predetermined radius but not smaller by incorporating portions in the sheathing tube walls that are spaced apart when the tube is straight (**18**), but that bump up against one another at (**20**) when the tube is bent to the minimum radius, thus preventing further bending.

4 Asymmetry

Make object asymmetrical or increase asymmetry in the object.

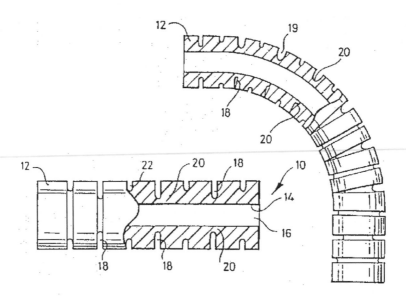

Figure 10.11 – *Design principle 3, Local quality: Schematic of optical fiber sheathing tube (Stockman and Benton, 1995)*

Example 6: Design of asymmetric cross-sectional diameter helical springs. Developed from Lin et al. (1993). *Design Problem:* With reference to Figure 10.12(a), helical springs (**1**) are often used to maintain the contact of a follower (**2**) on the cam (**3**) in a cam-follower system. Problems occur at high speeds when the cam angular velocity, and therefore the oscillation frequency of the follower, approaches the natural frequency of the spring resulting in a drop in the spring force near the highest follower position (Figure 10.12(b)). This position is where the maximum spring force is required to keep the follower in contact (**4**) with the cam as the follower begins to move downward. The force reduction can lead to loss of contact between the follower and the cam (**5**). A possible solution is to use a stronger spring, but the additional force increases the wear on the system and requires additional power. *Design Solution:* Use helical springs with asymmetrical (non-uniform) cross-sectional areas. Diameter variation is calculated to suppress the effect of the resonance on the force supplied by the spring within its operating frequencies.

5 Joining

Merge homogeneous objects or those intended for adjacent operations.

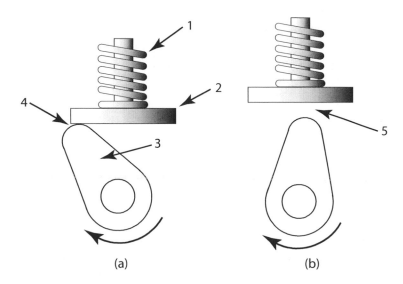

(a)   (b)

Figure 10.12 – *Design principle 4, Asymmetry: (a) Schematic of cam-follower system showing the use of a helical spring to keep the follower on the cam surface. (b) Near maximum lift position of follower illustrating 'follower-toss,' where the follower loses contact with the cam surface.*

6 Universality

Allow an object to perform several different functions. Remove redundant objects.

7 Nesting

Place an object in another, which in turn is placed within a third object, and so on. Let one object pass through the cavity into another.

8 Counterweight

Attach an object with lifting power or use interactions with the environment. *Examples:* Aerodynamic lift (environment).

9 Preliminary counteraction

Carry out a counter-action to the desired function before the desired-function is performed.

10 Preliminary action

Carry out the required action before it is needed, or set-up objects to be able to carry out their actions as soon as required.

11 Introducing protection in advance

Compensate for the low reliability of an object by introducing protections against failure before the action is performed.

Example 7: Casement window operator. Developed from Sheets et al. (1998). *Design problem:* Casement windows that have a sash which pivots outward along a vertical edge are well known. A crank is typically used to actuate a linkage for pushing the window open and pulling it shut. One problem with these casement window operator linkages is that they often suffer from window creep, the inability of the window to remain fixed at a given position. Creep can be caused by wind or other forces that may act against the window. *Design solution:* With reference to Figure 10.13, a friction bearing (**32**) interacting with the threaded screw drive member (**66**) - used to open and close the window - with a self-locking thread design holds the window at any position in either static load conditions or cyclical load conditions from wind and other forces eliminating window creep.

Example 8: Miniature flashlight. Developed from Maglica (1996). *Design problem:* High quality flashlights are commonly sealed for protection from moisture and other harmful environmental elements. Proper sealing is achievable with machined metallic flashlights that employ nonpermeable materials and can be constructed with reliable sealed joints. Occasionally, however, pressure builds up within the flashlight barrels due to outgassing from defective batteries. Designs have included vent holes or simple manufacturing imperfections have unintentionally created vent passages. Where moisture is considered to be a problem such vent holes may include a moisture impervious diaphragm to allow the passage of air but not moisture into and out of the internal chamber of the flashlight. Such devices are believed to be less than optimum in that various harmful elements in gaseous form can be drawn into the internal chamber of the flashlight. Further, such devices cannot resist substantial overpressure resulting from deep submersion or other equivalent conditions. *Design solution:* With reference to Figure 10.14, a one-way valve (**62**) is placed in the seal (**33**) between the tail cap (**22**) and the barrel (**21**). The valve prevents air flow from the outside into the interior of the flashlight, but allows overpressure within the flashlight to escape.

212

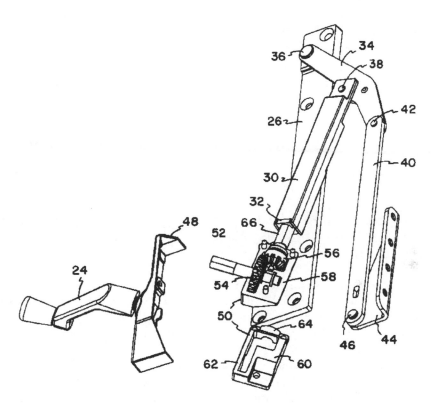

Figure 10.13 – *Design principle 11, Introducing protection in advance: Use of a friction bearing in a casement window operator (Sheets et al., 1998)*

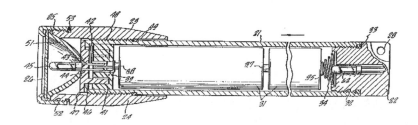

Figure 10.14 – *Design principle 11, Introducing protection in advance: Miniature flashlight (Maglica, 1996)*

213

12 Equipotentiality

Alter conditions such that the object does not need to be moved up or down in the potential field.

13 Opposite solution

Implement opposite action of that specified. Make a moving part fixed and a fixed part moving. Turn the object upside-down.

14 Spheroidality

Switch from linear to curvilinear paths, from flat to spherical paths, and so on. Make use of rollers, ball bearings and spirals. Switch from direct to rotating motion. Use centrifugal force.

Example 9: Shock absorbing damper device. Developed from Axathammer (1990). *Design Problem:* Vehicle suspension struts have both a shock absorbing and a wheel guiding function. They are therefore subject to axial and traverse loads, as well as bending moments. With reference to Figure 10.15, the piston rod (**3**) slides through a guiding sleeve (**4**) that provides friction (shock absorbing function) and restricts the piston to axial motion. The presence of traverse forces and bending moments, however, considerably increases the friction between the sleeve and the piston rod reducing driving comfort in the vehicle and resulting in high wear of the sleeve. *Design solution:* Add a set of guiding rollers (**6**) whose axis of rotation is perpendicular to the axis of the piston rod (**3**). The rollers resist traverse forces and bending moments with minimal resistance to axial motion. Further, the rollers are concave shaped increasing the contact area between the rollers and the piston rod, decreasing the engagement pressure. Lower pressure reduces the wear on both the piston rod and the rollers.

15 Dynamism

Allow the object or environment to become optimal at any stage of work. Make the object consist of parts that can move relative to each other. Make a fixed object moveable.

16 Partial or excessive action

If action not wholly attainable, try for slightly less or slightly more.

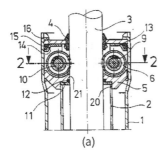 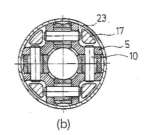

(a)　　　　　　　　　　　(b)

Figure 10.15 – *Design principle 14, Spheroidality: Schematic of shock absorbing or oscillation damper device. Figure (b) is the cross-section through the cutting plane 2-2 (Axathammer, 1990).*

17 Moving into a new dimension

Increase the object's degree of freedom. Use a multi-layered assembly instead of a single layer. Incline the object or turn it on its side. Use the other side of an area.

18 Use of mechanical vibrations

Make the object vibrate. Increase the frequency of vibration. Use resonance, piezovibrations, ultrasonic, or electromagnetic vibrations. *Examples:* Mechanical vibrators used to separate items.

19 Periodic actions

Use periodic actions, change periodicity, use pauses between impulses to change the effect.

20 Uninterrupted useful effect

Keep all parts of the object constantly operating at full power. Remove test or set-up runs.

21 Rushing through

Perform a process or individual stages of a process at high speed.

22 Turning harm into good

Obtain a positive effect from harmful factors. Remove a harmful factor by combining it with other harmful factors. Strengthen a harmful factor until it ceases to be harmful.

23 Feedback

Introduce feedback. If feedback is present, change it. *See* Example 1 on page 222.

24 Go between

Use an intermediate object to transfer or transmit the action. Merge the object temporarily with another object that can easily be taken away.

25 Self service

The system should service and repair itself. Use waste products from the system to produce the desired actions.

Example 10: Ash melting furnace. Developed from Fukusaki (1996). *Design problem:* In areas where municipal and industrial waste are disposed of by incineration, the residual ashes are buried in landfills. This has become increasingly difficult due to scarcity of suitable areas for landfills and as the environmental regulations applicable to both incineration and ash-burying become more stringent. As a result methods have been implemented to make slag out of the ash by melting the ash at a high temperature, followed by solidification. Slag is non-polluting and can be used in the construction, packaging and insulation industries. Incineration plants that adopted this approach run into high operation costs as electricity or oil was used as the main heat source for melting the ash. *Design solution:* Control the waste incineration so as to leave enough unburnt matter (e.g.,unburnt carbon) in the ash. Use the unburnt carbon as the major fuel (heat source) in the ash melting process, significantly reducing the use (and therefore cost) of electricity or oil.

Example 11: Remote control holder and illuminator. Developed from Galvin(1996). *Design problem:* TV, VCR and DVD remote controls are typically operated in low light environments making the selection of the appropriate key on the remote control difficult. Numerous remote control holders have since been developed but all require the use of an electric light source and batteries, both of which add significant cost to the device. *Design solution:* With reference to Figure 10.16, a magnifying mirror (**12**) reflects the low ambient light from the television (waste product in the system) onto the keypad (**101**) of the remote control (**100**) illuminating the keys.

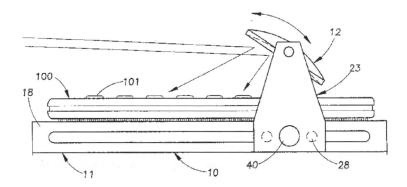

Figure 10.16 – *Design principle 25, Self service: Use of ambient light to view the buttons on a remote control in low-light conditions (Galvin, 1996)*

26 Copying principle

Use simplified, inexpensive copies instead of unavailable, complicated or fragile objects. Replace an object by its optical copy, make use of scale effects. If visible copies are used, switch to infra-red or ultra-violet copies.

27 Inexpensive short life instead of expensive longevity

Replace an expensive object that has a long-life with many inexpensive short-life objects. *Examples:* Paper plates, disposable diapers.

28 Replacement of a mechanical pattern

Replace a mechanical pattern with an optical, acoustical or odor pattern. Use electrical, magnetic or electromagnetic fields to interact with the object. Switch from fixed to movable fields that change over time. Go from structured to unstructured fields. *See* Example 14 on page 226.

29 Hydraulic or pneumatic solutions

Replace solid parts with gas or liquid parts. *Examples:* Air mattress, water-bed, water-jet massage machines.

30 Use of flexible and fine membranes

31 Use of porous materials

Make the object porous or use porous elements. For objects already porous, fill pores in advance with some useful substance.

Example 12: Tape head cleaning device. Developed from Nouchi et al. (1996). *Design problem:* Tape recorders record signals onto magnetic tape that can later be played back. After long periods of use, however, constituents of the magnetic tape (for example, magnetic particles, binder and resin) collect on the tape player head surface degrading the recording and playback quality. Under low humidity conditions, there is a sizable increase in the deposits from the tape onto the recording head that cannot be removed with the typical low force head cleaners that introduce a solvent to the tape head followed by a wiping action. A low force is required to prevent damage to the tape head. *Design solution:* With reference to Figure 10.17, the head cleaning device (**5**) consists of a cleaning roller (**3**) that is pressed against the magnetic head surface (**2**) of the cylindrical tape head (**1**). The cleaning roller (**3**) is made from a soft, porous material (**3z**), for example felt fibers, that are impregnated with abrasive particles (**3c**) that are attached to the felt using a polymer resin (**3b**). The abrasive particles grind off the deposits formed on the head surface without requiring an increase in the contact force.

32 Use color

Change the color or translucency of an object or its surroundings. Use color additives to observe certain objects or processes. If such additives are already used, employ luminescence traces.

33 Homogeneity

Interacting objects should be made with objects of the same material, or materials with the same properties.

34 Discarding and regenerating parts

An object that has fulfilled its task and is no longer required should be automatically discarded or made to disappear. Objects that become useful later should be automatically generated. *Examples:* Solid-rocket boosters and main fuel tanks on space shuttle (discarded).

35 Altering an object's aggregate state

Change an object's state.

36 Use of phase changes

Use of phenomena that occur during phase changes.

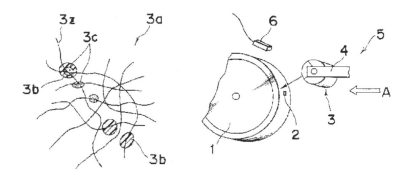

Figure 10.17 – *Design principle 31, Use of porous materials: Impregnation of porous felt tape head cleaning device with abrasive materials to increase effectiveness (Nouchi et al., 1996)*

37 Application of thermal expansion

Use of expansion or contraction of materials by heat. Use of materials with different thermal expansion coefficients.

38 Using strong oxidation agents

Replace air with enriched air, or enriched air with oxygen. Treat the air or oxygen with ionizing radiation. Use ionized oxygen. Use ozone.

39 Using an inert atmosphere

Replace normal atmosphere with inert one or vacuum. *Examples:* Light bulb.

40 Using composite materials

Change homogeneous materials with composites.

Example 13: Noble metal watch case. Developed from Marthe (1994). *Design Problem:* Watch cases formed from noble (precious) metals, for example gold, silver or platinum, tend to be very expensive and heavy. A design solution to reduce both the weight and cost is to hollow out as much of the case as possible (Design principle 1 Removal). Although this solution maintains the structural integrity of the watch case, the thin case wall is susceptible to denting. *Design solution:* With reference to Figure 10.18 a frame composed of an alternate material (**25**), for example brass, is placed within the circumference of the hollowed out noble metal case (**22**). The brass frame is designed to fit tightly in the hollowed space. With this design, the noble metal case can be

219

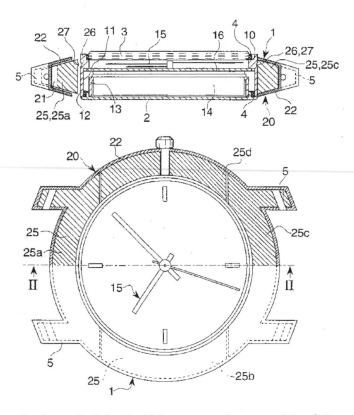

Figure 10.18 – *Design principle 40, Using composite materials: Schematic of watch case composed of composite materials (Marthe, 1994)*

made even thinner (further reducing cost) than the hollowed case without the frame, as the combined use of the noble metal and brass creates a composite case stronger than one made purely of the noble metal either before or after being hollowed out.

10.6 Technical Contradictions and the Contradiction Matrix

Technical contradictions refer to the standard engineering trade-off, i.e., changing one parameter to make an aspect of the system better makes another aspect of the system worse. Examples include

1. Increasing the stiffness of an airplane's wings to reduce vibration during flight (good) increases the weight of the plane (bad).

2. Reducing the engine size in an automobile to decrease fuel consumption (good) decreases available horse-power (bad).

3. Adding more windows to a house to improve passive lighting (good) increases heat loss from the house during winter (bad).

The general parameters used to describe system metrics can be used to formulate the *technical contradictions* within a system. Elimination of the technical contradiction may yield the desired final design.

How can one eliminate the technical contradictions easily? Do certain contradictions lend themselves to a particular solution irrespective of the actual problem at hand? These are the questions Althsuller set out to answer. In his patent analysis, Altshuller reformulated the problems in terms of the general parameters listed in Table 10.3 and noted their solutions. He soon realized certain technical contradictions were associated more frequently with particular design principles than others. He tabulated these observations to create the contradiction matrices.

The *contradiction matrix*[2] lists the most probable design principles for the solution of a particular technical contradiction. A portion of the matrix is shown in Table 10.5. A complete listing of the tables can be found in Appendix C. Note that the column headings represent the worsening general parameter in the contradiction and the row headings the improving one. The recommended design principles are listed at the intersection of a particular row and column. For example, if the technical contradiction was defined with parameter **2 weight of a moving object** as the improving feature and parameter **10 force** as the worsening feature, the contradiction matrix recommends design principles **8 counterweight**, **10 preliminary action**, **19 periodic actions**, and **35 altering an objects aggregate state**. An interactive form of the contradiction tables with numerous examples for each of the design principles can be found on the website *http://cede.psu.edu/~ogot/triz/*.

At first glance, there may appear to be no connection between the technical contradiction and the proposed solutions. This is because your mind may still be subject to "psychological inertia" where it is looking for the obvious or common solution. In the framework of the recommended design principles, however, the design team should use concept generation techniques (for example, *brainstorming*) to come up with new concepts to address the design problem. Teams must remain disciplined and restrict themselves only to ideas that fall within the current design principle under consideration.

In the event that a solution cannot be found from the recommended design principles (recall that these are simply the most probable), the design team should consider all forty design principles one by one.

---

[2]The contradiction matrix presented in this text was developed by Altshuller and his colleagues in 1969. The principles within the matrix are slanted towards mechanical systems, but are still applicable. Other researchers since have updated or developed new contradiction

Table 10.5 – *Contradiction matrix fragment: Parameters 1-2 vs. Parameters 1-11. The columns are the generalized performance parameters that deteriorate as the generalized parameters in the corresponding rows are improved.*

	1	2	3	4	5	6	7	8	9	10	11
1			15 8 29 34		29 17 38 34		29 2 40 28		2 8 15 38	8 10 18 37	10 36 37 40
2				10 1 29 35		35 30 13 7		5 35 14 2		8 10 19 35	13 29 10 18

Revisiting Example 1: Ice Dispenser Example

Revisiting the ice-maker example (*see* Page 199) and consulting the general parameters used to describe engineering system metrics listed in Table 10.3, the following technical contradiction can be formulated:

> In an effort to *improve* the quality of the ice and the ice dispensing process (**39 Productivity**), there is an increase *(worsening effect)* in **17 Temperature** and fusion of ice (**23 Substance loss**).

The contradiction matrix suggests the following general design principles as the most probable course of action to solve the problem.

- *35 Altering an objects aggregate state.*
- *21 Rushing through.* Perform a process or individual stages of a process at high speed.
- *28 Replacement of a mechanical pattern.* Replace a mechanical pattern with an optical, acoustical or odor pattern. Use electrical, magnetic or electromagnetic fields to interact with the object. Switch from fixed to movable fields changing over time. Go from structured to unstructured fields.
- *10 Preliminary action.* Carry out required action before it is needed, or set-up objects to be able to carry out their actions as soon as required.
- *23 Feedback*

Based on these suggestions, focused brain-storming is performed to generate concepts to remove the technical contradiction and thus solve the problem. In this example, the inventors used **feedback** for their solution. Recall that the IFR stated that,

> "**A resource** will eliminate the **frictional heat generated by the agitator fins** within the **ice storage chamber** during the **time the agitator is in operation** without complicating the system

---

matrices.

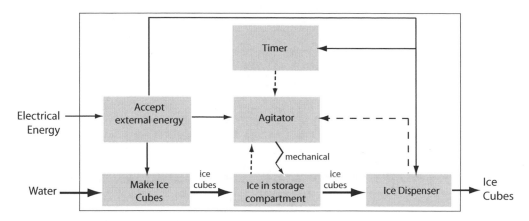

Figure 10.19 – *Energy-Material-Signal model of ice-maker showing* **harmful effect (mechanical energy)** *of agitator on the ice stored in the ice storage compartment*

while performing the **task of preventing ice-cubes from fusing together**."

One must therefore first look at the EMS model (Figure 10.19) to see if there are any resources that can be used in the context of **feedback** to solve the problem. A close examination of the figure indicates the presence of signals emanating from each of the three events that trigger the agitator. The inventor decided that instead of having the agitator independently controlled by three different events, all three events would be monitored and their actions **fed back** to a single controller that ensured that the agitator did not consecutively activate two or three times (Figure 10.20). For example, if the agitator had just been activated by the timer, and ice was dispensed, the controller would ignore the latter action and prevent the agitator from activating again.

Example 14: Rain Sensor for a Motor Vehicle

Developed from Petzold (1991). Several motor vehicles have rain sensors embedded within their windshields, automatically turning on the windshield wipers when a certain level of moisture is detected. Once the level of moisture is sufficiently low, the sensors automatically stop the wipers. The problem with this design was that the sensor was integrated in or on the windshield resulting in a cost-intensive manufacturing process. In addition, retrofitting existing vehicles with the sensor required the replacement of the entire windshield. The EMS model of the original rain sensor assembly is shown in Figure 10.21(a). The IFR can be written as:

"**A resource** will eliminate the **high cost** within the **sensor and windshield integration process** during the **time of the manu-**

223

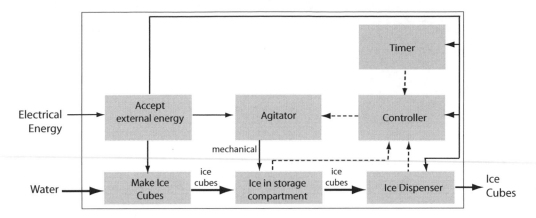

Figure 10.20 – *Energy-Material-Signal model of ice-maker showing removal of* **harmful effect** *by introducing* **feedback** *through a controller*

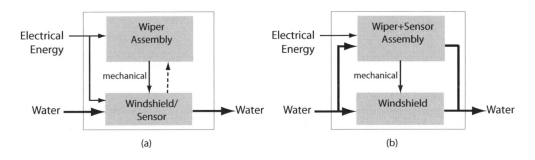

Figure 10.21 – *EMS models of rain sensor assemblies*

facture of new vehicles or when retrofitting old vehicles without complicating the system while performing the **task of sensing rain and automatically activating the windshield wipers.**"

With reference to the general parameters used to describe engineering systems listed in Table 10.3, one can formulate the following technical contradiction:

In order to *improve* the **32 manufacturability** of the sensor/windshield assembly, the **35 flexibility** of the sensor application is *diminished.*

The contradiction matrices suggest the following design principles as the most likely to solve the problem:

- *2 Removal.* Remove from the object the disturbing or the necessary part.

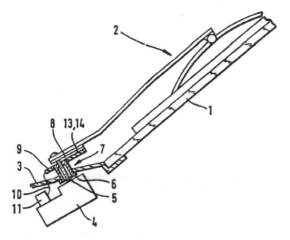

Figure 10.22 – *Windshield wiper assembly showing location of rain sensor (Petzold, 1991)*

- *13 Opposite solution.* Implement opposite action of that specified. Make a moving part fixed and a fixed part moving. Turn the object upside-down.

- *15 Dynamism.* Allow the object or environment to become optimal at any stage of work. Make the object consist of parts that can move relative to each other. Make a fixed object moveable.

Focused brainstorming, restricted to these suggestions, is used to generate concepts. In this example, the inventor chose to use **2 removal**. His aim was to create a rain sensor that could be used at minimal expense as original equipment on a new vehicle or as an add-on to an existing vehicle. The first step was to see if resources already present in the system could be used in some way to solve the problem. In this case the only other resource is the windshield wiper assembly (Figure 10.21 (a)). With reference to Figure 10.22, the rain sensor was no longer integrated with the windshield (**removal**), but developed as a separate component **(9)** that is added to the windshield wiper assembly and has a separate wiper blade **(8)**. With this arrangement, the sensor is no longer placed directly on the windshield. As a result, a special manufacturing process is not required for new vehicles, and existing vehicles can easily be retrofitted with the device. Finally, the additional wiper removes any moisture present on the sensor at the same time the main wiper removes moisture from the windshield. The EMS model for the new design is illustrated in Figure 10.21(b).

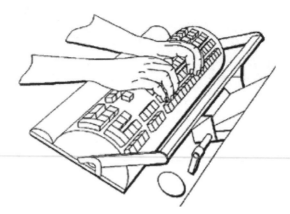

Figure 10.23 – *Schematic of an alternate cylindrical keyboard design (Kolsrud, 2000)*

Example 15: Keyball

Developed from Kolsrud (2000). Keyboards are a common data entry device for computers. As computers (and by extension the use of keyboards) become pervasive in everyday life, more time is spent using them for numerous tasks. The need to enter data quickly, coupled with the length of time spent using the keyboard, has led to numerous users developing carpal tunnel syndrome in their wrists, and the possibility of arthritic complications. One way to alleviate the problem is to reduce the speed of typing and lessen the force or pressure experienced by the wrist.

Numerous devices have been proposed to alleviate the problem. One is to shape the keyboard in the form of a cylinder (Figure 10.23) to minimize the bending of the wrist. An alternative design modifies the locations of the keys to correspond to the natural position of the fingertips. Although these designs somewhat relieve the stress placed on the forearm and wrist, the typist's wrists are still subject to the weight of the hands and the force from the depression of the keyboard keys. The EMS model for the design problem is illustrated in Figure 10.24. The IFR can be written as

> "**A resource** will eliminate the **stress** within the **wrist** during **typing at moderate speeds and long periods of time** without complicating the system while performing the **entering of data into the computer**.

Based on the general parameters used to describe engineering system metrics (Table 10.3), one can formulate the following technical contradiction:

> In order to reduce (*improving feature*) the stress on the wrist (**31 Harmful action generated by design**) the wrist must be com-

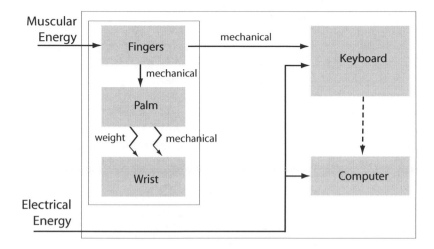

Figure 10.24 – *EMS model of interaction between the human hand and a typical keyboard showing the* **harmful effect (mechanical energy and palm weight)** *on the wrist*

pletely supported and the rate of data entry (**9 Velocity**) must be reduced (*worsening feature*).

The contradiction matrices suggest using the following design principles to overcome the contradiction:

- *1 Segmentation.* Similar to sub-division. Divide the object into as many sub-divisions as is practical.
- *28 Replacement of a mechanical pattern.* Replace a mechanical pattern with an optical, acoustical or odor pattern. Use electrical, magnetic or electromagnetic fields to interact with the object. Switch from fixed to movable fields changing over time.
- *3 Local quality.* Change the object's environment structure from homogeneous to non-homogeneous, or let different parts of the object perform different functions.
- *25 Self service.* The object should service and repair itself. Use waste products from the object to produce the desired actions.

As previously mentioned, brainstorming initially restricted to these solutions is performed to generate concepts. In this case, the inventor chose **replacement of mechanical pattern** for his solution using a device called a keyball. Figure 10.25 illustrates the top view of the device. The system includes a forearm rest (**11**) to support the typist's forearms. The device has two keyballs (**15 16**) upon which the typist rests her palms. Each keyball has electronic sensors

((23) in Figure 10.26) to detect pressure exerted by the finger or thumb tips - significantly lower than required to depress a key on a conventional keyboard. As the palm is fully supported at all times, less force is transferred back to the typist's wrist as compared to a conventional keyboard. With reference to Figure 10.27, the keyballs can swivel (similar to a joystick) on a pivoted link (24) whose position is detected by an electronic sensor (25). A typist could then move the keyballs foward, backward, left and right, and depress their fingers corresponding to positions on a traditional keyboard. The combination of detecting of which finger is depressed and the location of the keyball pivot generates the appropriate character that is then transferred to the computer. The EMS model of the keyball is shown in Figure 10.28.

10.7 Physical Contradictions

Physical contradictions are situations where an object is required to be in two contradictory states or have opposite requirements at the same time. Examples include

1. An automobile airbag should inflate rapidly to protect occupants from high speed impacts, but should inflate slowly to avoid harm to shorter passengers.
2. Software should be easy to use but should be able to perform complex functions.
3. Sport-utility vehicles should have high clearance to traverse rugged terrain, but should have low clearance to avoid rollovers.

Once an object has been formulated as a physical contradiction, one of the TRIZ Condensed Standards (described in the next section) can be used to generate concepts to solve the problem.

If a solution cannot be found, the physical contradiction must be *separated* using one of the four separation principles. These are

1. Separation in time
2. Separation in space
3. Phase transition: for example, solid - liquid - gas - plasma, paramagnetic - ferromagnetic.
4. Separation in structure

The separation principles are related to certain design principles as displayed in Table 10.6. These design principles can therefore be used to achieve separation. After separation, the same TRIZ Condensed Standards are consulted again for solution concepts.

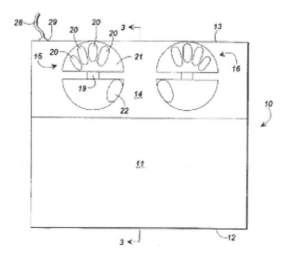

Figure 10.25 – *Top view of keyball system (Kolsrud, 2000)*

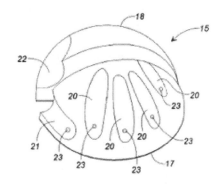

Figure 10.26 – *Close up view of a single keyball (Kolsrud, 2000)*

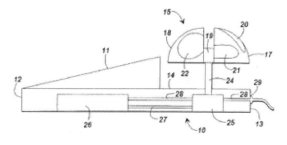

Figure 10.27 – *Cross-sectional view of keyball system (Kolsrud, 2000)*

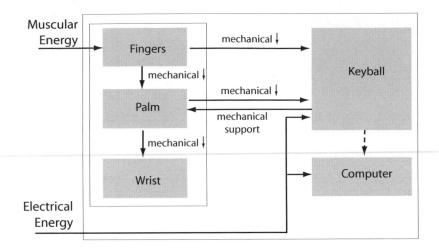

Figure 10.28 – *EMS model of interaction between the hand and the keyball*

Table 10.6 – *Relationship between separation and design principles*

Separation in	Related Design Principles
Time	15 Dynamism 20 Uninterrupted useful effect 24 Go between
Space	2 Removal 7 Nesting 17 Moving to a new dimension
Structure	1 Segmentation 3 Local quality 5 Joining 13 Opposite Solution

The TRIZ Condensed Standards

Once a physical contradiction has been formulated, the designer then attempts to seek a solution(s) from the set of TRIZ Condensed Standards. The latter have been developed from the 76 classical TRIZ standard solutions. One of the complaints that has been leveled against the traditional 76 standard TRIZ solutions is the significant degree of repetition among solutions. Several authors have presented reduced versions of the 76 solutions. Soderlin (2002), preferring to use 'rules' as opposed to 'standards,' reduced the number of solutions to 16 rules. Orloff (2003) renames the standard solutions as 'compact standards,' and reduces the number to 35. In this text, we have used suggestions from Soderlin and Orloff in generating our **Condensed Standards** composed of 27 solutions. In addition to reducing the number of solutions from 76, these Condensed Standards (a) use the language and jargon typical in engineering design and (b) replace the substance-field models found in the original 76 solutions with the energy-material-signal (EMS) models.

The classical 76 standard solutions fall into five classes:

1. Class I: Improving the system with little or no change
2. Class II: Improving the system by changing the solution
3. Class III: System transitions
4. Class IV: Detection and measurement
5. Class V: Strategies for simplification

The Condensed Standards reduce these five classes to three sets of standards:

1. Condensed Standards I: Improving the system with little or no change
2. Condensed Standards II: Improving the system by changing the solution
3. Condensed Standards III: Detection and measurement

The Condensed Standards are presented in Tables 10.7 - 10.9. Within the tables, the numbers in italics refer to the classic TRIZ standard solutions on which the condensed sets are based. Where applicable, solution fragments based on the EMS model are included. Note that more than one Condensed Standard is often incorporated into the final design solution. The use of the Condensed Standards is best explained through example presented in the following sections.

Example 16: Air-leakage Through Mating Pipes

Consider high pressure air running through two mating pipes that leaks through the joint to the lower pressure environment (*see* the EMS model in Figure 10.29(a)). The IFR can be written as:

> "**A resource** will eliminate the **air leakage** within the **joint between the two mating pipes** during the **time the air within the pipe is at a high pressure** without complicating the system while performing the **task of transferring the high pressure air from one location to the other**."

Searching for the appropriate **resource** using the TRIZ tools would solve the problem that has been clearly identified in the IFR. This problem can also be defined as a physical contradiction: *"The air has to be high pressure for the design operation; the air has to be low pressure to avoid leakage."* The IFR can therefore be stated in terms of the contradiction as

Table 10.7 – Condensed Standards I (9 solutions): Improving the System with Little or no Change. *This class looks at ways to modify a system in order to produce a desired outcome or eliminate an undesired one. An additive can be material, energy, voids, systems, sub-systems or super-systems.*

	Problem	Solution
1.1		Without changing the system, add a temporary or permanent, internal or external additive. The additive may or may not be present in the environment *(1.1.2-1.1.4, 1.2.1-1.2.2, 1.2.4, 5.1.1.1-5.1.1.3, 5.1.1.6, 5.1.4, 5.2.2,5.2.3)*
1.2		Change the environment. *(1.1.5)*
1.3		If a moderate energy is insufficient, but higher energy is damaging, apply higher energy to an additive that acts on the original system. *(1.1.7)*
1.4		Both low and high energy levels are required. Use an additive to protect those sub-systems that require low energy. *(1.1.8, 1.2.3)*
1.5		Heat a material above its Curie point to neutralize harmful magnetic effects. The Curie point is the temperature above which a ferromagnetic material loses its ferromagnetism. *(1.2.5)*
1.6		Use a small amount of a very active additive *(5.1.1.4)*.
1.7		Add additives to a copy or model of the object if it is not possible to add to the original *(5.1.1.7)*.
1.8		Desired additives can be obtained by decomposition of other materials *(5.5.1)*, such as hydrogen from water decomposition.
1.9		Desired additives can be obtained by combining other materials *(5.5.2)*.

Table 10.8 – Condensed Standards II (11 solutions): Ways to improve the system by changing it. *An* additive *can be material, energy, voids, systems, sub-systems or super-systems.*

	Solutions
2.1	Apply an additional energy source to the system *(2.1.1, 2.1.2)*. Example: Use of water and detergent (chemical energy) is not very effective for washing clothes. Agitating the system (mechanical energy) improves the cleaning.
2.2	Replace or add to energy existing in the system that is difficult to control with energy that is easier to control *(2.2.1)*. From the Laws of Evolution, in order of improved controllability: mechanical to thermal to chemical to electric to magnetic to electromagnetic energy.
2.3	Replace uncontrolled energy with energy that has predetermined patterns *(2.2.5)*.
2.4	Replace a uniform or uncontrolled system with a non-uniform system having a predetermined structure *(2.2.6)*.
2.5	If it is difficult to accurately control small quantities, use large quantities and remove the extra *(1.1.6)*.
2.6	Match or mismatch frequencies of elements within the system *(2.3.1,2.3.2)*. Example: *Noise canceling headphones* introduce a second signal with the same frequency but 180° out of phase.
2.7	A pair of incompatible or independent actions can be accomplished by running one during the down time of the other *(2.3.3)*.
2.8	Add ferromagnetic materials (objects or liquids) and/or electric generated magnetic fields (dynamic, variable or self-adjusting) *(2.4.1-2.4.11)*. For example: *Maglev trains* use dynamic magnetic fields to levitate and propel trains at high speeds.
2.9	Transition to the super-system. Simplify, improve the links between or create bi- and poly-systems *(3.1.1-3.1.4)*. Example: *Modern traffic lights* have replaced the use of a single light bulb with a large array of light emitting diodes (LEDs). LEDs have a longer life and use significantly less energy than regular bulbs.
2.10	Transition to the micro-level by dividing the system into smaller and smaller units *(3.2,5.1.2)*.
2.11	Make use of a material's phase transitions *(5.3.1-5.3.5)*.

Table 10.9 – Condensed Standards III (7 solutions): Detection and measurement. *Mainly used for control. Often the best designs are those with automatic control that do not require detection or measurement, but which utilize physical, chemical or geometrical effects within the system. An additive can be material, energy, voids, systems, sub-systems or super-systems.*

	EMS Model	Class III Problem
		Current measurements in the system are insufficient
	EMS Model	**Solutions**
3.1	S'	Modify the system to make detection and measurement unnecessary *(4.1.1)*
3.2	S → Sᶜ →	Measure a copy or image of the system. *(4.1.2)*.
3.3	S ⊣□─□⊢	Make two detections instead of continuous measurement *(4.1.3)*.
3.4	S → A →	Introduce an additive (into the system or its environment) that reacts to changes in the system. Measure changes in the additive or changes in the energy from the additive. *(4.2.2-4.2.4)*. Example: *Wind tunnels:* Adding smoke particles (additive) into the air flow in wind tunnels (original system) makes it easy to observe (measurement) the flow of air around objects.
3.5		Determine the state of a system by measuring the changes in scientific effects known to occur in the system. This could include the system's natural frequency *(4.3.1-4.3.3)*.
3.6		Add ferromagnetic materials (objects, particles, liquids) to the system or its environment and measure changes to the magnetic field *(4.4.1-4.4.5)*.
3.7		Measure the first or second derivatives in time or space *(4.5.2)*. For example: *Ground-based radar systems* measure *changes in the frequency* (second derivative of displacement) to accurately determine position, velocity and acceleration of an aircraft.

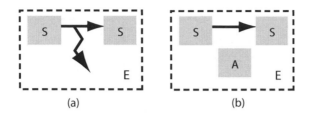

Figure 10.29 – *EMS model for air-leakage through two mating pipes*

"During **the time the air in the pipe is under high pressure, the resource** must provide on its own the **transfer of high pressure air through the pipes** and has to provide the **desired action to prevent the leakage of air to the atmosphere.**"

From the Table 10.7 of the Condensed Standards I, solution 1.1 could be used (Figure 10.29 (b)), where teflon would be the additive (the resource) that when placed between the threads of the mating pipes prevents leakage.

Revisiting Example 2: Fluidized Bed Combustion Furnace Erosion

An analysis of the fluidized bed combustion example previously described on page 199 and whose EMS model is illustrated in Figure 10.30, reveals that the high pressure air flow into the furnace necessary for combustion is causing the unburnt coal to collide with and erode the furnace walls. The problem can be defined as a physical contradiction: *"The high pressure air flow is required for combustion; the air flow should have a low pressure to reduce erosion on the furnace walls."* The IFR can therefore be stated in terms of the contradiction as

"During **the time the furnace is in operation with the forced air feed on,** forced air must provide its own **high pressure to facilitate the rapid combustion of the coal in the furnace** and provide **low pressure to avoid erosion at the walls.**"

A solution is sought from the Condensed Standards, starting from the first set of standards to the third until a set of acceptable solutions is found. An applicable Condensed Standard I solution (Table 10.7) could be *1.1 Without changing the system, add a temporary or permanent, internal or external additive. The additive may or may not be present in the environment.* Recall that an "additive" can be material, energy, system, or voids. Brainstorming and other concept-generation techniques would now be employed to generate designs based on this standard solution. An example design concept is illustrated in Figure

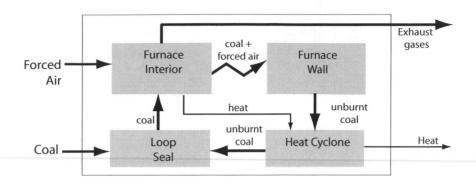

Figure 10.30 – *EMS model of fluidized bed combustion furnace showing* **harmful effects (coal+forced air)** *on the furnace wall*

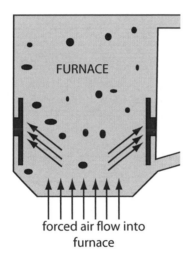

forced air flow into
furnace

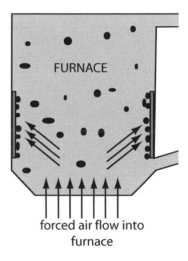

forced air flow into
furnace

Figure 10.31 – *Concept 1 for FBC furnace wall erosion problem based on standard solution 1.1*

Figure 10.32 – *Concept 2 for FBC furnace wall erosion problem based on standard solution 1.1*

10.31. An internal protective wall made of a more resistant material could be added in the corrosion-prone areas of the furnace wall.

A second concept based on this solution would make the inner wall of the furnace in erosion-prone areas sticky, where coal particles would attach and form a barrier between other coal particles and the furnace wall (Figure 10.32). A third concept would add a forced air stream at the walls (mechanical energy) as illustrated in Figure 10.33. The second air stream could reduce the impact of unburnt coal particles on the furnace wall, reducing the erosion.

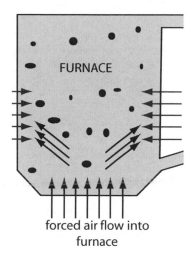

forced air flow into
furnace

Figure 10.33 – *Concept 3 for FBC furnace wall erosion problem based on standard solution 1.1*

These examples illustrate how the creation of the EMS model and understanding the type of design problem, followed by concept generation based on the Condensed Solutions, can lead to innovative designs.

10.8   Summary

Through numerous examples, this chapter has introduced three of the several tools used in TRIZ for systematic concept generation. Along with the tools, a simple algorithm was also presented to maximize the benefits from using the tools. The steps in the algorithm may seem tedious at first, as design teams have an urge to rush to begin concept generation before the problem is fully understood. But it is only after a structured analysis of the problem is performed and the right problem defined that the correct solution can be found. With practice, the steps become increasingly intuitive and the quality of the solutions generated by design teams will significantly increase.

Finally, there are several good online sources for tutorials, case studies and scholarly work on TRIZ. These include

1. The TRIZ Journal, www.triz-journal.com

2. The TRIZ Experts, www.trizexperts.net

3. The Altshuller Institute for TRIZ studies, www.aitriz.org

4. Ideation TRIZ, www.ideationtriz.com

References

Altshuller, G.S., *Creativity as an Exact Science: The Theory of the Solution of Inventive Problems*, New York: Gordon and Breach, 1988

Axathammer, L., "Shock Absorbing or Oscillation Damper Device", US Patent No. 4,966,257, October 30, 1990.

Bright, M., "Seeing the Forest for the TRIZ", *Profit Magazine*, Oracle Publishing, pp 43, November 2003.

Domb, E., "The Ideal Final Result: Tutorial", TRIZ Journal, no. 2, 1997.

Domb, E., "Finding the Zones of Conflict: Tutorial", TRIZ Journal, no. 6, 1996.

Domb, E., "How to Help TRIZ Beginners Succeed", TRIZ Journal, no. 4, 1997a.

Domb, E., "Contradictions: Air Bag Applications", TRIZ Journal, no. 7, 1997b.

Fukusaki, M., Hagihara, H., and Inoue, S., "Ash Melting Furnace", US Patent No. 5,493,578, February, 1996.

Galvin, W., "Remote Control Holder and Illuminator", US Patent No. 5,485,359, January 16, 1996.

Ganguly, D. Y., "TRIZ of the Trade", The Economic Times, March 21, 2003.

Kolsrud, A., "Keyball", US Patent No. 6,053,646, April 25, 2000.

Kowalick, J., "Tutorial: Use of Functional Analysis and Pruning, with TRIZ and ARIZ, to solve 'Impossible-to-Solve' Problems.", TRIZ Journal, no. 12, 1996.

Kowalick, J., " 'No-compromise' Design Solutions to the Air-Bag Fatalities Problem", Triz Journal, no. 7, 1997.

Lee, J-G, Lee, S-B, and Oh, J-M, "Case Studies in TRIZ: FBC (Fluidized Bed Combustion) Boiler's Tube Erosion", TRIZ Journal, no. 7, 2002.

Lin, Y., Hodges, P.H., and Pisano, A.P., "Optimal Design of Resonance Suppression Helical Springs", Journal of Mechanical Design, Vol. 115, September 1993, pp. 380-384.

Maglica, A., "Miniature Flashlight", US Patent No. 5.485,360, January 16, 1996.

Mann, D., "Case Studies in TRIZ: A Better Wrench", TRIZ Journal, no. 7, 2000.

Mann, D., "Assessing the Accuracy of the Contradiction Matrix for Recent Mechanical Inventions", TRIZ Journal, no. 2, 2002.

Marconi, J., "ARIZ: The Algorithm for Inventive Problem Solving: An Americanized Learning Framework", TRIZ Journal, no. 4, 1998.

Marthe, C.-A, "Nobel Metal Watch Case", US Patent No. 5,493,544, March 27, 1994.

McGraw, D., "Expanding the Mind", *ASEE Prism*, Summer 2004.

Nakamura, Y., "The Effective Use of TRIZ with Brainstorming", TRIZ Journal, no. 6, 2001.

Nouchi, N., Sakaguchi, M., and Yoda, H., "Head Cleaning Device", US Patent No. 5,543,179, August 6, 1996.

Orloff, M., *Inventive Thinking through TRIZ: A Practical Guide*, Berlin:Springer, 2003.

Petzold, S., "Rain Sensor for a Motor Vehicle", US Patent No. 5,900,821, May 4, 1991.

Pickel, H., "Injection assembly for an injection molding machine", US Patent No. 5,540,495, December 20, 1994.

Raskin, A., "A Higher Plane of Problem-Solving", Business 2.0, pp. 54-56, June 2003.

Royzen, Z., "Tool, Object, Product (TOP) Function Analysis", TRIZ Journal, no. 9, 1999.

Smith, E., "From Russia with TRIZ: An Evolving Design Methodology Defined in Terms of a Contradiction", ASME Mechanical Engineering Magazine, March 2003.

Soderlin, P., "TRIZ The Simple Way", TRIZ Journal, no. 2, 2002.

Sheets, B.J., Wintrone, L.L. and Kondratuk, M.W., "Window Casement Operator", US Patent No. 5,815,984, October 6, 1998.

Stockman, A.J. and Benton, S.C., "Optical Fibre Sheathing tube", US Patent No. 5,473,723, December 5, 1995.

Tasi, C-C and Tseng, C-H., "Using TRIZ for an Engineering Design Methodology Course at NCTU in Taiwan", TRIZ Journal, no. 3, 2000.

Terninko, J., Domb, E., and Miller, J., "The Seventy-six Standard Solutions, with Examples. Section one." TRIZ Journal, No. 2, 2000.

Terninko, J., Domb, E., and Miller, J., "The Seventy-six Standard Solutions, with Examples-Class 2." TRIZ Journal, No. 3, 2000.

Terninko, J., Domb, E., and Miller, J., "The Seventy-six Standard Solutions, with Examples-Class 3." TRIZ Journal, no. 5, 2000.

Terninko, J., Domb, E., and Miller, J., "The Seventy-six Standard Solutions, with Examples-Class 4." TRIZ Journal, no. 6, 2000.

Terninko, J., Domb, E., and Miller, J., "The Seventy-six Standard Solutions, with Examples-Class 5." TRIZ Journal, no. 7, 2000.

Tsuchikawa, Koji, "Ice Dispenser", US Patent No. 6,257,009, July 10, 2001.

Exercises

1. Many outdoors sports played on natural grass require lines be drawn to indicate boundaries and other field markings. Conventionally, such lines are marked by chalk or paint to create the line on the grass. The drawback with these techniques is that the line markings are temporary and must be reapplied from time to time. Also, contestants may run over the lines and obliterate them during the game. Alternatively, the lines may be created by using diesel or other chemicals to kill the grass, a process that has become environmentally unacceptable. (i) Create an

EMS model, (ii) define the zones of contradiction, (iii) determine which general engineering parameters define the problem, (iv) state the technical contradiction, and (v) use the suggested principles to develop several concepts.

2. A number of devices have been patented to accomplish the task of removing the cork from a champagne bottle. Most of them attempt to pry or pull the stopper out by exerting a force between the top of the bottle and the enlarged, exposed part of the cork. Champagne cork are tightest before the initial displacement, which breaks the bond to the bottle created over weeks or months of storage. Teeth forced into the uncompressed cork can tear or fracture the cap, leaving even less to work with. (i) Create an EMS model, (ii) define the zones of contradiction, (iii) determine which general engineering parameters define the problem, (iv) state the technical contradiction, and (v) use the suggested principles to develop several concepts.

3. Commercial cooking ovens are designed to cook food quickly and efficiently. The heat they generate, however, causes the exterior to become very hot, posing a hazard to both employees and adjacent equipment. In addition, the heat radiated from the exterior raises the ambient temperature of the kitchen resulting in higher energy costs through wasted heat and increased air conditioning. (i) Create an EMS model, (ii) define the zones of contradiction, (iii) compose an IFR, and (iv) state the physical contradiction.

4. From the US patent office website (www.uspto.gov), find a patent whose problem can be defined as a physical contradiction. The problem being solved by the patent is usually identified in the 'Background of Invention' section. (i) Create an EMS model, (ii) define the zones of contradiction, (iii) compose an IFR, (iv) state the physical contradiction, and (v) describe the solution used by the inventor. Relate the solution, if possible, to one of the standard solutions. Note that the separation principles may have been used prior to use of the standard solutions.

5. From the US patent office website (www.uspto.gov), find a patent whose problem can be defined as a technical contradiction. The problem being solved by the patent is usually identified in the 'Background of Invention' section. (i) Create an EMS model, (ii) define the zones of contradiction, (iv) determine which general engineering parameters define the problem, (v) state the technical contradiction, and (vi) determine which of the 40 design principles (could be more than one) were used by the inventor. Are the principle(s) used the same as those recommended by the contradiction matrix?

Part III

Overview of Materials and Manufacturing

Chapter 11

Materials and Material Selection

11.1 Introduction

Selecting materials usually begins in the third stage of the design process, **preliminary design** (Figure 11.1). Although this text concentrates on the first two stages, problem definition and conceptualization, a few general ideas about the preliminary design stage are given. The preliminary design stage is iterative, meaning the design becomes successively more detailed based on continuous, increasingly sophisticated analyses, drawings and discussions among team members. It involves numerous activities being conducted in parallel, making it essential that there is effective communication between all team members to ensure success.

Preliminary design essentially involves the infusion of discipline-specific knowledge into the previously generated concepts. It involves

1. Determination of overall system layout
2. Determination of shapes and assemblies for all components
3. Materials selection
4. Construction and testing of prototypes (Paul and Beitz, 1996)

Pahl and Beitz (1996) suggest a 14 point check list for preliminary design. An abbreviated nine point check list is shown in Table 11.1. As the design team

Figure 11.1 – *The third stage of the design process: preliminary design*

Table 11.1 – *Checklist for embodiment design. Adapted from (Paul and Beitz, 1996).*

	Areas	**Examples of issues to be addressed**
1	Function	Are the stated functions met? To what extent are the customer needs meet?
2	Working principle	Are the desired outcomes obtained from the selected working principles?
3	Layout	Selected layout, component shapes and sizes and materials should provide the desired strength, stiffness, stability, etc.
4	Safety	Ensure that the safety of all components in the product, it's operation and its effects on the environment are accounted for.
5	Ergonomics	Take into account the human-machine interface issues and the aesthetics of the final product.
6	Assembly	Is the product easy to assemble?
7	Recycling	Can the product be reused or its components easily recycled?
8	Maintenance	Is the product easy to maintain? Does it require frequent maintenance?
9	Costs	Is the final product within the specified cost constraints?

goes through this stage, it should ensure that the questions raised within the table, if appropriate, are addressed. The remainder of the chapter will provide an overview of materials and material selection.

11.2 Materials and Material Selection

One of the key areas in the design process is the selection of the appropriate materials. A poor material choice can lead to failure of a part or system or to unnecessary cost. In selecting the appropriate material one must consider

1. Material properties. Affects the part performance.
2. Material processing. Affects the manufacturing cost and therefore the final part cost.
3. Material cost.
4. Availability. Is the material available in the desired quantity and time frame? The availability of the required material can affect the production timelines. Often for small design projects, the minimum purchase size or quantity may far exceed the team's need.

The relative importance of the above four factors varies depending on the application. For example in military and aerospace applications, pushing the material properties to their limits normally takes precedence over cost. For consumer products, lowering cost typically plays the leading role. For projects on

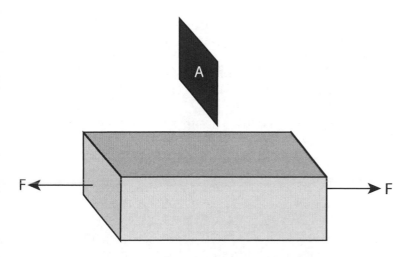

Figure 11.2 – *Schematic illustrating concept of stress*

a tight schedule, design teams should plan ahead because the required material may not be immediately available when needed. A delivery delay of a couple of weeks or months, for example, could derail a project's schedule if the order is made at the last minute.

The process of material selection is a difficult one. The designer becomes a decision-maker and is typically faced with multiple and often conflicting material characteristics, as well as a large number of constraints.

11.3   Mechanical Properties of Materials: Stress-Strain

When considering the mechanical properties of materials, the two most common parameters are *stress* and *strain*. A brief definition of each is given in the following sections.

Stress

Stress is a measure of the intensity of a force acting on a material. Stress in its simplest form can be defined as the ratio of the applied force to the area over which it is applied. Figure 11.2 illustrates a beam of uniform cross-section, A, subject to a tensile force, F. The average normal stress, σ, can be defined as

$$\sigma = \frac{F}{A} \tag{11.1}$$

Typically units are pounds per square inch (*psi*) or pascals (*Pa*), where a pascal is a Newton per square meter. Prefixes representing steps of 1,000 are frequently used to reduce the space taken up by numbers presented in tables and reports. For example, the prefixes kilo (k), mega (M) and giga (G) represent 10^3, 10^6 and 10^9, respectively. Examples of usage include 1 kPa $= 10^3$ Pa, 1 MPa $= 10^6$ Pa, and 1 GPa $= 10^9$ Pa. Stresses can be a result of tensile, compressive or shear forces. For shear stress, τ, the force is applied parallel to the area and has the same units as normal stress.

Why is it important to calculate stresses for a particular design? Every material has two important parameters related to stress that define its suitability for different applications. The first is the material's *yield stress*. This is the stress value which if exceeded results in *plastic (permanent) deformation* or yielding of the material, i.e., the material deforms but does not return to its original shape. If the material is subject to stress values below the yield stress, it undergoes *elastic deformation*, i.e., once the load is released it returns to its original shape. The second important parameter related to stress is the *ultimate strength*. This is the maximum stress value that the material can endure before it breaks.

Strain

A material subject to an applied force undergoes deformation proportional to the magnitude of the force. The deformation as a fraction of the original object dimension defines the elongation per unit length or *strain*. With reference to Figure 11.3, an object with uniform cross-section and original length, L, is subject to a tensile force, F, resulting in a deformation, ΔL. The strain, ϵ, can be defined as

$$\epsilon = \frac{\Delta L}{L} \qquad (11.2)$$

where *e* is a dimensionless parameter. Except for a few materials such as rubber, strain tends to be *very* small. Strain is calculated for a material under tensile, compression or bending forces.

Stress-Strain Diagrams: Hooke's Law

When studying materials it is common to plot diagrams that relate stress and strain for a particular material. One should note that stress-strain diagrams vary widely between materials and even for the same material, the latter due to variations in test conditions (for example, temperature or speed of test). A 'typical' stress-strain curve for a *ductile*[1] material such as mild steel is illustrated in Figure 11.4.

---

[1]Ductility is the ability of a material to *plastically* strain without fracture. Conversely, a *brittle* material does not.

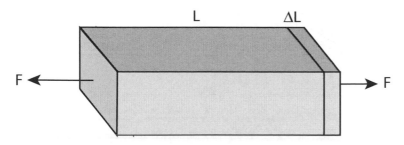

Figure 11.3 – *Schematic illustrating concept of strain*

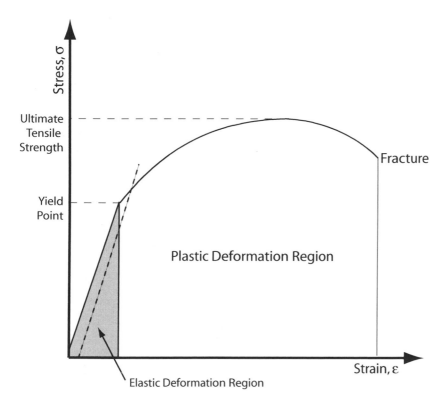

Figure 11.4 – *General stress-strain curve for a metal under tension*

With reference to Figure 11.4, the following observations can be made and generally hold true for all materials:

1. From the origin (no load condition) to the *yield point*[2], the relationship between stress and strain is practically linear. This relationship is known as **Hooke's Law** and is defined as

$$\sigma = E\epsilon \quad or \quad E = \frac{\sigma}{\epsilon} \qquad (11.3)$$

where the proportionality constant, E, is called **Young's Modulus of Elasticity** and has the units of stress. As strain, ϵ, is typically a very small quantity, E is generally a large value. The larger the value of E, the steeper the slope on the stress-strain curve. Physically, this can be interpreted as meaning that a larger stress is required to achieve the same strain, or put a different way, that the material is more resistant to deformation. Young's Modulus can therefore be used as a measure of a material's *stiffness*. The larger the value of E, the stiffer the material.

2. For applied forces (and corresponding stress values) that lie within the elastic region, the material returns to its original dimension once the load is removed.

3. In the plastic deformation region, the material undergoes permanent deformation.

4. For most materials, similar curves are obtained for tension or compressive loads. Notable exceptions are cast iron and concrete which are both very weak in tension but very strong in compression.

In deciding which material to use for a particular application, a large number of material characteristics must be considered. The following sections discuss common mechanical, thermal, and manufacturing characteristics.

11.4 Typical Mechanical Properties for Material Selection

Hardness

Hardness is a measure of the resistance to permanent indentation. Hard materials are resistant to scratch and wear. Several tests are available to measure hardness, the most common ones being the Brinell hardness test (used for testing metals and non-metals of low to medium hardness), the Rockwell hardness test and the Vickers hardness test.

---

[2]Technically, it should be to the *proportionality limit*. The two points have been found to be very close, however, and for practical purposes can be treated as the same point.

Young's modulus of elasticity (E)

Proportionality constant between stress (force resisted per unit area) and strain (fractional elongation/compression due to an applied load) of a material within the linear elastic region. Typical units are Newtons per meter squared (N/m^2) or pounds per square inch (psi).

Yield strength (σ_y)

Maximum stress a material can resist before yielding (undergoing permanent deformation). Yield strength can be evaluated for shear, tension or compression loads. Typical units are Newtons per meter squared (N/m^2) or pounds per square inch (psi).

Ultimate strength (σ_{ult})

Maximum stress levels that a material can resist before breaking. Ultimate strength is also evaluated for shear, tension or compression loads. Typical units are Newtons per meter squared (N/m^2) or pounds per square inch (psi).

11.5   Typical Thermal Properties for Material Selection

Thermal conductivity

Measure of the rate at which heat can be conducted through a material. Conduction is the heat transfer process where thermal energy is transferred within a material purely by thermal motion, without the transfer of mass. Thermal conductivity is measured with the *coefficient of thermal conductivity, k*, and with units $Btu/in. - hr -^\circ F$ or $J/sec - mm -^\circ C$. Btu (British thermal unit) and J (joule) are the English and SI units of thermal energy. The higher the coefficient, the better the thermal conductivity. For cases where thermal insulation is required, such as insulation in the walls of buildings, materials with low thermal conductivity are used.

Specific heat

Specific heat, C, is the amount of thermal energy required to increase a unit mass of a material's temperature by 1 degree. English and SI units are $Btu/lb -^\circ F$ and $J/Kg -^\circ C$, respectively.

Coefficient of Thermal Expansion

When a material is heated, its density (mass per unit volume) decreases. This decrease is equivalent to a unit mass of the material occupying a larger volume or the material undergoing thermal expansion. The coefficient of thermal expansion, α, gives a measure of an object's change in length per degree change in temperature. It can be defined as

$$\alpha = \frac{\Delta L}{L \Delta T} \tag{11.4}$$

where L, ΔL and ΔT are the length, change in length and change in temperature, respectively. Units are in/in/°F or simply $°F^{-1}$ (mm/mm/°C or $°C^{-1}$). The higher a material's coefficient of thermal expansion, the more it expands (and contracts) from temperature changes.

For designs that are exposed to relatively large temperature variations, design teams should select materials whose expansion and contraction do not negatively impact the design application. Alternatively, measures should be incorporated in the design to account for the dimensional changes. For example, bridges are typically made of steel with average values of $\alpha = 6.7 \times 10^{-6} \, °F^{-1}$. If a bridge had a 100 ft span (at 30 °F) and was exposed to temperature variations from 30 °F in the winter to 100 °F in the summer, the bridge would expand by

$$\Delta L = \alpha L \Delta T = 6.7 \times 10^{-6} \times 100 (100 - 30) = 0.0469 \quad ft \quad (0.563 \quad in.) \tag{11.5}$$

Bridges are therefore designed to allow portions of the span to slide relative to each other thereby accommodating the expansion.

11.6   Typical Electrical Properties for Material Selection

Resistivity

Electricity flows through solid materials and liquids via electrons and ions, respectively. For a given voltage difference, V, the amount of electricity or current that flows through the material can be calculated from Ohm's law:

$$I = \frac{V}{R} \tag{11.6}$$

where I and R are the current and resistance, respectively. Units for V, I and R are volts (V), amperes (A) and Ohms (Ω). Consider a material of length, L, and cross-sectional area, A. Its resistance can be calculated from

$$R = \rho \frac{L}{A} \tag{11.7}$$

where ρ is the material's *resistivity*. Units are inch-ohm or meter-ohm. Resistivity measures a material's ability to resist electricity; the higher its value the higher the resistance of the material. Note, however, that resistivity changes with temperature. For example, as the temperature of a metal is raised, its resistivity goes up.

Dielectric Strength

Materials can be categorized in terms of their electrical properties as *conductors* (low resistivity), *semiconductors* (moderate resistivity) or *insulators* (high resistivity). Conductors include metals, semiconductors include silicon (widely used in the electronics industry), and insulators include most ceramics and polymers. For an insulator, the *dielectric strength* is the voltage required to break down the insulation (i.e., allow electrical conduction) through a unit thickness of the material. Typical units are $V/in.$ or V/m. The higher the dielectric strength, the greater the voltage the material can insulate against.

11.7 Typical Manufacturing Properties for Material Selection

Several properties can be considered during the material selection process. These include

1. **Availability**. Multiple supply sources, future availability, available in form and sizes needed.
2. **Processing properties**. Ability to machine, mold, forge, form, cast, harden, weld, or heat treat.
3. **Variability in properties**. How consistent are the material properties from order to order?
4. **Joining techniques**. For the given material what are the fastening options? Do they meet the requirements for the design?

11.8 General Material Categories

Materials available for most engineering applications can be classified into five broad categories: ceramics, polymers, composite materials, metals and alloys, and liquid and gases. Two widely used online resources for material properties are

1. **Matweb**, http://www.matweb.com. Provides material properties for over 2800 metals, plastics, ceramics and composites.
2. **Plastics USA**, http://www.plasticsusa.com. Extensive database of polymer material properties.

Ceramics

Formed through the fusion of powders under high pressure and temperature. Example powders include silica (SiO_2), aluminum oxide, beryllium oxide or silicon carbide. Materials in this category include glass, clay products (for example, tiles, bricks, chinaware), cement and concrete[3] Typical properties include high hardness, corrosion and high temperature resistance, and low electrical conductivity. Ceramics have high strength but are brittle.

Polymers

Includes plastics, rubbers (elastomers), fibers and coatings. General properties include easy fabrication, low weight, low cost, very low electrical conductivity (insulators), resistance to corrosion, low temperature resistance (can withstand temperatures between 200° - 300° C), weak, compliant and durable. They can be easily formed and molded.

Composite Materials

Combination of different materials, usually a reinforcing component and a binder. General properties include brittle, low weight, high stiffness, low thermal conductivity, high fatigue resistance and high strength to weight ratios (Figure 11.5). Reinforcing components include glass, carbon, ceramics, metal or boron fibers.

Metals and Alloys

Traditionally popular for engineering systems due to strength, toughness, high electrical and thermal conductivity, and stiffness. They are relatively easy to process through machining, casting, forming, stamping, and welding but are susceptible to corrosion.

Liquids and Gases

Play a major role in thermal, hydraulic and pneumatic systems. Used for heat transfer, material flow, power/pressure transmission, and lubrication. General properties include low electrical and thermal conductivity, and typically low cost.

---

[3]Although concrete is technically a composite, both its components are ceramics (Groover, 1996).

Figure 11.5 – *Experimental all-composite aircraft to be used for low earth orbit flights. Engineers choose composites to take advantage of the high strength to weight ratio (Courtesy www.xprize.org).*

11.9 Properties of Common Metals

This section will discuss the properties and industry-standard naming conventions for common metals used for design projects. These naming conventions are required when ordering the materials or when looking up their characteristics. Note that a *metal alloy* is a metal composed of two or more elements.

Steel and Alloys

Steel and alloys find applications where strength, hardness and stiffness are more important than weight. Steel is formed by adding low percentages (0.77% to 2.11%) of carbon to iron (Groover, 1996). Carbon percentages between 2.11% to 5%, in addition to 1% to 3% silicon yields *cast iron.*

Different materials can further be added to steel to create steel alloys with enhanced properties. Steel alloys follow designations from the American Iron and Steel Institute and the Society of Automotive Engineers (AISI-SAE) system. Table 11.2 lists the common designations and the corresponding attributes. Note that the last two digits indicate the fractional percentage of carbon, by weight. For example a 1320 steel alloy contains 0.20% carbon.

Table 11.2 – *Attributes of different steel alloys. Compiled from Groover (1996) and Rufe (2002).*

AISI-SAE No.	Added Elements	Characteristics
10XX	Carbon	Increased strength, hardenability, and wear resistance. Reduces ductility and weldability.
13XX	Manganese	Increased strength; increased ductility after heat treatment.
23XX - 25XX	Nickel	Increased strength toughness and corrosion resistance.
3XXX	Nickel, Chromium	Increased toughness and ductility (from nickel); increased wear and corrosion resistance (from chromium).
4XXX	Molybdenum	Increased toughness, wear and creep resistance. Creep is defined as plastic deformation under a sustained load.
5XXX	Chromium	Increased corrosion resistance.
6XXX	Chromium, vanadium	In addition to chromium properties, vanadium provides increased strength, toughness and abrasion resistance.
8XXX - 9XXX	Chromium, nickel, molybdenum	Increased attributes from each.

Aluminum and Alloys

Aluminum and alloys find wide use in engineering applications due to their finished appearance, resistance to corrosion, high strength-to-weight ratio, and relative ease to machine. In addition, aluminum has high thermal and electrical conductivity and is very ductile. Wrought aluminum alloys (what you typically purchase) follows an YXXX numeric designation, where Y represents the primary alloy material and XXX the alloy's purity. Table 11.3 lists the common designations and corresponding attributes. For the 1XXX series, the last two digits represent the purity of the alloy; for example, 1066 aluminum is 99.66% pure.

Copper and Alloys

Copper has very low electrical resistivity and, combined with its relatively low cost, results in a wide use as an electrical conductor. In addition, it has very good thermal conduction properties and is resistant to corrosion.

On the downside, its strength to weight ratio is very low. Common alloys of copper include *brass* (65% copper and 35% zinc) and *bronze* (90% copper and

Table 11.3 – *Attributes of different aluminum alloys. Compiled from Groover (1996) and Rufe (2002).*

Number	Alloy Elements	Characteristics
1XXX	Commercially pure ($> 99\%$)	Corrosion resistant, good workability, and low strength.
2XXX	Copper	High strength to weight ratio but lower corrosion resistance.
3XXX	Manganese	Good workability but moderate strength.
4XXX	Silicon	Low melting point.
5XXX	Magnesium	Good corrosion resistance and high strength.
6XXX	Magnesium and Silicon	Good weldability and machineability; corrosion resistant and medium strength.
7XXX	Zinc	High strength.

Table 11.4 – *Average material properties for common metals and alloys. Compiled from Rufe (2002) and Groover (2002).*

Material	Modulus of Elasticity psi (MPa)	Tensile Strength psi (MPa)	Melting Temperature ^{o}F (^{o}C)	Density lb/in (g/cm^3)
Aluminum	10e6 (0.69e5)		1220 (660)	0.098 (2.7)
Copper	16e6 (1.1e5)		1981 (1083)	0.327 (8.96)
Iron	30e6 (2.09e5)		2,802 (1,539)	0.287 (7.87)
Steel				
1010 (hot rolled)		44,000 (304)		
1020 (hot rolled)		55,000 (380)		
1040 (hot rolled)		75,000 (517)		

10% tin).

Average material properties for common metals and allows are presented in Table 11.4.

11.10   An Overview of Polymers

Polymers, commonly referred to as plastics, find wide use in design projects due to their relatively low cost and good machineability. Polymers are essentially chain-like structures composed of repetitive hydrocarbon units. Carbon makes up the backbone of the molecule with hydrogen atoms bonded along the backbone. Polyethylene $(C_2H_4)_n$, the simplest polymer structure, for example, is illustrated in Figure 11.6.

Some polymers consist of carbon and hydrogen chains only, for example polypropylene $(C_3H_6)_n$, polybutylene, polystyrene $(C_8H_8)_n$, and polymethylpen-

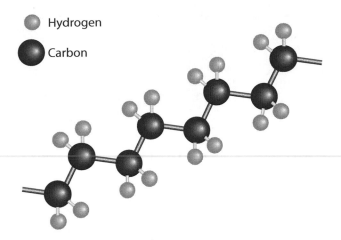

Figure 11.6 – *Polyethylene hydrocarbon chain*

tene. Other polymers add additional elements, such as, oxygen, chlorine, fluorine, nitrogen, silicon, phosphorous, and sulfur, to give the polymers added material properties. Examples include polyester (contains oxygen), nylon (contains nitrogen), polyvinyl chloride $(C_2H_3Cl)_n$ (commonly referred to as PVC and contains chlorine), and teflon (polytetrafluoroethylene $(C_2F_4)_n$ and contains fluorine).

Molecular Arrangement of Polymers

Polymers can be arranged in an *amorphous organization*, that is, the chains lack a specific form, by controlling and quenching the polymerization process. An amorphous arrangement of molecules is similar to a plate of spaghetti, making this group of polymers generally transparent. Their applications include food wrap, plastic windows, headlights, and contact lenses. Translucent and opaque polymers have a *crystalline arrangement*, that is, the chains form a distinct pattern. The degree of translucence or opaqueness of the polymer is directly affected by the extent of crystallinity: the more crystalline the structure, the more opaque the polymer.

General Characteristics of Polymers

Polymers can be very resistant to chemicals, making them ideal materials for chemical storage containers. Polymers can be both thermal and electrical insulators, finding applications in, for example, household appliances, electrical cords, and electrical outlets. Thermal resistance is useful in the kitchen, as the handles of pots and pans are made from polymers. Other examples of the use of polymers for thermal insulation include thermal underwear worn by skiers (polypropylene) and fiberfill in winter jackets (acrylic).

Polymers are very light in weight with varying degrees of strength. They can be processed in numerous ways to produce thin fibers or very intricate parts. They can be molded into bottles or the bodies of cars or mixed with solvents to become an adhesive or a paint. A brief overview of the manufacturing processes for polymers is given in Section 13.2 on page 295.

Polymers can be categorized into three broad categories: *thermosets, thermoplastics* and *elastomers.*

1. **Thermosets.** Undergo an irreversible curing process during heating and shaping resulting in a permanent chemical change in their molecular structure. As a result, once cured they cannot be remelted and reshaped, making them generally not recyclable. Examples of thermosets include polyurethanes found in mattresses, cushions, insulation, ski boots, and toys; unsaturated polyesters used in varnishes, boat hulls and furniture; and expoxies used in glues and coating on electrical circuits.

2. **Thermoplastics.** Comprise approximately 80% of commercial plastic. Their chemical composition remains the same during heating and melting, making them recyclable. Examples include polyethylene used in packaging, electrical insulation, and milk and water bottles; polypropylene used in carpet fibers, automotive bumpers, microwave containers and external prostheses; and polyvinyl chloride (PVC) used in siding, piping, credit cards, automobile instrument panels, floor and wall coverings, and sheathing for electrical cables.

3. **Elastomers.** Commonly referred to as rubbers, elastomers undergo large *elastic* deformation when subject to relatively low loads.

Plastics deteriorate but never decompose completely. In 2000, plastics made up 9.5% of United States' trash by weight compared to paper, which constituted 38.9%. Glass and metals made up 13.9% (APC, 2000). Recycled plastics can be blended with virgin plastic (plastic that has not been processed before) without sacrificing properties in many applications. These include polymeric timbers for use in fences, picnic tables and outdoor toys. Plastic from 2-liter bottles can be spun into fiber for the production of carpet. Plastics that cannot be recycled can be converted into energy through controlled combustion.

Plastics can provide several advantages over traditional materials (for example, metals, glass). These include

1. Being lighter and safer than glass for domestic applications, for example, storage containers, bottles, etc.

2. Inexpensive.

Table 11.5 – *Average material properties for common commercial polymers. Compiled from Rufe (2002) and Groover (2002).*

Material	Modulus of Elasticity psi (MPa)	Tensile Strength psi (MPa)	Melting Temperature ^{o}F (^{o}C)	Density lb/in (g/cm^3)
ABS	300,000 (2,100)	7,000 (50)		0.039 (1.06)
Acrylic	400,000 (2,800)	8,000 (55)	392 (200)	0.044 (1.2)
Polycarbonate	350,000 (2,500)	9,500 (65)	446 (230)	0.044 (1.2)
Polyethylene				
Low density	20,000 (150)	2,000 (15)	240 (115)	0.33 (0.92)
High density	100,000 (700)	4,000 (30)	275 (135)	0.035 (0.96)
PVC	400,000 (2,800)	6,000 (40)	414 (212)	0.051 (1.4)

3. Manufacturing processes that require less energy than glass or metals due to lower melting points.

4. Finishing by painting or plating is generally not required as colors can be blended right in during the forming of the plastics themselves.

5. Low melting points that make in creating intricate shapes easier.

11.11 Properties of Common Polymers

This section will discuss the properties associated with five common commercial polymers (all thermoplastics): polycarbonate, polyethylene, acrylics (trade name: plexiglass), polyvinyl chloride (PVC), and Acrylonitrile-Butadiene-Styrene (ABS). Their material properties are summarized in Table 11.5.

Polycarbonate

Is a transparent polymer with excellent mechanical properties and relatively high heat resistance. It finds wide applications, such as in windows and windshields.

Polyethylene

Is the most widely used polymer. It is inexpensive, chemically inert and easy to process. Polyethylene comes in many forms, the most common being *low-density polyethylene (LDPE)*, lower crystallinity and density, and *high-density polyethylene (HDPE)*, higher crystallinity and density resulting in a stiffer, stronger polymer with a higher melting point.

Acrylics

A key polymer in the acrylic family is *polymethylmethacrylate* (trade name *plexiglass*). It has excellent transparency and is used instead of glass in numerous optical applications like automotive tail lights and aircraft windows. Unlike glass, however, it scratches easily.

PVC

Has largely varying properties depending on the additives it is combined with. For example adding plasticizers can transform the typically rigid PVC (used in pipes, flooring, and toys) to flexible PVC (used in film, sheets, and wire and cable insulation).

Acrylonitrile-Butadiene-Styrene (ABS)

Has very good mechanical properties (strength, ductility and impact resistance). It finds applications in automobile and appliance components, as well as pipes and fittings.

11.12 Steps in Material Selection

The material selection process can be broken down into three basic steps (Figure 11.7):

1. **Determination of required critical material properties from the design operating conditions and environment**. Note that material selection occurs at every step of the design process. At the conceptual stage a wider spectrum of materials should be considered to inspire more innovative designs. But which material properties should one consider using in the material screening process? These will depend on the possible failure modes likely to be encountered during the service of the parts of the design in question, as well as on other desired characteristics. By establishing all the possible failure modes for each particular component and matching them with the associated material properties that those that failure modes, a list of material properties for the screening process can be established. Table 11.6 lists some of the common failure modes and the associated influencing material properties.

2. **Screening of large material databases for candidate materials that meet the critical material properties determined in step 1.** These critical properties can typically be divided into three groups: non-discriminating parameters, go/no-go parameters and discriminating parameters.

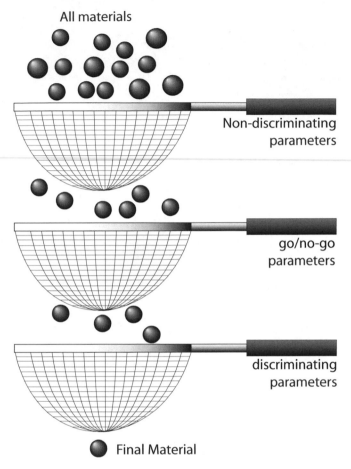

Figure 11.7 – *Three general steps in material selection*

a) *Non-discriminating parameters* are those that must be met if the material is to be used at all. Examples include availability (if the material is not available it cannot be used) or corrosion resistance (if the material will corrode in its operating environment it cannot be used).

b) *Go/no-go parameters* are minimum (or maximum) properties values which candidate materials must meet, where any excess over (or under) these fixed values does not make up for deficiencies in other properties. Examples may include cost and strength.

c) *Discriminating parameters* are those where a minimum (or maximum) property value must be met by a candidate material, and where any excess over (or under) these values *can* make up for deficiencies

Table 11.6 – *Common failure modes and associated influencing material properties*

Mode of Failure	Influencing Material Property								
	US	YS	CS	SS	FP	E	CR	HD	CE
Fatigue (High Cycle)	•				•				
Fatigue (Low Cycle) [1]					•	•			
Gross Yielding		•		•					
Buckling			•			•			
Wear								•	
Thermal Fatigue							•		•
Creep							•		
Gross Deformation						•			

[1] Low Cycle Fatigue is influenced by fatigue properties and the ductility of a material. The latter, however, is determined by the modulus of elasticity.

US - Ultimate Strength	YS - Yield Strength
CS - Compressive Yield Strength	SS - Shear Yield Strength
FP - Fatigue Properties	E - Modulus of Elasticity
CR - Creep Rate	HD - Hardness
CE - Coefficient of Expansion	

in other areas. Examples include cost, density and strength.

Note that depending on the material application, a characteristic that is considered a go/no-go parameter for one application may be considered a discriminating or non-discriminating parameter in another. For example in most military and space applications, cost is treated as a discriminating parameter, whereas for most consumer products, cost is a go/no-go parameter. In this step first non-discriminating parameters and then go/no-go parameters are used to narrow down the candidates materials to a smaller list.

3. **Selecting the final material based on a trade-off of discriminating parameters**. Decision tools discussed in Chapter 6 can be used for this task.

Case Study: Design of a Remote Control Robot for Battlebot Competition

The following case study is taken/comes from the final report of one of the capstone senior design projects at Rutgers University in 2002. It illustrates the use of material parameters and decision matrices for material selection. The student group members were Michael J. Brown, Stephanie A. David, Juan B. Melli-Huber, Jason J. Nikitczuk, Thomas G. Recchia and Justin G. Philpott.

Table 11.7 – *AHP pair-wise comparison chart used to obtain normalized weights for material property criteria in material selection of battlebot armor*

	Stiffness	**Strength**	**Density**	**Sum**	**Weights**
Stiffness	1.00	2.00	0.33	3.33	0.27
Strength	0.50	1.00	0.33	1.83	0.15
Density	3.00	3.00	1.00	7.00	0.58

Introduction. The design team's primary goal was to design, build, and test a robot capable of competing in the annual Battlebots competition[4]. In order to compete in this competition, each participating robot must meet certain guidelines that include constraints on size, weight, control, materials, and drive systems.

After completion of an external search, several design concepts were generated through brainstorming, with the design team finally selecting a four-wheel drive system. The next task was to select the appropriate material to be used as armor on the machine. Due to weight restrictions (the machine had to weigh ≤ 120 *lb*), the design team required a light material, with high stiffness and strength. The latter two properties to resist attacks from and collisions with competitors.

Material Selection. In seeking the appropriate material to use, the battlebot team was primarily concerned about weight, strength and stiffness, with corresponding material properties of density (ρ), yield (σ_u) and ultimate (σ_y) stress, and Young's modulus of elasticity (E), respectively. These would be used as the discriminating parameters during material selection. An AHP pairwise comparison chart (Figure 11.7) was used to rank the three properties and provided normalized weights that were then used in the weighted decision matrix to select the final material.

After considering non-discriminating and go/no-go material characteristics, the team narrowed down their choices to three potential battlebot armor materials: aluminum 6061, steel 1010 and Lexan (a bullet-proof polycarbonate manufactured by General Electric). Their relevant material properties are listed in Table 11.8.

A weighted decision matrix, shown in Table 11.9, was used to select the final material. Although steel had superior strength and stiffness characteristics, Lexan was extremely light with adequate strength and stiffness characteristics. As density had the highest weight (0.61 out of 1), Lexan topped the decision matrix. The top two materials, Lexan and steel, were then compared against machineability. Lexan is much easier to machine than steel. Based on the results from the weighted decision matrix and the secondary material characteristic of machineability, Lexan was therefore chosen as the material for the battlebot armor. Cost was not a factor, as both were relatively inexpensive for the quantities required. A picture of the 'Lexan Armoured' battlebot under construction is shown in Figure 11.8.

---

[4]More information on the competition can be found at http://www.battlebot.org.

Table 11.8 – *Typical material properties for the three candidate materials*

	units	Steel (1010)	Aluminum (6061)	Lexan (9034)
Ult. stress	σ_u, ksi	52.9	18	9.5
Yield stress	σ_y, ksi	44.2	8	9
Density	ρ, lb/in^3	0.28	0.1	0.043
Young's modulus	$E\ ksi$	30,000	10,000	330

Table 11.9 – *Weighted decision matrix for final material selection of battlebot armor*

	σy (ksi)	Norm.	E (ksi)	Norm.	r lb/in3	Norm.	Total
Weights		0.15		0.27		0.58	
Steel	44.2	1.00	29000	1.00	0.28	0.15	**0.51**
Lexan	9	0.20	330	0.01	0.043	1.00	**0.61**
Aluminum	8	0.18	10000	0.34	0.1	0.43	**0.37**

Figure 11.8 – *Lexan battlebot under construction*

11.13   Summary

An overview of the key elements in the material selection process are presented in this chapter. In addition, a discussion of two of the major types of materials, metals and alloys and polymers, is given to show that design teams will be faced with a wide array of choices from which to select their materials. Care should therefore be taken to ensure that the 'best' materials, considering all the evaluation criteria, are selected for the job.

References

American Plastics Council, http://www.apc.org, viewed June 2002.

Groover, M.P., *Fundamentals of Modern Manufacturing: Materials, Processes, and Systems*, Upper Saddle River: Prentice-Hall, 1996.

Pahl, G. and Beitz, *Engineering Design. A Systematic Approach*, 2nd Edition, London: Springer-Verlag, 1996.

Popov, E., *Mechanics of Materials*, Upper Saddle River: Prentice-Hall, 1976.

Rufe, P.D., *Fundamentals of Manufacturing*, Dearborn: Society of Manufacturing Engineers, 2002.

Bibliography

Groover, M.P., *Fundamentals of Modern Manufacturing: Materials, Processes, and Systems*, Upper Saddle River: Prentice-Hall, 1996.

Matweb.com, http://matweb.com, viewed May 2004.

Pahl, G. and Beitz, *Engineering Design. A Systematic Approach*, 2nd Edition, London: Springer-Verlag, 1996.

Popov, E., *Mechanics of Materials*, Upper Saddle River: Prentice-Hall, 1976.

Rufe, P.D., *Fundamentals of Manufacturing*, Dearborn: Society of Manufacturing Engineers, 2002.

Chapter 12

Physical Models and Prototypes

12.1 Introduction

The construction of three dimensional physical *models* or *prototypes* of a design allows teams to evaluate their designs and provides better visualization of the idea than an equivalent 3D computer model.

A physical model is similar to the actual design. It may be scaled up or down or made of different materials than the actual design. The model allows design teams to test form and fit, as well as conduct physical and performance tests whose results are scaleable to the actual design. For example, architects and structural engineers make models of buildings, bridges, road systems, etc. prior to construction. These models are primarily used for three dimensional visualization of the final design. Engineers in the aerospace and automobile industry frequently use scaled-down models of airplanes and cars, respectively, to perform wind-tunnel testing. The tests allow the engineers to come up with shapes that reduce drag (air resistance) without compromising structural integrity. For both vehicles, as drag goes up, so does fuel consumption. Electrical and computer engineers frequently use bench top models of circuits for testing purposes, before production of the final circuit boards or integrated circuits chips.

Physical models can generally be divided into four broad categories (Dieter, 1991):

1. *Proof-of-concept model.* Used to show the basic design principle of the concept. It is very elementary and constructed from readily available materials.
2. *Scale model.* Primarily used to determine form and fit as well as visually communicate what the design will look like. It is typically non-operational.
3. *Experimental model.* This is a functional model, but may not look exactly like the final design. It is used to perform extensive testing to ensure the

design will meet specified performance requirements.

4. *Prototypes.* Are an exact replica of the final design, including the materials used. Prototypes are usually constructed for cases where test results from models would be inadequate and tests would need to be conducted on the final design. Examples include automobile crash testing and airplane test flights.

The decision as to which type of physical model to build will depend, amongst other factors, on (1) the purpose of the model or prototype, (2) the cost and time to build, and (3) the required number of copies of the final design. An example of the latter criterion would be the design and construction of buildings. As a single copy of a building is generally required, it would not make sense to build an exact replica (prototype) but rather a scaled-down model. On the other hand, systems and assemblies that go into buildings are usually prototyped for safety, as hundreds or thousands of units for each model may be required. For example, hundreds of the same HVAC or plumbing systems are found in numerous buildings. As a result these systems are prototyped prior to commercial production and installation.

This chapter provides a general overview of manufacturing methods that could be used by design teams to create crude physical models.

12.2   Rapid Prototyping - An Overview

Rapid-prototyping or rapid-modeling allows the 3D visualization of part geometry, form, fit and function, as well as aesthetics and manufacturability. Rapid prototyped parts are also for the manufacture of molds in the casting industry. These 'masters' allow the easy inclusion of any eleventh hour changes to the mold.

This text concentrates on three common rapid-prototyping techniques: 3D printing, Stereolithography (SLA) and Fused Deposition Modeling (FDM). All are classified as Solid Freeform Fabrication (SFF) methods. As the materials used do not have the final structural strength or mechanical and thermal properties as the final products, they only provide a physical model before the final parts are manufactured using other processes and materials.

Rapid-prototyping machines convert the 3D computer-aided design (CAD) models into the 3D physical models. For all three methods, this is accomplished through the following general steps:

1. **Creation of a CAD solid model.** Most popular solid-modeling programs (for example Alibre, AutoCAD, IDEAS, Pro/Engineer and Solidworks) support the creation of files in a format understood by rapid-prototyping machines. Using any of these programs, therefore, you can create the solid model of the part to be manufactured.

2. **Conversion of CAD solid model files to .STL format**. The .STL file format, originally developed for Stereolithography, has evolved to become the industry *de facto* standard. The format approximates the solid-model surfaces via a series of triangles. Most CAD software can convert the solid model files to .STL files simply by performing a *'Save as...'* and selecting the .STL format. Some programs will list the format as 'Rapid Prototyping'. Irrespective of the program used, it is advisable to save the .STL as a binary file, as it saves on time and file size.

3. **Addition of support structures.** If necessary, support structures are added to the part (more on this in the next section).

4. **Creation of the layers.** Prior to building the actual part, separate software specific to the rapid-prototyping machine is used to slice the part into a series of thin consecutive layers. Each layer has the same thickness.

5. **Part construction.** The rapid-prototyping machine then forms each layer from the bottom up, building the next layer on top of the previous. In this manner, the part is 'built', layer by layer.

Stereolithography

Developed in 1984 by Charles Hull, stereolithography was the first commercialized Solid Free Form (SFF) technique. Hull joined forces in 1986 with Raymond Freed and established 3D Systems, Inc. for the development and marketing of stereolithography-based rapid-prototyping equipment (Jacobs, 1996).

Stereolithography produces solid models by systematically curing a liquid photopolymer resin with an ultraviolet laser beam. The resin turns from liquid to solid when struck by ultraviolet light. Figures 12.1 and 12.2 illustrate a stereolithography machine produced by 3D Systems Inc. The PC pictured controls the build process.

The part to be produced is 'built' layer by layer on the build platform, which rests below the surface of the liquid resin. On initiation of the build process (refer to Figure 12.3(a)), the platform is lowered precisely one layer thickness below the surface of the liquid resin. A Neon-Helium laser beam then accurately traces out the cross-sectional area of the first layer of the part. Using a series of mirrors mounted on high resolution stepper motors, the laser beam can be accurately moved to any X-Y location on the resin surface. To create voids or holes the laser is selectively turned off and on as needed. Everywhere the laser beam makes contact with the resin, the liquid solidifies. Layer thickness ranges from 0.002" to 0.02". In general, the thinner the layer, the smoother and higher tolerance of the produced parts, but the longer the build time.

On completion of the first layer, the build platform lowers by another layer thickness, allowing fresh liquid resin to flow over the previously solidified layer (refer to Figure 12.3(b)). In older stereolithography machines, a wiper passes over the solidified layer to even out the liquid level and reduce the effects of

Figure 12.1 – *External view of a 3D systems stereolithography machine*

Figure 12.2 – *Stereolithography machine with the build chamber door open, displaying the build platform and the vat containing the resin. The build chamber is enclosed to contain the resin fumes.*

surface tension on the part edges. Recent machines, however, utilize pump-driven recoating systems to achieve the same effect. The laser beam then traces out the next layer. The resin cures and bonds to the previously 'built' layer as shown in Figure 12.3(c). In this manner, the part is built from the bottom up, layer by layer.

Geometries with overhangs or undercuts must be given special attention during the build process. Overhangs in the object are not self-supporting during the layer-by-layer build process. Consequently, a series of support structures are designed as part of the object to give support to the overhang. The support structures are generally very thin and incorporated either manually or automatically into the object design. They are later removed once the build process is complete.

When the build is complete, the platform raises the part out of the liquid resin allowing excess resin to drain back into the vat. After manually swabbing any left over resin, the part undergoes a final cure in a post-curing apparatus (looks like an oven) that bathes it in intense UV light for a period of time ensuring that all the resin has cured.

Any support structures that were created are now removed and surfaces undergo finishing processes, for example sanding, as necessary. Figure 12.4 displays several examples of parts made from stereolithography.

Fused Deposition Modeling (FDM)

Fused Deposition Modeling is similar to Stereolithography in that the three dimensional part is built layer by layer based on the 'slices' generated from the CAD model. But unlike stereolithography, the FDM systems build their parts by sequentially depositing a polymer that is reeled out from a spool. As the polymer passes through a heated head, it melts to a liquid state and is then deposited during the build process. The polymer rapidly solidifies once deposited, bonding to any previous layers. Once the first layer is built, the head moves up the equivalent of one layer thickness and begins to deposit the next layer. In this manner the 3D part is quickly constructed layer by layer. Material deposit rates can be up to 900 in. per minute, making the build process quite rapid (Crump, 1992).

Several polymers can be used for the FDM process. These include Acrylonitrile Butadiende Styrene (ABS), investment casting wax and E20 (a thermoplastic polyester-based elastomer). These materials are typically interchangeable on the same machine depending on the application. All the materials are inert, non-toxic thermoplastics with melting points less than 180^oF. FDM machines do not require any special rooms similar to stereolithography where the toxic resin needs special environmental considerations.

The FDM process was developed by Scott Crump, who founded the Stratasys Corp. to commercialize FDM-based machines in 1988. In 2004 the Stratasys Corp. was still the industry leader in FDM machines. Compared to Stereolithography, FDM machines are generally cheaper and provide faster build times. Stereolithography, however, generally provides higher accuracy, better dimensional tolerance, and better surface finish.

3D Printing

Like all other rapid prototyping machines, 3D printers build solid parts layer by layer from the bottom up. Z Corp's (http://www.zcorp.com) printers work by first spreading a thin layer of powder. Next a binder or glue is sprayed onto the powder corresponding to the solid portions of the part being manufactured. Areas that are not sprayed remain as powder that are later removed and reused. The liquid resin is sprayed on using a print head similar to that found in conventional ink-jet printers (hence the name 3D printing). The process is repeated, thus building the part layer after layer. The technology was first developed at the Massachusetts Institute of Technology.

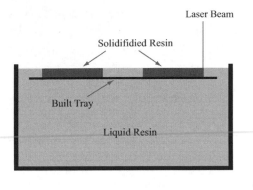

(a) Building the first layer by systematic curing of the resin with a UV laser beam

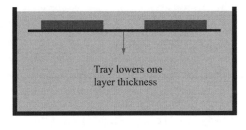

(b) Once a layer is built, the tray lowers the equivalent of one layer thickness allowing liquid resin to flow over the previously built layer

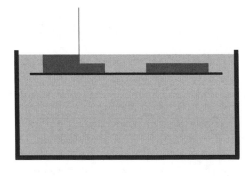

(c) The laser beam begins to selectively cure the resin and build the next layer

Figure 12.3 – *Steps in the Stereolithography build process*

 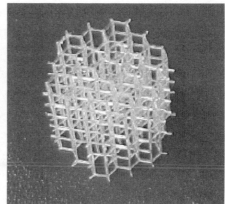

Figure 12.4 – *Sample parts made from the stereolithography process* (Courtesy, Dr. Noshir Langrana, Rutgers University)

The Z-Corp 3D printers can use three basic types of materials depending on the application of the manufactured part (Z Corp, 2004):

- **Plaster-based material:** provides high strength and accurate representation of design details. Of the three, it is the best for thin-walled parts.

- **Composite-based material:** cannot obtain as accurate design details as plaster-based materials due to a slightly larger minimum layer thickness. Great for snap-fit and assembly applications.

- **Starch-based material:** is the weakest and cheapest of the three. Great for high speed printing or the manufacture of large bulky parts.

Irrespective of the material used, all completed parts can be impregnated with resins and waxes to improve the strength durability and temperature resistance of the parts. In addition, the parts can be readily sanded and painted for a finished appearance.

3D Systems, Inc. (http://www.3dsystems.com) manufactures printers that deposit very thin layers of a thermoplastic/wax material to build the part. Unlike FDM machines where the polymer is reeled out from a spool, these printers place small 'dots' of material in much the same way as conventional ink-jet printers. The series of dots placed in the desired locations form the current layer. Once a layer is complete, the next layer is built in a similar way. Resolutions of up to 600 dots per inch (dpi) can be obtained.

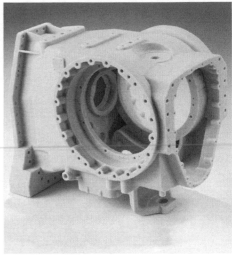

Figure 12.5 – *Sample parts made using 3D printing* (Courtesy, Z-corp, Inc.)

12.3 Machining

Machining is the creation of a final desired shape using cutting tools to remove excess material from a work piece. Examples include milling, turning and
drilling. Machining plays a key role in the manufacturing process as it

- Is applicable to a wide variety of materials: nearly all solid metals and
  most plastics.
- Allows the generation of regular geometries - for example flat planes, holes,
  and arcs - whose combinations can produce a wide variety of shapes.
- Can achieve very tight tolerances as compared to other processes ($>$ 0.001
  in).
- Can yield very smooth surfaces as compared to other processes ($>$ 0.16
  μin)

Machining is therefore typically used to obtain the final geometry dimension
and finish of a part after another process has been used to get the part to the
rough shape. For example, a band saw could be used to obtain an approximate
dimension and shape of the part followed by machining to get the final form and
finish.

The Machining Decision Process

The machining process can be loosely divided into the following steps:

1. **Analyze the part geometry** to decide which machining operations are required to obtain the desired part geometry. Examples of operations include creating holes, removing material on a flat plane, removing material about an axis, cutting the material to a rough shape, etc.

2. **Select the machining process(es)** based on the required machining operations determined in (1). Machining operations include cutting, drilling, turning, milling, etc.

3. **Analyze the material.** What is the part to be made of? Material properties will dictate, among other things, (a) the tools to use; (b) speed at which the machining process is performed; (c) if lubrication is required; (d) steps to be taken if heat is a problem, for example when working with steels; and (e) steps to be taken if cracking is a possibility, for example when working with brittle materials.

4. **Determine tools necessary** to perform required machining operations based on the conditions determined above.

5. **Write down each step of the entire machining process.** This forces you to think through the entire machining process, eliminating mistakes and identifying potential problems early. Once you are in the shop, you <u>must</u> be ready to go, not still deciding what step comes next.

6. **Estimate manufacturing time for each part.** All machining processes can typically be divided into four sequential parts:

   a) **Set up.** Getting together the required tools and jigs and fastening the material to the appropriate machine.

   b) **Rough cuts.** Removes fairly large amounts of material at a time. During this phase, the cutting speed is typically slow with a large depth of cut and feed rate. This results in relatively high material removal rates, quickly arriving close to the final dimensions. The disadvantage, however, is that the surface finish is very poor.

   c) **Finishing cuts.** Removes very small amounts of material. A relatively high cutting speed coupled with very low depth of cut and feed rate results in very low material removal rates but a good final finish.

   d) **Tear-down and clean up.** Returning tools to their proper location, removing the complete part and jigs off the machine, and then cleaning the machine and the floor around it.

   Keeping these four aspects of the machining process in mind helps in estimating the time it takes to complete each part. This assures completion within allotted time slots.

7. **Manufacture the part.**

Figure 12.6 – *Dangerous situation: Chuck key left in drill press chuck*

Machine Shop Safety Overview

The machine shop, by its very nature, is a dangerous environment. Many people have been cut, blinded, paralyzed and even killed in machine shops doing seemingly benign tasks. Safety must be your **primary** concern when using the shop. Good safety practices will ensure a productive and safe project for everyone. The main safety issues are listed below.[1]

1. **Wear Safety Glasses**. Safety glasses must to be worn any time you enter the machine shop. If you need to take them off to rest your eyes or clean the lenses, leave the room. Glasses protect you not only from the hazards associated with your work, but also from those of other people in the shop. This is by far the most important safety rule.

2. **Tie up long hair**. Hair caught in the spindle of rotating machinery has disastrous results.

3. **Avoid wearing any loose clothing**. For the same reasons as in (2), loose clothing (e.g. ties, scarves) can get caught in the spindle of rotating machinery resulting in serious injury.

4. **Never leave the chuck key in the chuck**. This rule applies to the drill press (see Figure 12.6), milling machines and the lathe (see Figure 12.7).

---

[1]Compiled by Prof. Nick Glumac, Department of Mechanical and Industrial Engineering, University of Illinois.

Figure 12.7 – *Dangerous situation: Chuck key left in lathe chuck*

Figure 12.8 – *Dangerous situation: Wrench left in milling machine drawbar*

If a machine is switched on with the chuck key in the chuck, the key will be ejected from the machine at a high speed and can cause serious injury or death.

5. **Never leave the wrench in the milling machine drawbar**. The same logic as in (4) applies here (see Figure 12.8). If the wrench is left on the drawbar, it will be flung when the machine is turned on. Do not let go of the wrench unless it is sitting in its resting place off to the side of the machine.

6. **Sufficiently tighten tools and workpieces in vices, holders and chucks.** Once you place a tool in a holder (e.g., on the lathe) or in a collet (e.g., on the milling machine or lathe), make sure you tighten it sufficiently. A rapidly spinning tool, with razor sharp edges, that drops out of a holder can have disastrous consequences. Similarly make sure your work piece is securely fastened in a vice (e.g., on the mill) or in a chuck or

Figure 12.9 – *Using scrap material to push work piece through the band saw. Note that the student's fingers are on the side and not in the direct path of the blade*

collet (e.g., on the lathe).

7. **Watch for clearances between tools and vices/chucks**. Driving a tool into a spinning chuck or driving a spinning tool into a stationary vice will likely cause the tool to shatter, spewing shrapnel over a wide area. Pay special attention to clearance issues, especially on the lathe. If operating close to the chuck, check for clearances **with the machine off** by bringing in the tool as close to the chuck as you need to. Spin the chuck by hand, checking all around the tool and chuck for critical clearances. If it appears too close, rearrange your set up so that clearance is not an issue. A similar problem can happen if too deep a cut is taken. If at the start of a pass the machine makes excessive noise or vibration, stop immediately and recheck your spindle speed and depth of cut.

8. **Keep your hands far from the band saw blade**. More injuries occur on the band saw than perhaps any other machine. This is because your hands are in close proximity to the blade. Always use a piece of wood or metal to push on your work piece. Further, keep your hands to the side of the blade so that if they slip, they do not encounter the blade (Figure 12.9).

9. **Remove the chips on the edges of machined pieces with pliers or a brush, NOT your fingers.** This rule should be quiet obvious, yet this type of injury is quite common. The chips on the edges are very sharp and can result in deep cuts.

10. **During setup or clean up, look out for sharp tools.** As you setup or cleanup, the sharp edges of a cutting tool can result in deep cuts. If

276

possible, move the tool away or keep it within its protective casing.

11. **Always wear closed shoes in the shop**. Frequently tools and work pieces fall to the floor and can cause serious damage to exposed toes. Consequently, sandals or other forms of open shoes should not be worn in the shop.

12.4   An Overview of Fastening Methods

Adhesives

Adhesives find wide use in joining a large variety of similar and dissimilar materials including metals, plastics, ceramics and wood. In general, all adhesives, once applied in a liquid state, undergo a chemical change, solidifying and adhering to the mating parts. This process, referred to as curing, may require heat and or pressure as a catalyst. The time it takes to cure, *the curing or setting time*, can range from a couple of seconds to a few days. The strength of the bond can be quantified in terms of its shear strength and its tensile strength. The shear strength is the force that would be required to separate the joint as the mating parts slide over one another. The tensile strength, on the other hand, is the force required to pull the parts apart. There are numerous types of adhesives available on the market. The following sections provide a broad description of the common *types of adhesives* (Groover, 1996; McMaster, 2002).

EPOXIES

One-part and two-part epoxies provide high-strength bonds to metal, rubber, fiberglass, urethane, wood, glass, ceramics, concrete, and some plastics. Two-part epoxies require mixing to initiate the curing process, often resulting in a limited application time. Applications include bonding, laminating, and filling spaces between surfaces. The curing time for epoxies can range from a few minutes to several hours. They have good resistance to high temperatures, solvents, and impact. They exhibit a shear strength to 5,800 *psi* (40 MPa) and a tensile strength to 7,200 *psi* (50 MPa).

Metal-Filled Epoxies, often referred to as liquid metals, have the addition of powdered metal fillers. Applications include bonding, sealing, plugging, patching, repairing, and rebuilding worn and eroded metal back to original dimensions. They are typically nonrusting, resistant to most chemicals, water, and temperature and can be machined after they cure.

ACRYLICS

The primary application of acrylics is bonding. Available in one-part and two-part formulations, acrylics begin to harden within minutes (sometimes seconds) yielding high-strength bonds. They are especially good for plastic to metal, rubber to metal, and plastic to plastic bonding. Two-part acrylics work with a catalyst and require mixing. One-part acrylics don't need mixing but are used with a primer/activator (usually a liquid) that's sprayed or brushed onto the surface being bonded. Acrylics are more tolerant of moisture as well as oily, dirty, and unprepared surfaces than super glues. Although they can be used for filling spaces between surfaces, they are not as good as epoxies. They exhibit a shear strength to 4,900 *psi* (34 MPa) and a tensile strength to 3,600 *psi* (25 MPa).

URETHANES

Like acrylics, the primary application of urethanes is bonding. Because they form more flexible bonds than epoxies and acrylics, they are often used on films, foils, and elastomers. They bond well to metal, rubber, polyvinyl chloride (PVC), polycarbonate, and especially wood. Curing time ranges from a few minutes to 48 hours. They exhibit a shear strength to 2,800 *psi* (19 MPa) and a tensile strength to 7,250 *psi* (50 MPa).

SUPER GLUES

Commonly referred to as cyanoacrylates, crazy glues, or super bonders, their primary application is the bonding of closely mated parts with minimal weather exposure. Their curing times are relatively short. They adhere to a wide range of substances, including rubber, plastic, and metal, but are not recommended for use on glass. Super glues are poor at filling spaces between surfaces and give inferior resistance to impact, temperature extremes, moisture, and solvents. They exhibit a tensile strength to 2,500 *psi* (17 MPa).

ANAEROBICS

Commonly referred to as thread lockers or retaining compounds, anaerobics are acrylic-based adhesives that reach full strength in the absence of air and the presence of metal, such as when confined between the threads of a nut and bolt assembly. Applications include locking threads, sealing, gasketing, bearing retention, and bonding/retaining cylindrical assemblies. When combined with other materials, they can be used as a sealant for pipe joints and threads. Depending on the anaerobic picked, the bonds may or may not be permanent. Anaerobics have relatively short curing times and provide excellent resistance to solvents, water, weather, and temperatures up to 400°F. They exhibit a shear strength to 2,900 *psi* (20 MPa) and a tensile strength to 5,800 *psi* (40 MPa).

SILICONES

Applications include low-strength bonding, sealing, filling spaces between surfaces, encasing items in adhesive, and gasketing. Most are one-component RTV (room-temperature vulcanizing) liquid rubbers that keep their rubber properties under almost any conditions. They cure through exposure to moisture in the air in 24-72 hours. They provide excellent resistance to temperature extremes (-60° to +450°F), as well as chemicals, UV radiation, ozone, and weather. Silicones make great gap fillers and sealants for low-stress applications. They adhere to rigid and flexible substrates, including metal, glass, fiberglass, cement, canvas, rubber, plastic, ceramics, and wood. Their high viscosity make them good for vertical and overhead applications. They exhibit a shear strength to 450 *psi* (3 MPa) and a tensile strength to 725 *psi* (5 MPa).

CONSTRUCTION ADHESIVES

Commonly referred to as panel adhesives, construction adhesives are made of generally viscous plastics formulated of natural or synthetic rubber in a solvent or water carrier. Among other uses/applications, they are used for installation of paneling, drywall, foam, and flooring. They cure quickly, are excellent for filling spaces between surfaces, and remain flexible when dry.

CONTACT CEMENTS

Manufactured from synthetic rubber, typically neoprene dispersed in a solvent or water, contact cements form permanent bonds. Applying the cement to the two surfaces to be joined, letting them air dry, and then bringing the surfaces together results in them sticking instantly. Applications are primarily laminating and bonding.

HOT MELTS

Applications include bonding, sealing, filling spaces between surfaces, caulking, packaging, and parts-holding for nailing and other final assembly. Hot melts are manufactured from thermoplastics that melt when heated and solidify as they cool. Polymers used include ethylene vinyl acetate, polyethylene, butyl rubber, polyamide, polyurethane and polyester. They are ideal for rigid-to-flexible, low-strength bonds that begin to harden quickly. They exhibit a shear strength to 1,000 *psi* (7 MPa) and a tensile strength to 1,300 *psi* (9 MPa). They're excellent for filling spaces between surfaces and provide good adhesion to most materials, especially porous surfaces such as wood, paper, and leather. Hot melts have low solvent resistance, limited temperature resistance, and require dispensing equipment such as glue guns.

Brazing and Soldering

For both processes, a heat source is used to melt a bonding agent (typically a metal alloy) that flows between the parts to be mated. The melting point of the bonding agent must be lower than that of the parts to be joined. The agent

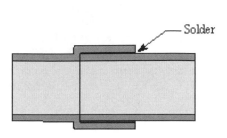

(a) Schematic of overlapped tubing to give soldered joint additional strength

(b) Soldered copper tubing showing overlapped soldered joint

Figure 12.10 – *Overlapping segments to achieve structural strength in soldered joints*

bonds to both parts and solidifies as it cools, creating a strong mechanical joint. In *Brazing* the bonding agents are typically alloys of copper, silver, aluminum, silicon, or zinc, necessitating relatively high temperatures, above 840°F. Metals joined by brazing include copper, aluminum, and nickel alloys, as well as plain carbon, and alloy and stainless steels.

Soldering, on the other hand, uses bonding agents that have lower melting points, 392° - 572^oF (200° - 300^oC). Soldering alloys include lead-tin, tin-zinc, tin-silver and lead-sliver. Due to the weaker mechanical strength of solder alloys (as compared to brazing alloys), soldered joints should be created with overlapping fittings or interlocking lap joints to give additional mechanical strength (Figure 12.10). Common applications of soldering include piping and tubing joints and the attachment of components to circuit boards and wires to components in the electronics industry.

Soldering is used on copper and brass parts, typically with a lead(50%)-tin(50%) alloy solder. The lead-tin alloy is best suited for low strength, low temperature applications where cleanliness is not required. At temperatures below 212°F (100^oC), lead-tin soldered joints have a shear strength of $\sim$ 5000 *psi* (35 MPa).

For engineering applications flames are the main heat source. In the electronics industry electric solder irons are the norm. Table E.3 in Appendix E lists the composition of common solders and their applications.

Threaded Fasteners

Threaded fasteners, commonly referred to as bolts or screws, are mostly used for applications that require frequent disassembly. Figure 12.11 illustrates the physical meaning of the common terminology used for threaded fasteners. A description of each term follows.

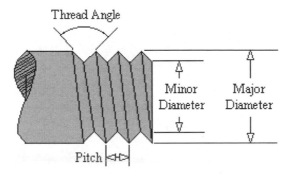

Figure 12.11 – *Screw thread terminology*

- **Thread Angle.** Typically 60°.
- **Pitch.** The distance between two successive crests. It represents the distance the screw would travel lengthwise corresponding to one revolution of the screw.
- **Minor Diameter.** The smallest diameter on the thread.
- **Major Diameter.** The largest diameter on the thread.

Screw threads are categorized by various standards. In the United States threads are specified according to the *American Standard Unified* and *American National* threads standards. The two main thread series under both standards are the *coarse thread series* and the *fine thread series*. Coarse threads, referred to as United National Coarse (UNC) threads, are for general use and provide maximum strength. Fine threads, referred to as United National Fine (UNF) threads, are primarily used where small adjustments are required or for parts under shock and vibration that may loosen the screws. UNF threads are more difficult to shake loose than UNC threads. Tables E.1 and Table E.2 in Appendix E lists the primary characteristics for different size fasteners from UNC and UNF standards, respectively.

In addition to the type of thread, the thread tolerance must also be specified. There are four broad categorizations of thread tolerances:

- **1A and 1B.** The A and B correspond to external and internal thread, respectively. In general the lower the number, the larger the thread clearance. Thread clearance is the distance between the screw and its mating parts, for example between a bolt and a nut. 1A/1B threads have the largest clearance and are primarily used where dirty or scratched parts need to be assembled.

Table 12.1 – *Examples of threaded fastener designation*

Specification	Description
$\frac{1}{2}$-13 UNC-2A	External threaded fastener, 1/2 in. major diameter, 13 threads per inch and a 2A tolerance.
$\frac{1}{4}$-20 UNC-1B	Internal threaded fastener, 1/4 in. major diameter, 20 threads per inch and a 1A tolerance.
$\frac{1}{4}$-13 UNF-3A	External threaded fastener, 1/4 in. major diameter, 28 threads per inch and a 3A tolerance.
6-32 UNC-3A	External threaded fastener, gauge number 6 (0.138 in.) major diameter, 32 threads per inch and a 3A tolerance.

- **2A and 2B.** Best suited for most general applications and usually supplied if a thread tolerance is not specified.
- **3A and 3B.** Provide a very precise fit.
- **5A and 5B.** Give an interference fit and are therefore used for semi-permanent applications. An interference fit is where there is negative clearance between the screw and its mating part.

Threaded fasteners are specified using a four-part designation: *'dia-tpi series-tol'*, where *dia* specifies the major diameter; *tpi*, the number of threads per inch; *series*, the thread series; and *tol*, the tolerance. For major diameters under 1/4 in., the gauge number and not the actual diameter is specified. Several examples are presented in Table 12.1 to illustrate its use.

Finally, in addition to the selection of screw types and sizes, material choice based on the application of the fastener must be made. Common fastener materials include

- **Aluminum.** Resistant to weather corrosion. Has the same strength as mild steel but one third the weight. Aluminum also has good electrical and thermal conductivity. Most common alloys used are 2024-T4, 2011-T3, and 6061-T6.
- **Brass.** Softer than steel and stainless steel. It is nonmagnetic and provides very good corrosion resistance.
- **Nickel and alloys.** Provide good corrosion resistance and superior performance at elevated temperatures.
- **Nylon.** A nonconductive material that resists chemicals and solvents.
- **Steel.** Several alloys of steel are commonly used. These include

  - **Type 316 Stainless Steel.** Nonmagnetic. Contains molybdenum giving it better resistance to corrosion than 18-8 stainless steel.

- **18-8 Stainless Steel.** Can be mildly magnetic. Provides superior corrosion resistance in chemical, marine and generally corrosive applications.
- **Black Oxide Finish 18-8 Stainless Steel.** An oxidizing treatment that creates a black oxide coating giving improved corrosion resistance and lubrication.
- **Zinc-Plated Steel.** Provides good to excellent rust resistance. Manufactured from 1006-10038 carbon steel.

- **Silicon Bronze.** Composed of 95-98% copper and silicon for strength. Non-magnetic and provides superior corrosion resistance to salt water, gases and sewage.
- **Titanium.** As strong as steel with only 60% of the weight giving them a high-strength to weight ratio. As a result it is widely used in the aerospace industry. Provides better resistance to chemicals, salt water and acids than stainless steel.

Pipe Tubings and Fittings

Piping finds infinite uses in engineering from conduits of liquids and gases to structural members. This section gives a brief overview of piping materials, pipe sizes, threads and thread conventions.

COMMON MATERIALS AND APPLICATIONS

The following list provides the properties and applications of common materials used for manufacturing pipes. Metallic materials are listed fist, followed by polymers.

- **Aluminum** is strong, ductile, malleable and lightweight. It can be anodized for better corrosion resistance. For low-pressure systems with water-based fluids in agricultural and some food-processing applications.
- **Brass** is soft, provides tight seals and is easier to install than other metals. It can be used interchangeably with copper where heavier walls are required. It resists corrosion from salt water as well as fresh water polluted with waste from mineral acids and peaty soils. Primarily used with water for plumbing and heating. Also good for use in pneumatic and marine applications.
- **Cast iron** is a harder, more brittle iron. It is used with steam, water (heating and cooling), and is good for fire protection applications. **Malleable Iron** is a softer and more ductile iron and is used with gas, oil, and water. It is also good for industrial plumbing.

- **Steel** is used for high-pressure systems in hydraulic, pneumatic and petro-chemical applications. **Stainless Steel** is corrosion-resistant, retains a lustrous appearance, provides high strength at high temperatures, and helps prevent contamination of products being transported. Type 304 stainless steel is a low-carbon chromium-nickel stainless steel. Type 316 stainless steel is similar to Type 304 but has a higher nickel content, as well as molybdenum content for stronger resistance to heat and corrosion. Stainless steel pipes are used with gas, oil and water. They are sanitary and noncontaminating making it a good choice for use in food, dairy, brewery, beverage and pharmaceutical industries.
- **Acrylonitrile-Butadiene-Styrene (ABS).** Good strength, ductility and impact resistance. Generally has good corrosion resistance.
- **Polyethylene.** Is a soft, flexible and ductile plastic with good corrosion resistance.
- **Polypropylene.** Flexible and resistant to acids, bases and numerous solvents.
- **Polyvinyl Chloride (PVC).** Rigid material with good strength and corrosion resistance. **Chlorinated Polyvinyl Chloride (CPVC).** Added chlorine makes it applicable at higher temperatures than PVC.

PIPE SIZE

Pipe size refers to the industry accepted size designation and is not the actual dimensions of the pipe. For example, a male pipe fitting with a 1/2 in. pipe size actually has an outer diameter (OD) of 0.840 in. Tables E.4 and E.5 in Appendix E give the correspondence between industry standard 'pipe sizes' and the actual OD of the pipe. The thickness of a pipe, and hence the inner diameter (ID), is specified by the *pipe schedule*.

- **Schedule 10** is the thinnest-wall pipe and therefore the lightest.
- **Schedule 40** is the most common pipe wall thickness and great for all-around use.
- **Schedule 80** is a thicker, heavier pipe appropriate for high-pressure applications where greater mechanical strength is required.
- **Schedule 160** is the thickest, heaviest pipe of the common piping schedule. It finds use in the most demanding applications where high pressures and temperatures are present. Tables E.4 and E.5 in Appendix E relate the OD, ID and pipe thickness for various schedules. Note that the terms 'IPS', 'size', 'pipe size' and 'nominal pipe size' all have the same meaning.

PIPE THREADS

Pipe threads come in two forms (Figure 12.12): tapered or straight (parallel). Tapered threads, mated together and tightened, compress to form a seal. The

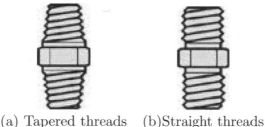

(a) Tapered threads   (b)Straight threads

Figure 12.12 – *Schematics of (a) Tapered and (b) Straight threads*

Figure 12.13 – *Use of Teflon tape as sealant*

National Pipe Taper (NPT) threads is the common pipe thread standard used in North America. They are often more specifically referred to as FPT for female threads (internal) and MPT for male threads (external). NPTF (Dryseal) threads are modified NPT threads that provide a better seal than NPT threads. However, it is recommended that all threads (NPT and NPTF) a sealant compound or Teflon tape to assure a leak-free seal (Figure 12.13).

Straight threads, on the other hand, are used solely for mechanical joining to hold a fitting in place, making it necessary to have an O-ring (elastomer), hard metal seal, or soft metal seal. Straight pipe threads standards include National Pipe Straight Mechanical (NPSM), National Pipe Straight Locknut (NPSL) and National Pipe Straight Hose (NPSH) threads. Irrespective of the type of thread used, make sure that the pipes are of the same size when matching pipe OD for male threads and pipe ID for female threads.

References

Ahlers, R. J., and Reinhart, G., "Rapid Prototyping," SPIE Proceedings Series, vol. 2787, 1996.

American Plastics Council, 'Plastics Resource: Information on plastics and the environment', http://www.plasticsresource.com, June 2002.

Crump,S.S., "Rapid Prototyping Using FDM", Modern Casting, pp. 36-37, April 1992.

Dieter, G.E., *Engineering Design: A Materials and Processing Approach*, 2nd Edition, New York:McGraw-Hill, Inc., 1991.

Groover, Mikell, *Fundamentals of Modern Manufactuing: Materials, Processes and Systems*, Upper Saddle River: Prentice-Hall, 1996.

Jacobs, F. Paul, *Stereolithography and Other RP&M Technologies*. New York:ASME Press, 1996.

McMaster-Carr Inc., Online Catalog, http://www.mcmaster.com, June 2002.

Moore, J.H., C.C. Davis and M. Coplan, *Building Scientific Apparatus - A Practical Guide to Design and Construction,* Second edition, Reading, Massachussetts: Perseus Books, 1991.

Mott, Robert, *Machine Elements in Mechanical Design*, 3rd edition, Upper Saddle River:Prentice Hall, 1999.

Stereolithography.com, http://www.stereolithography.com, June 2002.

Wood, L., *Rapid Automated Prototyping: An Introduction*. New York: Industrial Press Inc., 1993.

Z Corp, Official Website, http://www.zcorp.com, viewed April 2004.

Bibliography

Ahlers, R. J., and Reinhart, G., "Rapid Prototyping" SPIE Proceedings Series, vol. 2787, 1996.

American Plastics Council, "Plastics Resource: Information on plastics and the environment", http://www.plasticsresource.com, June 2002.

Crump,S.S., "Rapid Prototyping Using FDM", Modern Casting, pp. 36-37, April 1992.

Groover, Mikell, *Fundamentals of Modern Manufactuing: Materials, Processes and Systems*, Upper Saddle River: Prentice-Hall, 1996.

Jacobs, F. Paul, *Stereolithography and Other RP&M Technologies*. New York:ASME Press, 1996.

McMaster-Carr Inc., Online Catalog, http://www.mcmaster.com, June 2002.

Moore, J.H., C.C. Davis and M. Coplan, *Building Scientific Apparatus - A Practical Guide to Design and Construction,* Second edition, Reading, Massachussetts: Perseus Books, 1991.

Mott, Robert, *Machine Elements in Mechanical Design*, 3rd edition, Upper Saddle River:Prentice Hall, 1999.

Stereolithography.com, http://www.stereolithography.com, June 2002.

Wood, L., *Rapid Automated Prototyping: An Introduction*. New York:Industrial Press Inc., 1993.

Chapter 13

Commercial Manufacturing Processes

13.1 Manufacturing Processes for Metals - An Overview

Metal manufacturing processes occur either at or close to room temperature (cold-working) or at an elevated temperature (hot-working). Advantages of cold working include better accuracy and surface finish, heating of work not required and the resulting strain hardening increases the strength and hardness of parts. Disadvantages include residual stresses in parts, larger forces required for operations, and limited deformation from ductility and strain hardening of the metal.

In *hot working*, manufacturing processes operate at the re-crystallization temperature (approximately half the melting point). Advantages include lower forces and power required to deform metal, general freedom of part from residual stresses, no part strengthening from work hardening, and substantially more plastic deformation can be produced than from cold working. Disadvantages include the inability to hold tight tolerances, the power required to heat the metal, and the material surface has a characteristic oxide scale.

Casting

Metal casting is the process by which a metal or metal alloy is poured into a mold and hardened in the shape of the mold cavity. In principle (but not quite so simple in practice)

1. The metal is melted
2. It is poured into the mold
3. It is allowed to cool and solidify

Figure 13.1 – *Internal combustion engine intake manifold produced via casting*

4. The finished part is removed from the mold

Metal casting allows the creation of complex parts both with internal and external shapes. It can be used to make small or large parts, and is generally applicable to any metal that can be heated to a liquid state. Some casting processes can produce the finished product without the need for further machining processes. The casting process is also well suited for mass production. An example of an intake manifold of an internal combustion engine produced by casting is shown in Figure 13.1.

The casting process has several drawbacks that vary in seriousness depending on the type of method used. They include:

1. Poor dimensional accuracy and surface finish
2. Limitations on mechanical properties
3. Porosity (may have voids)
4. Hazardous working conditions

The *mold* contains the cavity whose shape and size conform to the part to be manufactured. It is designed to be slightly oversized to account for shrinkage that occurs during the solidification and cooling of the molten metal. The mold can be made from numerous materials including sand, metals, plaster and ceramics.

Molds can be categorized as either *expendabe* or *permanent*. Expendable molds are destroyed to remove the part once the casting is complete. These molds allow the manufacture of more intricate geometries. Permanent molds, on the other hand, are used over and over again. They are more difficult to design because the mold must be able to open, to allow the removal of the part, and close again to cast the next part.

The metal to be used for casting is first heated to raise its temperature to a point suitable for pouring. The pouring temperature, or temperature, above

Figure 13.2 – *A rolling mill opened to expose the rollers*

the metals melting point, is important because it must be sufficiently high to allow the metal to flow entirely into the mold before solidification occurs. In addition, the pouring rate must be just right. If it is too slow, the molten metal may begin to solidify before it has filled the entire mold cavity. If it is too fast, turbulent flow may create air pockets that may cause voids in the cast part.

Rolling

Rolling uses two opposing rolls to draw material into the gap between them, thereby reducing the material's thickness (Figure 13.2 and 13.3). Rolling is a capital-intensive process as it requires massive pieces of equipment, i.e., the rolling mills. The high cost can therefore only be made up by large production quantities.

The basic process can be explained with an example from steel mills. With reference to Figure 13.3, a cast steel ingot that has just solidified and therefore is still relatively hot is placed in an oven to bring the entire piece to a uniform temperature. The ingot is then passed through the rolls.

In *flat rolling*, the final product has a rectangular cross-section where the width is greater than the height. The thickness is reduced by a small amount referred to as the *draft*, d. The latter is defined as

$$d = t_{en} - t_{ex} \qquad (13.1)$$

289

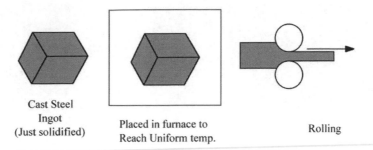

Cast Steel
Ingot
(Just solidified)

Placed in furnace to
Reach Uniform temp.

Rolling

Figure 13.3 – *Steel mill example showing the basic steps in the rolling process*

where t_{en} and t_{ex} are the entrance and exit thicknesses, respectively. The *reduction*, r, can therefore be defined as

$$r = \frac{d}{t_{en}} \qquad (13.2)$$

If a series of rolling operations are used, the overall reduction is the sum of the individual drafts divided by the original thickness. As the thickness of the material is reduced, *spreading* or an increase in width also occurs. Spreading is more pronounced with lower width-to-thickness ratios and low coefficients of friction.

Forging

Forging uses compressive forces to generate distinct shapes using die hammers and hydraulic presses. Die hammers deform the material with a series of high velocity impacts. Hydraulic presses deform the material through controlled high-pressure motion. All metals are forgeable to different degrees. *Forgability* is a subjective measure of the extent to which the material can be deformed. It depends on the metal's composition, crystalline structure, and mechanical properties.

The forging process involves several sequential steps:

1. The material is cut to the approximate size from ingot stock.
2. The work piece is then heated in an oven until it reaches the appropriate forging temperature. An oven is used to ensure a uniform temperature throughout the work piece.
3. The heated work piece is placed between the dies and deformed into the desired shape.

4. Depending on the desired material properties, the deformed shape is either air-cooled or quenched to harden the parts.

5. If necessary, the deformed shape is finished using machining processes.

Forging dies are manufactured from hot-worked steel or medium carbon alloy steels to withstand the high forging temperatures and the abrasion and impacts between the dies and the materials. Forging can broadly be divided into two categories, open-die and impression-die forging.

In *open-die forging*, the work piece is deformed between flat or shaped dies that do not completely restrict the metal flow. Typically open-die forging is used to create simple shapes that then require significant additional machining to achieve final shape. Figure 13.4 illustrates several impacts in the deformation of a circular stock to reduce height and increase radius.

Impression or closed-die forging achieves deformation of the metal in one or more die impressions. The deformation is either through impacts or controlled compression. Unlike open-die forging, impression die forging restricts all material flow to the die. In this manner complex shapes requiring close to net shape can be produced. These require significantly less machining than parts produced from open-die forging. In addition, these parts tend to be of small to moderate size.

Lubricants are frequently used to minimize friction, abrasion and heat loss, and to enhance metal flow and permit release of forged parts from dies. The lubricants generally consist of graphite in oil or water.

Extrusion

Extrusion is a process that reduces the cross section of a block of metal by forcing it through a die orifice under high pressure (Figure 13.5). As a result, extruded products have uniform cross-section along their entire length. This process is generally used to produce cylindrical bars or hollow tubes. Irregular cross-sections are possible for more readily extrudable metals, such as aluminum. Extrusion is primarily a hot-working process, achieving considerable shape changes in a single operation and making it possible to form complex sections that cannot be produced in other ways. It offers economic advantages because dies are relatively inexpensive and interchangeable, allowing one machine to be used for the production of a wide variety of sections. Several samples of extruded products are shown in Figure 13.6.

With reference to Figure 13.5, extrusion essentially involves heating a metal billet to an appropriate temperature and then feeding it into the extrusion press. The temperature depends on, amongst other things, the type of metal being extruded and the extent of deformation required to achieve the final cross-sectional shape. The heated metal is forced by the action of a hydraulic ram through a steel die whose orifice has the desired cross-sectional shape of the final part. The

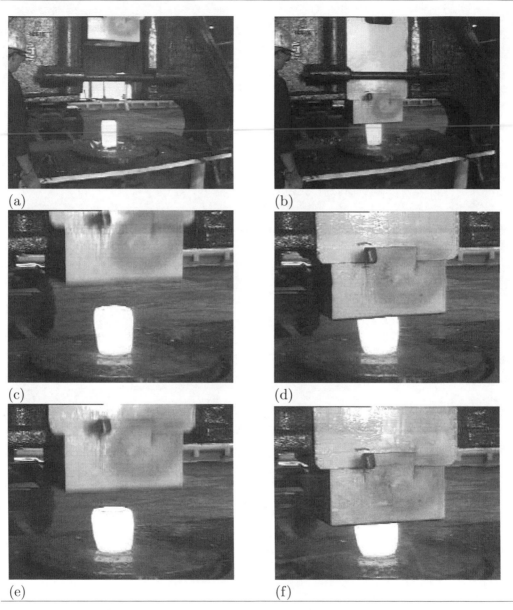

Figure 13.4 – *Multiple impacts are used to reduce the height and increase the radius of the work piece in this example of open die forging.* Images courtesy of Texas Metal Works, Inc.

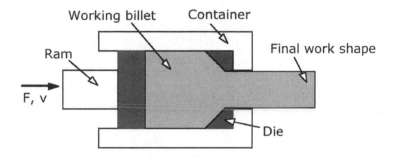

Figure 13.5 – *Schematic of basic extrusion process*

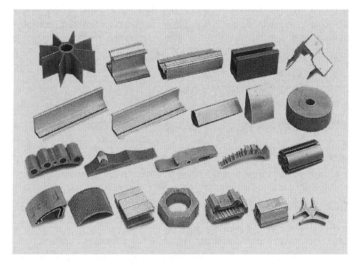

Figure 13.6 – *Sample extruded parts. Notice all have a uniform cross-section.*

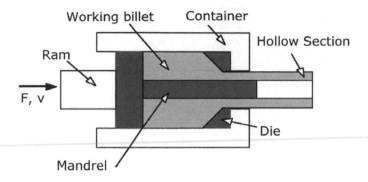

Figure 13.7 – *Extrusion of hollow sections by direct extrusion with the aid of a mandrel*

metal emerges from the die as a continuous bar, which is cut to the required lengths. Most metals and alloys can be shaped by extrusion.

In *direct or forward extrusion* (Figure 13.5), the metal billet is loaded into the container and forced by a ram through a die with one or more openings at the opposite end of the container. At the end of the ram stroke (*the run*), a small portion of material (*the butt*) cannot be forced through the die and is separated from the extruded product by cutting. The butt is recycled and used in future extrusions. Significant frictional forces are created between the billet and the walls of the container as the ram forces the billet to slide towards the die. These frictional forces cause a significant increase in the required ram force. For hot extrusion, the friction problem is compounded by the presence of an oxide layer that can cause defects in the extruded part.

To make tubular sections, a mandrel is arranged in the die orifice as shown in Figure 13.7. During extrusion the metal flows through the annular space created by the mandrel. Semi-hollow cross-section shapes can be produced in the same way.

Indirect, backward or reverse extrusion (Figure 13.8) has the die mounted to the ram. This differs from direct extrusion where the die is stationary and mounted on the far side of the container. The metal is forced through the moving die as the ram penetrates the work. As the metal is no longer being forced to slide against the container wall, there is no friction at the container walls resulting in a significantly lower ram force. Indirect extrusion, however, is limited by the lower rigidity of the hollow ram.

Hot extrusion operates above the metal's recrystallization temperature so that the work metal recrystallizes as it deforms. Hot working takes advantage of a decrease in flow stress at higher temperatures, which lowers tool forces and

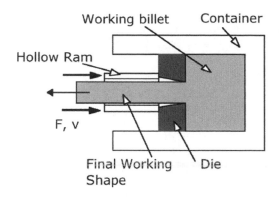

Figure 13.8 – *Schematic of indirect extrusion*

reduces equipment size and power requirements. *Cold extrusion*, performed at or about room temperature, is generally used to produce small machine parts. It can be used with any material that possesses adequate cold workability such as tin, zinc, copper and its alloys, and aluminum and its alloys. The punches and dies used in cold extrusion are subject to severe working conditions and are made of wear-resistant tool steels such as high-alloy chromium steels. To reduce friction, the tool surfaces are polished.

Precision cold forming can result in high speed production of parts with good dimensional control and good surface finish. Due to extensive strain hardening, it is possible to use cheaper materials with lower alloy content to get the same strength as using higher alloy content materials with hot extrusion. Extensive use is made of cold-formed, low-alloy steels in the automotive industry.

13.2 Manufacturing Processes for Plastics - An Overview

Plastics are shaped through numerous processes, including casting, extrusion, thermoforming, blow molding, transfer molding, injection molding, compression molding, and rotational molding. Two of the most popular, injection molding and extrusion, will be discussed.

Plastic Injection Molding

Plastic injection molding is the most common process for manufacturing plastic products. It involves

1. Heating a polymer to a molten state
2. Forcing the molten polymer to flow into a mold
3. Cooling and removing the molded part

Plastic injection machines come in all shapes and sizes depending on the size of the parts to be manufactured. The machines have two main systems: the injection system and the mold. The most common types of *injection systems* use a reciprocating screw mechanism (Figure 13.9). In the first step (Figure 13.9(a)) plastic pellets are loaded into a hopper which feeds them into the injection system. The screw mechanism within rotates, pushing the pellets forward. The root diameter of the screw increases, causing the pellets to melt due to friction and heat from the external heaters. Simultaneously, the screw mechanism moves backward accumulating molten plastic in the injection chamber. Once sufficient molten plastic has accumulated, the screw mechanism stops rotating and moves forward, injecting the molten plastic into the mold (Figure 13.9(b)). The rotary motion of the screw is either electric or hydraulic. The injection motion is primarily hydraulic.

The mold typically consists of two (sometimes more) pieces that automatically clamp together during the injection process and unclamp to release the formed part (refer to Figure 13.10). The mold has intricate internal water channels to rapidly cool the injected molten plastic. The water is recycled through a chiller, cooled, and passed through the mold again. Parts from plastic injection molding are typically net shape and produced in cycle times that range from 10 to 30 seconds.

Extrusion

Plastic extrusion, like metal extrusion, is a compressive process where the raw material is forced to flow through a die orifice with the desired cross-section. Samples of extruded plastics of various cross-sections are shown in Figure 13.11. The extrusion process consists of several stages (Figure 13.12):

1. **Feed.** Transports the high molecular weight polymer pellets away from the hopper into the barrel.
2. **Transition.** The root diameter of the screw increases, melting the polymer due to friction and heat from the external heaters surrounding the barrel[1]. The polymers are highly viscous in the molten state forcing the extruder pump to work under high temperature and pressure. The high shearing action from the screw feed mechanism divides, heats up and melts the polymer before being passed through the die.
3. **Metering.** The polymer is in a molten state and fed through the die.

---

[1]The screw is similar to that illustrated in Figure 13.9 of the plastic injection molding machines.

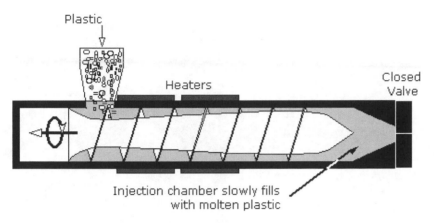

(a) Screw rotates and combines with heaters to melt plastic. At the same time the screw mechanism moves backward accumulating molten plastic in the injection chamber.

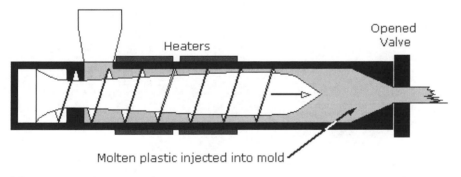

(b) Screw stops rotating and moves forward, injecting molten plastic into mold.

Figure 13.9 – *Injection system of plastic injection machine*

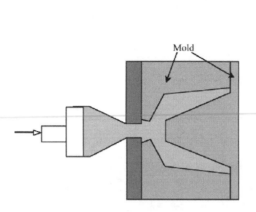

Figure 13.10 – *Schematic of a mold used in the plastic injection molding process*

Figure 13.11 – *Samples of extruded plastic parts*

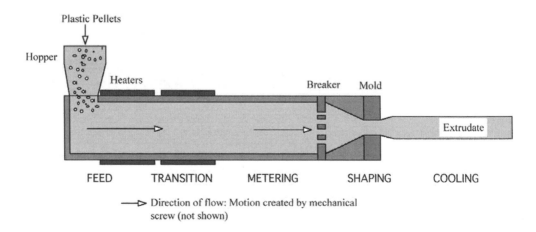

Figure 13.12 – *Schematic of the various stages in plastic extrusion*

4. **Shaping.** The polymer takes on the die's cross-sectional shape. The die must withstand high temperatures and pressures exerted on it as the polymer is forced through. The 'diehead pressure' depends upon several parameters, including

    a) The polymer properties. For example, the higher Young's modulus of elasticity, the higher the pressure. Recall that as Young's modulus of elasticity increases, the stiffer (resistance to deformation) the material becomes.

    b) The polymer temperature. Higher temperatures make the plastic more malleable, reducing required pressure.

    c) The shape of the die. The extent of deformation affects the required pressure.

    d) The flow rate through the die as the higher the flow rate, the higher the required pressure.

As contraction occurs when the molten plastic approaches the die orifice, there is an increase in the compressive stresses. As the molten plastic leaves the die orifice the compressive stresses are removed and the material expands *resulting in a cross-sectional area larger than that of the die orifice.* This process in which a molten polymer expands when imposed compressive stresses are removed is referred to as *viscoelasticity.* It must be taken into account when designing the die. Factors that affect the expansion include the fillers, temperature, shear rate, polymer type, recent shear history, and molecular weight distribution.

5. **Cooling.** After passing through the die, the plastic is cooled by blown air or water baths.

6. **Packaging.** For a continuous process, thin or very flexible extruded material is collected on rolls after cooling. Rigid materials are cut to lengths of up to 20 meters depending upon available transportation facilities.

Figure 13.12 illustrates an example of continuous extrusion. For batch extrusion, the material is forced through the die by a reciprocating piston with the process repeated for each part.

The extrusion process for plastics can result in some common defects. These include *melt fracture*, which occurs if constriction in the die is such that turbulent flow occurs, and *shark skin*, which is where the surface of the product is roughed as it exits the die.

Bibliography

Groover, Mikell, *Fundamentals of Modern Manufactuing: Materials, Processes and Systems*, Upper Saddle River: Prentice-Hall, 1996.

Moore, J.H., C.C. Davis and M. Coplan, *Building Scientific Apparatus - A Practical Guide to Design and Construction,* Second edition, Reading, Massachussetts: Perseus Books, 1991.

Rufe, P.D., *Fundamentals of Manufacturing,* Dearborn: Society of Manufacturing Engineers, 2002.

Part IV

General Design Considerations

Chapter 14

Green Design

By Andrew Lau

14.1   Introduction: What is Green Design?

Green design means *practicing engineering with the inclusion of natural systems, both as a model and as a fundamental consideration.* In the early stages of the Industrial Revolution, engineers consideration of the natural world primarily focused on natural resources, i.e., how to use the raw materials found in the world to make the products of industry. Attitudes and awareness began to change after World War II for many reasons. One was the incredible growth and industrial expansion that followed the war. Another was the expanded use of man-made chemicals and materials. The landmark publication that signaled the beginning of the environmental movement is *Silent Spring* by Rachel Carson in 1962.

In *Silent Spring*, Carson exposes the unanticipated consequences of using Dichloro-diphenyl-trichloroethane (DDT) for pest control. DDT was very effective at controlling insects like mosquitoes and black flies. Problems with it were first noticed by bird watchers who noted the unusual decline in the number of birds. DDT was found in extremely high concentrations in both fish and birds, where it had apparently *bioaccumulated* through the food chain. Bioaccumulation refers to an increase in the concentration of a chemical over time in a biological organism, as compared to the chemical's concentration in the environment. In birds, DDT caused eggs to become very fragile, leading to fewer infant birds.

A significant shift in perception resulted from Carson's book. The principle behind DDT's use is *linear*, that is the application of DDT in the environment causes the decline of a target organism, in this case, mosquitoes. End of story. The problem with linear thinking, however, is that it tends to neglect the complex connections and interactions of matter and energy in the environment, connections that in the case of DDT resulted in the unintended decline of many other living creatures. It is doubtful that if these impacts had been known ahead of time, that DDT use would have been allowed. Considering these connections requires thinking that is more *circular*, that is, the "life" of a product or material is considered from cradle-to-grave, or even better, from cradle-to-cradle. Furthermore, it requires thinking about *systems*, how one part interacts with other parts.

The field of ecology was founded around the beginning of the 20th century as a study of living organisms and their interactions with other living organisms and their surroundings. Ecologists study materials and energy flows within and between living systems. The problem made apparent in *Silent Spring* is that industries developing and applying DDT did not adequately consider ecological principles. What followed through the last half of the 20th century was the development of many new concepts and ideas that grew out of the attempt to connect seemingly disparate fields like economics, ecology, sociology, engineering, and philosophy. Nearly all of these attempts to further knowledge of human activity and its effect on the natural world recognize that the Earth and its living systems are a larger system within which all others function. Green engineering seeks to embed engineering practice in the larger realm of life on Earth. First one must understand the principles of ecology, and then consider how these principles can be followed in engineering practice.

14.2 Ecological Principles

Eco-effectiveness

The relationship of ecological principles to industrial practices has been the subject of several recent publications. One principle with the broadest scope is the concept of eco-effectiveness, as defined by McDonough and Braungart (2002). The authors list three principles of what they call "The Next Industrial Revolution," based on the ecological principles

1. Waste = Food
2. Use solar income
3. Respect diversity

They refer to this new goal of human activity as being grounded on *eco-effectiveness*. Eco-effectiveness is based upon modeling the impact of human activity and designs on nature. They point out that while efficiency has its place, it often boils down only to "being less bad." In some cases, efficiency merely sanctions continuing to use up resources and to pollute . Plus it provides no inspiring vision of good design. As a result of practicing design using the three above principles, researchers believe that we can take delight in products

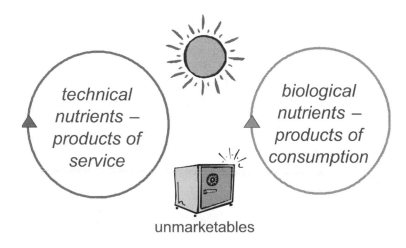

Figure 14.1 – *The two metabolisms involved in eco-effectiveness*

that can replenish, restore, and nourish the rest of the world, without sacrificing effectiveness.

Figure 14.1 captures some of the ideas in McDonough and Braungart (2002). This view puts all of matter ideally into two cycles, or metabolisms. Technical metabolism, on the left, includes materials that are recycled back into industrial goods. On the right is the biological metabolism, or the biosphere. The goal is for all of the products and materials manufactured by industry to safely feed these two metabolisms, thereby eliminating waste. Because this may never be perfectly achievable, particularly in the short term because some materials are too hazardous, there is a category of unmmarketables. These are materials that must be safely stored until ways are developed to detoxify them. Ultimately the goal is to have no unmarketables. Another principle illustrated in the figure is the use of current solar income as the only natural and sustainable energy source.

Sustainability

The concept of sustainable development was first defined in *Our Common Future* (WCED, 1987) as *the ability to meet the needs of the present without compromising the ability of future generations to meet their own needs.* The idea of sustainability is closely related to the ecological principle of *carrying capacity*, which refers to the maximum population of organisms that a given ecosystem can support. Carrying capacity, a natural limit, is determined by available resources and the complex interactions of species.

Carrying capacity can vary over time, as can populations of individual species. If a species is at its maximum population at a given time, diminished resources will cause a decline in that population, typically as a result of starvation and disease. Of course, decline in one species will affect other species in the ecosystem too. As a result, ecologists also describe an *optimum population* as an organism's population in an ecosystem that experiences only modest fluctuations over time. Naturally, the optimum population is smaller than the maximum, but it is more stable and less susceptible to decline. Sustainability is most closely related to the optimum population, which is always less than the maximum carrying capacity.

A question that arises is, *How close is humanity to the ecosystems carrying capacity?* Before trying to answer, it should be pointed out that humans have a highly developed ability to use technology to modify the apparent carrying capacity of a certain ecosystem. By using fossil fuels and advanced agricultural practices, humans are able to grow considerably more food than they could have without technology.

> "For every calorie of energy we get from our food, industrial agriculture puts between four and twenty calories of petroleum energy into fertilizer, equipment fuel, pesticides and herbicides, processing, and shipping. We are, in effect, eating oil." (Callenbach, 1998)

This apparent success of our agricultural system is predicated on the availability of cheap fossil fuels. Furthermore, the impact of these processes on soil quality, soil erosion, fertilizer runoff, and other species must be considered in the big picture.

Indeed, the concept of *environmental impact* attempts to account for the combination of key influences of human impact on the natural world. The IPAT equation for this is

$$I = PAT \qquad (14.1)$$

where I, P, A and T are environmental impact, population, affluence and technology, respectively. If we conclude that the impact of humanity is close to or exceeding the carrying capacity of the Earth, then Equation 14.1 can be used to consider options to reduce our impact. Below is a closer look at each of the three terms on the right-hand side of the equation.

1. **Population.** The world population increased rapidly after world war II (WWII), passing the 6 billion mark by 1999. The rapid increase after WWII was largely the result of longer lives from improved access to food, medicine, clean water and sanitation. The reduction in death rates, coupled with access by women to education and birth control, led to a reduction in birth rates. While some countries continue to have high population growth rates, the United Nations' projections for world population predict

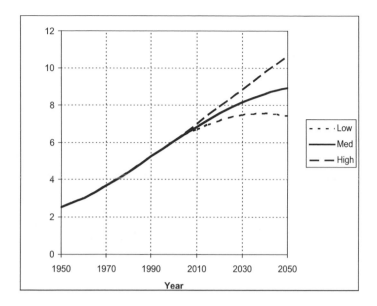

Figure 14.2 – *Historical and projected world population (UNPD, 2003)*

a leveling off at about 9 billion by 2050[1] (Figure 14.2). This is good news relative to impact – population is projected to stabilize by mid-century.

2. **Affluence.** Defines how much valuable stuff a person has. This is why it is often said that the average American has a much higher environmental impact than the average person in a developing country, and it is true. There is cause for concern here because even though population is likely to level off, the desire for more material goods seems insatiable. This aspect of impact may be the most challenging to control, as it depends mostly on psychological and sociological factors that are not directly affected by the work of engineers.

3. **Technology.** Unlike the first two, this factor is very much in the realm of engineers. With regard to environmental impact, technology refers to the level of resource use and waste generation associated with a certain level of technological development. In other words, technology attempts to account for the environmental impact of making our stuff (affluence). The historical trend of the Industrial Revolution holds at first for T to increase rapidly and then to increase less rapidly as more effort and resources are put to reducing impact through technology. Consider that over the next 50 years, P will increase by about a factor of 1.5, while A is estimated to increase by a factor of 3 to 5 (Graedel and Allenby, 2003). Just to

---

[1]Note that these are projections and the range of UN projections for 2050 is from 7.4 to 10.6 billion people.

maintain the same overall environmental impact, therefore, will require decreasing T by a factor of 4.5 to 7.5. This assumes that the present level is acceptable and sustainable. The next section will discuss one measure of sustainability, *ecological footprint.*

14.3   Sustainability Metric – Ecological Footprint

The concept of ecological footprint (EF) was developed to be an indicator and measure of the effective land area necessary to support human activity (Rees and Wackernagel, 1996; Chambers, 2001). It is in effect a measure of the amount of the Earth's productive surface required to supply our resources and assimilate our wastes. As such, it is comparable to environmental impact. The accounting process is based on six mutually exclusive uses of the Earth's biologically productive surface:

1. Growing crops
2. Grazing cattle
3. Harvesting timber
4. Catching fish
5. Accommodating infrastructure
6. Absorbing carbon dioxide emissions

EF was designed to be a conservative estimate; that is, it likely underestimates the actual amount of land needed. EF can be used in a number of ways. For example, it can be used by an individual to assess his own impact for comparison with others and for understanding the various contributors to his footprint, thus providing some insight into where changes may have the most effect. Several calculators are available on the Internet to calculate one's own EF. The author used the calculator on the organization *Redefining Progress'* website, www.rprogress.org, to calculate his own EF. The results are shown in Figure 14.3.

The author's EF was calculated to be 16 acres. While this seems reasonably large, it can be put into perspective in three ways. First, the average EF of a US resident is 24 acres per person. That's encouraging for the author since his EF is 33% less than average! Now the bad news. As shown in the figure, the Earth has only 4.5 acres of biologically productive acres per capita (if everyone had an equal share). Finally, if everyone on Earth lived like the author, it would require 3.7 Earths! One result not shown by the calculation is the current relationship of humanity's EF relative to the productive land and water available.

According to *Redefing Progress* (2004), in 2004 the average EF of each Earth citizen was 6 acres. But with only 4.5 acres available per person, the Earth's carrying capacity is currently exceed by 33%. Further, if everyone lived like the author, it would take 3.7 Earths, or the Earth's carrying capacity would be

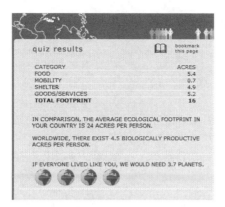

Figure 14.3 – *Estimated ecological footprint for the author*

exceed by 270%. If everyone on Earth lived like the average US citizen, it would take 5.6 Earths.

Recall from the previous IPAT analysis that just maintaining the same overall impact by the year 2050 would require decreasing T by a factor of 4.5 to 7.5. That was assuming maintaining the status quo. As the Earth's capacity is already exceed by 33%, the environmental impact of T must be decreased by a factor of 6 to 10.

That is a big challenge for this generation of engineers, and it is why green engineering must be practiced in all of the fields of engineering. The next section presents a valuable tool that serves as a guide in estimating environmental impact in engineering design and product development.

14.4   Life Cycle Assessment

A product's "life" begins when it is created from raw materials, continues through the product's use, and eventually ends with disposal. This is why life-cycle assessment (LCA) is often referred to as "cradle-to-grave" analysis. A new, broader term is also being used, "cradle-to-cradle," that recognizes that, consistent with natural cycles and the ideal of minimizing waste, a product should be designed to be returned to industry or to biodegrade. Rather than being thrown away at the end of the product life, the product would be disassembled, recycled, reused, and refurbished as feedstock for "new" products. In effect, the old product becomes reincarnated in new ones.

Consider the typical product life cycle shown in Figure 14.4, representing a company that manufactures a product with direct sales to customers. The product life is divided into five stages

1. Premanufacture

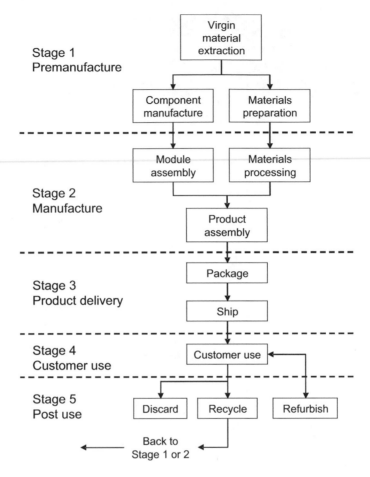

Figure 14.4 – *Stages in a product life cycle*

2. Manufacture
3. Product delivery
4. Customer use
5. Post use

The company that manufactures and sells the product may only be in direct management of stages 2 and 3. Yet the overall impact of the product is determined by all five stages. Considering the life and impact of a product at all stages is the basis of life-cycle assessment.

Even though the manufacturing company may only directly control the manufacture and product delivery, the product design directly affects all stages. For example, during material selection, greener materials could be selected for use

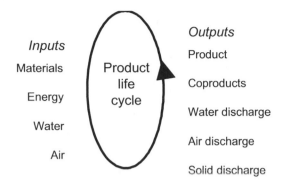

Figure 14.5 – *Inputs and outputs during product life cycle*

in the product. "Experts agree that roughly 80% of the environmental costs of a product are determined at the design stage, and that modifications at later stages of product development will have only modest effects" (Graedel and Allenby, 2003). Greener materials may come from renewable sources, have low environmental impact in extraction, be very energy-efficient, and/or be easily recycled. Specifications can be written to encourage or require subcontractors to deliver greener materials and components. Likewise, the product design directly impacts its use by the customer and its ease of being recycled back into technological nutrients, or its ability to benignly decompose.

The Structure of LCA

There is a lot of variation in the scope of LCA, but the general structure has three stages, with each stage being followed by an interpretation of results (Graedel and Allenby, 2003).

1. **Goal and scope definition.** This stage is crucial as it defines what will be considered, including the range of alternatives. Relevant questions include: How will the results be used and by whom? Do we have a target or mandate to achieve? How much detail is needed? How much can we afford to spend on this effort?

2. **Inventory analysis.** Attempts to quantify all of the inputs and outputs at each stage being considered (Figure 14.5). Here the scope, in space and in time, helps define the information to be considered. This stage is relatively straightforward once the details of what is to be inventoried are determined. A lot of data is available to support this effort.

3. **Impact analysis.** Involves translating the data from the inventory analysis into the environmental impact. As this step requires considerable judgment as to the seriousness of impacts, it is contentious and therefore

311

Table 14.1 – *Simplified life-cycle assessment matrix*

Life Stage	M	E	P
Pre-manufacture			
Manufacture			
Product delivery			
Customer use			
Post-use			

M-Materials E-Energy P-Pollution

requires explicit and consistent reasoning. Ranking the seriousness of impacts must be based on accurate information. In addition, as an LCA effort can have impact on the whole product life cycle, the LCA team should be diverse and represent the stakeholders, including material and component providers, design engineers, industrial engineers, marketing personnel, as well as customers.

Following these three stages, the LCA team must prepare a report with prioritized recommendations based on the LCA results. The general goal is to improve the condition of the environment while meeting customers' needs and providing a competitive product.

A Streamlined LCA Process

For many situations, including engineering design courses, a simpler LCA process called streamlined LCA (SLCA) can be used effectively. The process described here is based on the seminal work of Graedel and Allenby (2003).

The two important simplifications are (1) to lump the many impacts together and (2) to use a qualitative evaluation. A matrix similar to the decision matrices discussed in Chapter 6 is employed. For further simplicity, the environmental issues are lumped into three categories: *materials, energy,* and *pollution.* For each of these categories several aspects are considered. Establishing these categories and aspects is part of the scoping phase. A matrix is created that sets these three categories against the life stages of the product (Table 14.1).

The matrix is populated with scores using on an arbitrarily selected scale indicating the score's relative environmental impact. For example, consider materials aspects. One would consider whether the raw materials came from virgin, recycled, or renewable stock. In the green building industry, further consideration is currently given as to whether renewable sources are rapidly renewable, that is on the order of a year to a few years. One might also consider how far the materials need to be transported. For post-use, one would consider whether the materials could be readily reused or recycled or biodegraded. The toxicity of materials is also important.

A typical scale uses integer values ranging from 0 to 4, with 0 representing the lowest environmental impact and 4 representing the highest impact.[2] The assignment of environmental impact score depends a lot on the knowledge and experience of the evaluator. In most cases, there still needs to be research and investigation done in order to determine the values.

There are at least three ways to summarize the results of the SLCA.

1. **The completed matrix with no totals.** This allows options to be compared on a point-by-point basis. It does not readily indicate the overall impact.
2. **The completed matrix with totals.** Provides a simple one-value summary measure by summing the entries in each row, and totaling the row sums. The lower the total value the better.
3. **Target plot.** The matrix values are plotted on a target plot, sometimes called a web plot.

The following example illustrates the process of SLCA.

Example of a SLCA Analysis

To illustrate the SCLA analysis, consider choosing of framing materials for a building from the user's perspective, in this case an architect or an engineer. One option is wood (fir) studs, while the other option is steel, mainly from recycled stock. The steel is made locally (within 200 miles), while the wood studs are produced about 400 miles away and are sustainably harvested. When the building is eventually demolished, the steel studs could be readily recycled, while the wood studs may be landfilled, burned or could be reused with the appropriate infrastructure. In use, however, the steel studs conduct more heat, so the building uses more energy, and by extension, produces more pollution.

SLCA analyses for the wood and steel studs are shown in Tables 14.2 and 14.3, respectively. In both cases the results are summarized using completed matrices with totals. The totals for the wood and steel studs are 20 and 28, repsectively. Even though the steel stud is made mostly from recycled steel, it takes more energy to produce and in its use. It also produces more pollution from the mining and steel production processes. It therefore has a higher environmental impact than the use of wood studs.

The results from the SCLA analyses can also be displayed using target plots that can be plotted using a spreadsheet program like **Excel** (Figure 14.6 and 14.7). The closer the values are to the center, the less is the environmental impact. Another way of looking at it is the larger the enclosed area, the greater

---

[2]Note that this scale is the opposite of that proposed in Graedel and Allenby(2003), but it was chosen to be this way for two reasons. One is that it seems more logical if one is evaluating environmental impact, the worse the impact: the higher the numerical evaluation should be. Second is that this method facilitates plotting using target plots as shown in the next section.

Table 14.2 – *Simplified life-cycle assessment matrix for wood studs*

Life Stage	M	E	P	Sum
Pre-manufacture	2	1	0	3
Manufacture	0	1	1	2
Product delivery	1	3	2	6
Customer use	1	1	1	3
Post-use	3	1	2	6
Total				**20**

M-Materials E-Energy P-Pollution

Table 14.3 – *Simplified life-cycle assessment matrix for steel studs*

Life Stage	M	E	P	Sum
Pre-manufacture	2	2	2	6
Manufacture	2	3	2	7
Product delivery	1	2	1	4
Customer use	1	3	2	6
Post-use	1	2	2	5
Total				**28**

M-Materials E-Energy P-Pollution

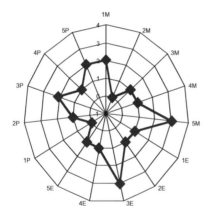

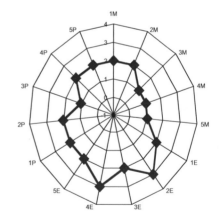

Figure 14.6 – *SLCA target plot for wood studs*

Figure 14.7 – *SLCA target plot for steel studs*

the overall impact. The plots also allows one to see where impact is greatest, suggesting opportunities for improvement. Table 14.4 lists the steps to create target plots using **Excel**.

The example of wood studs versus steel studs has been subjected to various levels of LCA by several different studies. One selected for comparison was commissioned by the Canada Wood Council and conducted by the Athena Insti-

Table 14.4 – *Steps to create target plots in* **Excel**

1. Define two columns in a **Excel** as shown in Table 14.5. The first column is the life stages, 1 to 5, corresponding to pre-manufacture through post-use, combined with the three categories, material (M), energy (E), and pollution (P). This results in fifteen row headings as shown.

2. Fill in the corresponding score relevant life stage and category from the SCLA matrix. The matrix for the wood studs recast for plotting is shown in Table 14.5.

3. In **Excel**, choose *Radar* for the chart type.

Table 14.5 – *The SCLA matrix for wood studs rearranged for generation of target plots in* **Excel**

1M	2
2M	0
3M	1
4M	1
5M	3
1E	1
2E	1
3E	3
4E	1
5E	1
1P	0
2P	1
3P	2
4P	1
5P	2

tute, developers and marketers of LCA software tools and databases. The study compared the framing options of wood versus steel for a 2400 square foot (floor area) house in Toronto (Trusty and Meil, 1999). The study considered the wall framing and the floors. Six key measures were evaluated in the LCA: initial embodied energy; ecologically weighted raw resource use; greenhouse gas emissions (both fuel and process generated); measures of emissions that contribute to air and water toxicity; and solid wastes. Embodied energy accounts for the direct and indirect energy associated with resource extraction, product manufacturing, on-site construction and all transportation with and between these three stages. The study did not account for differences in energy use associated with the higher conduction of heat by the steel studs.

The results from the study are shown in Table 14.6. Consistent with the SLCA results, steel framing was shown to have a significantly worse environmental impact than wood framing. The only measure where steel had a lower impact than wood was in the area of solid wastes. This comparison demonstrates that SLCA, if done thoughtfully and with adequate information, can provide valid results for decision making.

14.5 Summary

In green design, ecological principles are used to guide engineering design and decision making. Incorporating green design into all fields of engineering is

Table 14.6 – *LCA results from Athena Institute study (Trusty and Meil, 1999)*

	Wood Design	Steel Design
Embodied Energy (GJ)	255	389
Global Warming Potential (kg CO2 equivalent)	62,183	76,453
Air Toxicity (critical volume measure)	3,236	5,628
Water Toxicity (critical volume measure)	407,787	1,413,784
Weighted Resource Use (kg)	121,804	138,501
Solid Wastes (kg)	10,746	8,897

important today because the ecological footprint of humanity already exceeds the Earth's carrying capacity. By mid-century, the environmental impact of human endeavors needs to be reduced by a factor of 6 to 10, just to bring humanity's impact in balance with the Earth's natural limits. By considering the entire life cycle of a product, engineers can make informed decisions in the design of new products, as roughly 80% of the environmental impact of a product is determined at the design stage. In addition to the more complex life-cycle assessment (LCA), designers utilize streamlined versions, such as SCLA, to provide timely comparisons of the environmental impact of various design options.

References

Callenbach, E., *Ecology: A Pocket Guide*, Berkeley: University of California Press, 1998.

Carson, R., *Silent Spring*, 1962.

Chambers, N., Simmons, C., and Wackernagel, M., *Sharing Nature's Interest : Ecological Footprints as an Indicator of Sustainability*, London: Earthscan Publications, 2001.

Graedel, T. and B. Allenby, *Industrial Ecology*, 2nd edition, New Jersey: Pearson Education, 2003.

Mcdonough, W. and M. Braungart, *Cradle to Cradle: Remaking the Way We Make Things*, New York: North Point Press, 2002.

Rees, W. and Wackernagel, M., *Our Ecological Footprint: Reducing Human Impact on the Earth*, British Columbia: New Society Publishing, 1996.

Trusty, W., and Meil, J., "Building Life Cycle Assessment: Residential Case Study," Proceedings of Mainstreaming Green: Sustainable Design for Buildings and Communities, Chattanooga, TN, October 1999.

United Nations Population Division, World Population Prospects: The 2002 Revision Population Database, http://esa.un.org/unpp/, 2003.

World Commission on Environment and Development, *Our Common Future*, New York: Oxford University Press, 1987.

Exercises

1. Go to www.rprogress.org and complete the on-line assessment of your ecological footprint (EF). Base it on your current lifestyle. Print a copy of your results page (or use *Alt-PrtSc* to capture the image) and consider the following questions.

   a) How does your EF compare with the average American? Consider why yours might be different than average.
   b) Consider the components of EF based on the output categories (food, mobility, shelter, and goods/services) and on the individual questions in the survey. What changes could you make, and how might it impact your EF? For example, if you drive a car that gets 20 mpg, how would your EF change if it was 40 mpg? Or if you eat meat at every meal, how would it change if meat is part of only some of your, meals?
   c) From the various contributors to ones EF, make a list of some of the ways that engineers can make a difference in reducing impact.

2. Use the SLCA process to analyze a simple consumer product like a one-time use camera, electric toothbrush, or CD player. Use the internet to find information to help you assign values in the SLCA matrix. Plot the results in a target plot. Identify where you feel you need more information.

3. One of the issues in LCA is energy use during the customer-use stage of a product's life. Consider the energy used by two different lighting alternatives: a 75 W incandescent lamp versus a 20 W compact fluorescent lamp (CFL). The incandescent lamp costs $0.75 and lasts 750 hours. The CFL costs $8.00 and lasts 7,500 hours. Electricity costs $0.10/kilowatt-hour. Determine

   a) The energy used (kilowatt-hours) by each lamp over the longer CFL life of 7,500 hours.
   b) The number of lamps used over 7,500 hours.
   c) The cost of both lamps and energy over the 7,500 life for each alternative.
   d) Which alternative would you recommend and why?
   e) Is there any pollution issue with CFLs?

Chapter 15

Engineering Ethics

15.1 What is Engineering Ethics?

A dictionary defines ethics as "the philosophy of morals or the standard of character set by any nation or race." The definition implies an interchangeability between *ethics* and *morals*. The dictionary goes on to define morals as "pertaining to action with reference to right and wrong, and obligation of duty."

Based on these definitions, Engineering Ethics can be defined as the ethical or moral standard to which a particular professional engineering organization will hold you. These standards are embodied in a code of ethics. For example, for Mechanical Engineers, the relevant professional organizations are likely to be the American Society of Mechanical Engineers (ASME) and the National Society of Professional Engineers (NSPE). Other professional organizations aligned with the various engineering disciplines are listed in Table 15.2.

Ethics can also be defined as the understanding of moral values, the resolution of moral issues, and the justification of moral judgments. In this context, engineering ethics is the study of moral values, issues and decisions as they relate to engineering practice (Schinzinger and Martin, 2000).

One can distinguish between three kinds of ethics: personal ethics, common ethics and professional ethics.[1] Harris et al. (1995) state that

> "Professional ethics is the set of standards adopted by professionals insofar as they see themselves acting as professionals. Personal ethics is the set of ones' own ethical commitments, usually given in early home or religious training and often modified by later reflection. Common morality is the set of moral ideals shared by most members of a culture or society."

Despite the differences stated above, there is typically considerable overlap between the three types of ethics (refer to Figure 15.1). Overlap areas between all three could include moral concepts such as honesty and fairness, yet situations

---

[1] By considering engineering as work done by professionals, engineering ethics can be viewed as professional ethics applied to engineering.

do arise when differences occur. For example, consider the engineer who works for a medical device company who refuses to work on the development of new instruments to be used in abortion procedures because his *personal ethics* tells him that abortion is morally wrong. Yet from an *engineering ethics* perspective, it is perfectly ethical to work on the project, and in fact it may be considered unethical to refuse to do so. This example illustrates how personal ethics include morals and principles subscribed to by individuals that may not necessarily be accepted by others (common morality) or prescribed by engineering ethics. Engineers, therefore, occasionally need to resolve conflicts that arise between their personal ethics and engineering ethics.

Common morality is often viewed as analogous to common sense. For example, most people would agree that murder, stealing and lying are wrong. Yet situations do arise where all three can be justified and therefore excused by society. For example, killing during wars is not seen as immoral but often regarded as heroic! Engineering ethics, like common morality, should therefore be viewed as a set of guiding principles, not absolutes that must be adhered to in all situations. Engineers should assess each situation for its own merit, and only then decide on the ethical course of action to take.

15.2 Professional Societies and Codes of Ethics

Professional organizations' codes of ethics provide a common, agreed-upon standard for professional conduct (Harris et al., 2000). They provide a framework within which a professional can be judged. The codes themselves do not define new ethical principles, but rather embody principles and standards that are generally accepted as responsible engineering practice (Fleddermann, 1999). A sample code of ethics from the American Society of Mechanical Engineers is presented in Table 15.1 (ASME, 2004).

The full text for the National Society of Professional Engineers' code of ethics is given in Appendix D. Much longer than the ASME code, it breaks down how principles found in common morality apply to professional engineering practice. Despite the differences that exist between the codes of ethics of the various engineering societies, they all strive to achieve the same goal: to provide a set of guidelines for how engineers should behave with respect to clients, the profession, the public, and the law (Dym and Little, 2003).

Codes of ethics in of themselves are of no use unless engineers are aware of them and use them as a guide for their actions in various situations. Note, however, that ethical issues are seldom clear cut and often result in conflicting options. These conflicts may result from ambition, personal loss, personal gain, loyalty to workmates, loyalty to a company or concern for the public's welfare. In addition, ethical standards are usually relative and personal, seldom with an absolute standard.

Figure 15.1 – *Cross-over between common, personal and professional ethics*

For engineering students, codes of ethics used in conjunction with cases in engineering ethics can serve as an invaluable tool for becoming aware of the various ethical principles embodied in the codes, as well as stimulating the moral imagination by considering and discussing alternative courses of action and their implied ethical consequences. In addition, case studies serve to illustrate how codes do not always provide an answer in every situation, and how they occasionally suggest conflicting courses of action. Finally, cases illustrate how by their very nature codes of ethics are subject to interpretation with different individuals arriving at dissimilar conclusions when faced with the same ethical dilemma. Throughout this chapter, therefore, we will frequently draw upon ethics cases, most of them obtained from the Board of Ethical Review (BER) of the National Society of Professional Engineers (NSPE)[2].

Preventive ethics, analogous to preventive medicine, advocates that engineers should be proactive in their approach to professional ethics. To do so, they MUST think ahead to anticipate possible consequences of their professional actions, and in reviewing those consequences, decide what is ethically and professionally right. The core elements of preventive ethics are (1) stimulating the moral imagination, (2) recognition of ethical issues, (3) development of analytical skills, (4) eliciting a sense of responsibility, and (5) tolerance for disagreement and ambiguity (Baum, 1980, Harris et al., 2000). Each of these elements will be expanded upon in the following sections.

---

[2]The cases presented in this chapter are from the Board of Ethical Review of NSPE. The opinions expressed are based on data submitted to the board that do not necessarily represent all of the pertinent facts when applied to a specific case. These opinions are for educational purposes only and should not be construed as expressing any opinion on the ethics of specific individuals. The cases and opinions have been edited to reduce length.

Table 15.1 – *ASME Code of Ethics*

ASME Code of Ethics

The Fundamental Principles:

Engineers uphold and advance the integrity, honor, and dignity of the Engineering profession by:

1. Using their knowledge and skill for the enhancement of human welfare
2. Being honest and impartial, and serving with fidelity the public, their employers and clients
3. Striving to increase the competence and prestige of the engineering profession

The Fundamental Canons

1. Engineers shall hold paramount the safety, health and welfare of the public in the performance of their professional duties.
2. Engineers shall perform services only in areas of their competence.
3. Engineers shall continue their professional development throughout their careers and shall provide opportunities for the professional development of those engineers under their supervision.
4. Engineers shall act in professional matters for each employer or client as faithful agents or trustees, and shall avoid conflicts of interest.
5. Engineers shall build their professional reputation on the merit of their services and shall not compete unfairly with others.
6. Engineers shall associate only with reputable persons or organizations.
7. Engineers shall issue public statements only in an objective and truthful manner.
8. Engineers shall consider environmental impact in the performance of their professional duties.

Table 15.2 – *Professional organizations representing various engineering disciplines*

Organization	website
American Chemical Society (ACS)	www.acs.org
American Institute for Aeronautics and Astronautics (AIAA)	www.aiaa.org
American Society of Civil Engineers (ASCE)	www.asce.org
American Society of Mechanical Engineers (ASME)	www.asme.org
Biomedical Engineering Society (BMES)	www.bmes.org
Institute of Electrical and Electronics Engineers (IEEE)	www.ieee.org
Institute of Industrial Engineering (IIE)	www.iienet.org
National Society for Professional Engineers (NSPE)	www.nspe.org

15.3   Stimulating Moral Imagination

In order to reduce the chances of being taken by surprise, engineers should imagine possible design alternatives and their likely consequences. But to be able to do so, they must apply their technical knowledge and expertise in developing alternatives and use their best judgment to enumerate the likely consequences. This implies working only on projects that one is technically competent in, as embodied in Section II.2.a of the NSPE code of ethics that states "Engineers shall undertake assignments only when qualified by education or experience in the specific technical field involved." Only by being technically competent can an engineer stimulate their moral imagination.

But even when competent engineers formulate design alternatives and assess their consequences, they may not be able to account for all eventualities. For those situations, engineers must have the moral courage to make design changes or warn the public of possible dangers as soon as they are discovered. Let us consider two examples well documented in the literature, that illustrates these concepts.

On July 17, 1981, the Hyatt Regency Hotel in Kansas City, Missouri held a party in their atrium lobby. With many people standing or dancing on second, third and fourth floor walkways that were suspended over the first floor atrium, the second and fourth floor walkways collapsed, crashing to the crowded first floor (Figure 15.2). The collapse left 114 people dead and 200 injured.

In the ensuing investigation, it was discovered that the manner by which the second floor walkway was suspended had been altered during the construction phase of the project resulting in a structure more likely to fail. The original design (Figure 15.3 - *original design*) had both the second and fourth floor suspended walkways supported by the same rods fastened to the atrium roof. At both the second and fourth floor levels, nut/washer assemblies help keep the box beams and by extension the suspended walkways in place. During construction of the atrium, the builder decided to change the design and support the walkways by separate rods instead of a single one. The first rod was fastened to the atrium

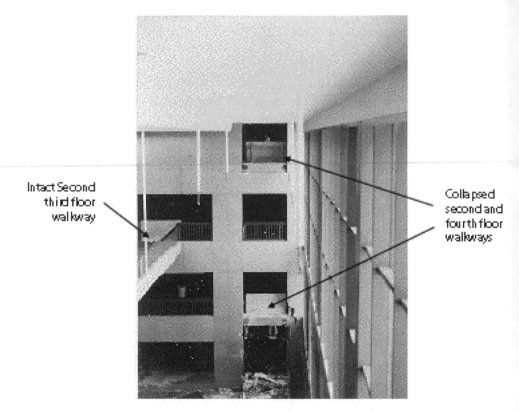

Figure 15.2 – *Photo of intact hanger rods from fourth floor walkway opening of the Hyatt Hotel*

roof and the fourth floor walkway, and the second was fastened to the fourth and second floor walkways (Figure 15.3 - *modified design*). The design change simplified the fabrication of the rods since they no longer needed to be threaded along most of their length as mandated in the first design.

A closer look at the redesign shows that the fourth floor nut/washer assembly had to now support a maximum load of 182 KN, twice the previous load capacity of 91 KN and far exceeding the load capacity it was originally designed for. Clearly, if the engineers had evaluated the implications of their design change - in this case a very simple calculation - an alternative design would have been sought and the tragic loss of life avoided.[3]

The second case involves the design of the Citibank Tower in New York City. In the mid 1970s, a structural engineer, William LeMessurier, was contracted to consult on the construction of a proposed new building for Citibank. At the

---

[3]Summary of Hyatt Regency case based on information at the Engineering Ethics website at Texas A&M University (ethics.tamu.edu).

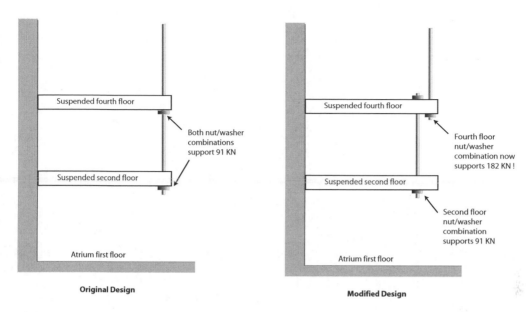

Figure 15.3 – *Illustration of design changes made to the support structure for the second floor suspended walkway*

corner of the proposed site stood an old church. A deal was struck between the church and Citibank, whereby the church would allow Citibank to build its building above the church in exchange for a new church building (Figure 15.4). To accomplish this, LeMessurier developed a structural frame for the new building with support columns at the center of each side, not at the corners as is typical. The corners could not be used due to the presence of the church. He then designed a series of eight story high diagonal braces that would channel the structural loads to the columns and brace the building against wind loading. In 1977, the new building was opened and hailed as a structural engineering marvel.

In 1978, while consulting on another building in Pittsburgh, LeMessurier was made aware of the fact that the diagonal steel girders in the Citibank building had been bolted into place, contrary to the original design that had specified much stronger welded joints. The contractors had switched to the use of bolts because it was cheaper and not as time-consuming. LeMessurier revisited his 90° wind-loading calculations for the building using bolts instead of welds and performed additional calculations for wind-loading at 45°. To his dismay and alarm, he found that for the 45° wind-loading there was a 160% increase in stress in the diagonal girders when compared to the original calculations. LeMessurier was thus faced with a choice. He could do nothing and face the potential collapse of the building during high sustained winds, for example during a hurricane, or

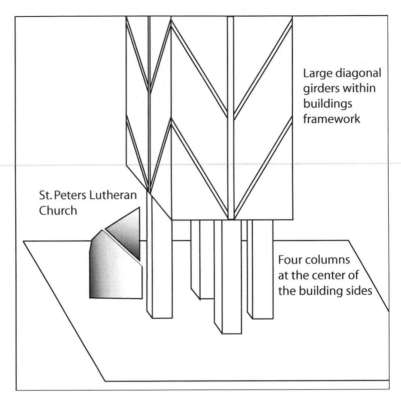

St. Peters Lutheran Church

Large diagonal girders within buildings framework

Four columns at the center of the building sides

Figure 15.4 – *Illustration of Citibank Building showing the internal diagonal bracing and the unique location of the support pillars*

he could make his findings known to the building owners and develop a solution.

LeMessurier took the latter option, at great risk to his firm and his reputation. A relatively simple solution was adopted to weld a series of metal strips over the bolted joint locations making the structure even stronger than in the original design. Although no major hurricane has hit New York City since 1978, if one did, at least the 'imminent' collapse of the Citibank Tower has been avoided[4]. This case illustrates that even highly competent and renowned engineers such as LeMessurier may not be able to anticipate all possible scenarios and consequences of their designs. But when those consequences are determined, they have the moral courage to act, find a solution and rectify the situation.

---

[4]Summary of the Citibank Building case is based on information at the Online Ethics Center for Engineering and Science at the CASE Western Reserve University (onlineethics.org).

15.4 Recognizing Ethical Issues

Ethical issues may not always be obvious. For example altering data to conform to a desired objective is an obvious unethical act. Consider the case of a team of engineers working on the next generation of hand guns. Should they integrate new safety features directly into the design, for example child safety locks, or should they omit them from the design leaving child safety in the hands of the buyer? If integrating the locks into the design is feasible, what responsibility does the engineering team have to do so? These are some of the ethical issues that can arise and should be recognized by engineers.

Engineers also need to recognize the ethical issues that may arise from their day to day activities. Consider the following two cases presented to the BER of the National Society of Professional Engineers.

Case 1: Joint Authorship of Paper (Beard et al., 1985)

Note: This case has direct relationship to situations where a student hardly participates in a team project, but still gets full credit due to the 'generosity' of other team members who are his friends.

Facts: Engineer A and Engineer B are faculty members at a major university. As part of the requirement for obtaining tenure at the university, both Engineer A and Engineer B are required to author articles for publication in scholarly and technical journals. During Engineer A's years as a graduate student, he had developed a paper which was never published and which forms the basis of what he thinks would be an excellent article for publication in a journal. Engineer A discusses his idea with Engineer B, and they agree to collaborate in developing the article. Engineer A, the principal author, rewrites the article, bringing it up to date. Engineer B's contributions are minimal. Engineer A agrees to include Engineer B's name as coauthor of the article as a favor in order to enhance Engineer B's chances of obtaining tenure. The article is ultimately accepted and published in a refereed journal. This case raises the following questions

1. Was it ethical for Engineer B to accept credit for development of the article?

2. Was it ethical for Engineer A to include Engineer B as coauthor of the article?

Discussion:[5] This case presents two distinct issues that, while not directly addressed by the Code of Ethics, are extremely important in regard to the integrity and honesty of intellectual work performed by university engineering faculty.

Turning to the first issue in this case, the BER is sensitive to the extremely difficult position in which many faculty members have been placed with regard to the so-called

---

[5]Reference Code of Ethics Section III.1. "Engineers shall be guided in all their professional relations by the highest standards of integrity." Section III.3.c. "Consistent with the foregoing, Engineers may prepare articles for the lay or technical press, but such articles shall not imply credit to the author for work performed by others."

rule of "publish or perish." This Board finds it extremely difficult to sanction a situation whereby Engineer A permits Engineer B, for whatever reason, to share joint authorship on an article when it is clear that Engineer B's contributions to the article are minimal. We think that Section III.3.c. speaks to this point. This Board cannot excuse the conduct of a faculty member who "takes the easy way out" and seeks credit for an article that he did not author. The only way a faculty tenure committee can effectively evaluate tenure candidates is to examine the candidates' qualifications and not the qualifications of someone else. For this Board to decide otherwise would be to sanction a practice entirely at odds with academic honesty and professional integrity. (See Section III.l.)

Secondly, the facts of the case raise the question of Engineer A's ethical conduct in agreeing to include Engineer B as coauthor of the article as a favor in order to enhance Engineer B's chances of obtaining tenure. However genuine Engineer A's motives may have been under the circumstances, we unqualifiedly reject the action of Engineer A. By permitting Engineer B to misrepresent his achievements in this way, Engineer A has compromised his honesty and forfeited his integrity. Engineer A is unquestionably diminished by this action.

Conclusions:

1. It was unethical for Engineer B to accept credit for development of the article.

2. It was unethical for Engineer A to include Engineer B as coauthor of the article.

Case 2: Engineer Misstating Professional Achievements on Resume (Beard, 1986)
Note: All students will be required to write several resumes while still in college. Resumes will be used to seek on campus employment, summer internships and ultimately a job after graduation. This case looks at the contents of resumes, specifically the notion that what you write in your resume should not only be true, but the implications or conclusions that a reader would draw from the contents should be as well.

Facts: Engineer A is seeking employment with Employer Y. As an employee for Employer X, Engineer A was a staff engineer along with five other staff engineers of equal rank. This team of six was responsible for the design of certain products. While working for Employer X, Engineer A along with five other engineers in his team participated in and was credited with the design of a series of patented products. Engineer A submits his resume to Employer Y and on it implies that he personally was responsible for the design of products which were actually designed through a joint effort of the members of the team. Was it ethical for Engineer A to imply on his resume that he was personally responsible for the design of the products which were actually designed through the joint efforts of the members of the design team?

Discussion:[6] In essence, the issue before the Board is to what extent, if any, may an individual engineer engage in less than total candor in the development of his professional resume. Although this issue may appear to be simple and easily resolved, a closer examination of the question reveals a number of difficult issues. An employment resume is in many senses a written description of the professional achievements and qualifications of the individual whose name appears at its heading. In the world of professional employment, it is the method by which prospective employees communicate their credentials, qualifications, experience, etc. to prospective employers. The resume is generally the first impression that a prospective employer gains of an individual. In fact, in the highly competitive employment environment that currently exists, the employment resume is the means through which the prospective employer "screens out" the less desirous, less qualified applicants. The importance of the employment resume cannot be overstated. It should therefore not be surprising that the contents of a resume can have an enormous impact upon the success of an employment applicant being considered for a position.

Today there is great pressure on the job applicant to stress those qualities and qualifications which will have the greatest impact and make the best impression upon those in a position of responsibility in the hiring process. Job seekers take great pains to stress those aspects of their educational and employment history which demonstrate their suitability for the particular employment position in question. In the context of the present case, we interpret Section II.5.a. to prohibit Engineer A to imply on his resume that he was personally responsible for the design of products which were actually designed through the joint efforts of the members of the design team. While we acknowledge that Engineer A did not in fact state that he was personally responsible for the work in question, we interpret the term "misrepresentation" in Section II.5.a. to include implications which are intended to obscure truth to a client, members of the public, or prospective employers for that matter. We stress, however, that we do not mean to suggest that unintentionally false or inaccurate statements would be unethical per se. Instead, we are referring to statements such as those made by Engineer A which are intentionally designed to mislead a potential employer by obscuring the truth.

Finally, we note that by his conduct, Engineer A appears to have been in violation of Section III.10.a. by failing to provide due credit to those five other staff engineers who worked with Engineer A in designing certain products. While we certainly are not suggesting that Engineer A indicate the names of the five other engineers on his employment resume, we do believe that Engineer A has an obligation to express the fact that the design work was performed as a result of a team effort as opposed to an individual effort. By noting such, we believe that in the context of the fact in this case, Engineer A would be in compliance with the spirit and intent of Section III.lO.a.

Conclusion: It was unethical for Engineer A to imply on his resume that he was personally responsible for the design of the products which were actually designed through the joint efforts of the members of the design team.

---

[6]Reference Code of Ethics Section II.5.a. "Engineers shall not falsify or permit misrepresentation of their, or their associates', academic or professional qualifications. They shall not misrepresent or exaggerate their degree of responsibility in or for the subject matter or prior assignments. Brochures or other presentations incident to the solicitation of employment shall not misrepresent pertinent facts concerning employers, employees, associates, joint venturers or past accomplishments with the intent and purpose of enhancing their qualifications and their work." Section III.1O.a. "Engineers shall, whenever possible, name the person or persons who may be individually responsible for designs, inventions, writings, or other accomplishments."

15.5   Developing Analytical Skills

Engineers need to develop analytical skills relating to ethical situations that are more qualitative than quantitative than most engineers are accustomed to. In these cases, the appropriate course of action is not typically apparent. Usually qualitative metrics such as utility, rights and duties must be blended with quantitative engineering data to come up with final decisions. Case studies provide an excellent mechanism for developing these skills.

15.6   Eliciting a Sense of Responsibility

Although engineers have ethics codes to guide their professional decisions, their application to certain moral problems may not always be obvious. Codes may need certain modifications for particular situations. One should therefore practice independent thought, and not succumb to indoctrination. Professional responsibility covers a wide array of areas. This introduction will focus on three: (1) public safety and health, (2) confidentiality and (3) conflicts of interest.

By the very nature of their work, engineers are in a position where the results of their actions can endanger *public safety and health*. They design products, processes, structures and systems that are used by the public with the implicit understanding that care has been taken to ensure that they are safe. This 'implicit contract' between engineers (designers) and the public (users) places a huge sense of legal and moral responsibility on engineers. According to Harris et al. (1995),

> "Engineers are morally responsible for harms they intentionally [knowingly and deliberately], negligently [unknowingly, but failing to exercise due care] or recklessly [aware that harm is likely to result] cause."

The responsibility not to cause harm does not only relate to safety but can take on other forms, for example, economic harm. Engineers are often privy to confidential information that is furnished to them by their clients, or developed by them on their clients' behalf. Engineers are obligated to exercise *confidentiality*, that is, not to divulge information in a manner that may cause 'harm' to their employers or clients. Breaking this confidentiality by divulging company A's trade secrets[7], for example, to a competitor while still an employee, or when the engineer changes employment to the competitor, would most probably cause economic harm to company A. This action would therefore be professionally unethical (and probably unlawful).

For situations like this one, the engineer must reflect and decide what is the 'best' course of action to take. The following case, from the National So-

---

[7] *Trade secrets* include information that a company does not want made publicly available and that typically gives it a competitive edge over their competitors. For example, the formula for Coca-Cola has been kept as a trade secret since the early 1900s.

ciety of Professional Engineers' Board of Ethical Review, looks at the issue of confidentiality.

 ## Case 3: Confidentiality of an Engineering Report (James et al., 1982)

Facts: Engineer A offers a homeowner inspection service, whereby he undertakes to perform an engineering inspection of residences by prospective purchasers. Following the inspection, Engineer A renders a written report to the prospective purchaser. Engineer A performed this service for a client (husband and wife) for a fee and prepared a one-page written report, concluding that the residence under consideration was in generally good condition requiring no major repairs, but noting several minor items needing attention. Engineer A submitted his report to the client showing that a carbon copy was sent to the real estate firm handling the sale of the residence. The client objected that such action prejudiced their interests by lessening their bargaining position with the owners of the residence. They also complained that Engineer A acted unethically in submitting a copy of the report to any others who had not been a party to the agreement for the inspection services. Did Engineer A act unethically in submitting a copy of the home inspection report to the real estate firm representing the owners?

Discussion:[8]

We note that this is not a case of an engineer allegedly violating the mandate of Section III.4. not to disclose confidential information concerning the business affairs of a client. That provision of the Code necessarily relates to confidential information given the engineer by the client in the course of providing services to the client. Here, however, there was no transmission of confidential information by the client to the engineer. Whether or not the client in this case actually suffered an economic disadvantage by the reduction of its bargaining power in negotiating the price of the residence through the owner having knowledge gained from the inspection report, the same principle should apply in any case where the engineer voluntarily provides a copy of a report commissioned by a client to a party with an actual or potential adverse interest.

It is a common concept among engineers that their role is to be open and above board and to deal in a straightforward way with the facts of a situation. This basic philosophy is found to a substantial degree throughout the Code (e.g., Sections II.3. and II.3.a). At the same time, Section II.l.c. recognizes the proprietary rights of clients to have exclusive benefit of facts, data, and information obtained by the engineer on behalf of the client. We read into this case an assumption that Engineer A acted without thought or consideration of any ulterior motive; that he, as a matter of course, considered it right and proper to make his findings known to all interested parties in order that the parties handle their negotiations for the property with both sides having the same factual data flowing from his services. Thus, although we tend to exonerate

---

[8]Reference Code of Ethics Section II.l.c. "Engineers shall not reveal facts, data, or information obtained in a professional capacity without the prior consent of the client or employer except as authorized or required by law or this Code." Section II.4. "Engineers shall act in professional matters for each employer or client as faithful agents or trustees."

Engineer A of substantial or deliberate wrongdoing, he was nevertheless incorrect in not recognizing the confidentiality of his relationship to the client. Even if the damage to the client, if any in fact, was slight, the principle of the right of confidentiality on behalf of the client predominates.

Conclusion: Engineer A acted unethically in submitting a copy of the home inspection to the real estate firm representing the owners.

An exception to maintaining *confidentiality* is if releasing the information may prevent grave harm to public safety and health (whistleblowing). In those situations, the engineer's ethical obligation to public safety and health supersedes the obligation to an employer or client. This is a clear example where different sections of the professional engineering organizations' codes of ethics may conflict - and in numerous other situations this conflict does arise. The 'definitions' of *whistleblowing* vary widely depending on the individual's point of view. Consider the following two views.

> 'Whistle-blowing' - the act of a man or woman who, believing that the public interest overrides the interest of the organization he[sic] serves, publicly "blows the whistle" if the organization is involved in corrupt, illegal, fraudulent, or harmful activity.' (Nader et al., 1972)

> 'Some of the enemies of business now encourage an employee to be disloyal to the enterprise. They want to create suspicion and disharmony and pry into the proprietary interests of the business. However this is labeled - industrial espionage, whistle-blowing or professional responsibility - it is another tactic for spreading disunity and creating conflict.' *Roche-GM chairman, 1971* (Martin and Schinzinger, 1989)

Whatever the definition, whistleblowing should be the action of last resort as a sign of serious corporate culture problems. Whistleblowing can be internal (skipping immediate supervisors to report to higher management) or external (reporting to authorities outside the company or to the public). In either case, whistleblowing is typically viewed as disloyalty to coworkers (internal) or coworkers and the organization (external). The sense of betrayal is viewed as less serious for internal whistleblowing as the problem is kept and possibly resolved within the organization, minimizing bad publicity typically associated with external whistleblowing.

According to Martin and Schinzinger (1989), it is morally permissible for engineers to engage in external whistle-blowing concerning safety if

1. The harm that will be done by the product to the public is serious and considerable.
2. They make their concerns known to their superiors.

3. Getting no satisfaction from their immediate superiors, they exhaust the channels available within the corporation, including going to the board of directors.

4. The employee has documented evidence that would convince a reasonable, impartial observer that her view of the situation is correct and the company policy wrong.

5. There is strong evidence that making the information public will in fact prevent the threatened serious harm.

On the other hand, one should not become a whistleblower as revenge against coworkers, supervisors or company, nor to exact future gains, for example from book contracts or speaking tours (Fleddermann, 1999). The US federal government, however, has been offering rewards to individuals who blow the whistle on companies and organizations who defraud the US government. The monetary rewards, a percentage of the money recovered, serves to overcome the stigma and anxiety typically associated with whistleblowing and to compensate the individual(s) for any possible loss of employment. Consider the case below that was presented to the Board of Ethical Review (BER) of the National Society of Professional Engineers. The BERs opinion on the case is based on NSPE's code of Ethics.

Case 4: City Engineer (Bechamps et al., 1989)

Facts: Engineer A is employed as the City Engineer/Director of Public Works for a medium-sized city and is the only licensed professional engineer in a position of responsibility in the city government. The city has several large food processing plants that discharge very large amounts of vegetable wastes into the city's sanitary system during the canning season. Part of the canning season coincides with the rainy season. Engineer A is responsible for the disposal plant and beds and is directly responsible to City Administrator C. Technician B answers to Engineer A. During the course of her employment, Engineer A notifies Administrator C of the inadequate capacity of the plant and beds to handle the potential overflow during the rainy season and offers possible solutions. Engineer A has also discussed the problem privately with certain members of the city council without the permission of City Administrator C. City Administrator C has told Engineer A that "we will face the problem when it comes." City Administrator C orders Engineer A to discuss the problems only with him and warns her that her job is in danger if she disobeys.

Engineer A again privately brings the problem up to other city officials. City Administrator C removes Engineer A from responsibility of the entire sanitary system and the chain of command by a letter instructing Technician B that he is to take responsible charge of the sanitary system and report directly to City Administrator C. Technician B asks for a clarification and is again instructed via memo by City Administrator C that he, Technician B, is completely responsible and is to report any interference by a third party to City Administrator C. Engineer A receives a copy of the memo. In addition, Engineer A is placed on probation and ordered not to discuss this matter further. If she does she will be terminated. Engineer A continues in her capacity as City Engineer/Director of Public Works, assumes no responsibility for the disposal plant and beds, but continues to advise Technician B without the knowledge of City Administrator C. That winter during the canning season, particularly heavy storms occur in the city. It becomes obvious to those involved that if waste water from the ponds containing the domestic waste is not released to the local river, the ponds will over flow the levees and dump all waste into the river. Under state law, this condition is required to be reported to the state water pollution control authority, the agency responsible for monitoring and overseeing water quality in state streams and rivers. Did Engineer A fulfill her ethical obligation by informing City Administrator C and certain members of the city council of her concerns?

Discussion:[9] The engineer's obligation to hold paramount the safety, health, and welfare of the public in the performance of his professional duties is probably among the most basic of duties. Clearly, its importance is evident by the fact that it is the very first obligation stated in the NSPE Code of Ethics. Moreover, the premise upon which professional engineering exists –the engineering registration process– is founded upon the proposition that in order to protect the public health and safety, the state has an interest in regulating by law the practice of the profession. While easily stated in the abstract, the breadth and scope of this fundamental obligation is far more difficult to fix.

As we have long known, ethics frequently involves a delicate balance between competing and, often times, conflicting obligations. However, it seems clear that where the conflict is between one important obligation or loyalty and the protection of the public, for the engineer the latter must be viewed as the higher obligation. Clearly, this case involves "endangerment to the public safety, health and welfare" by the contamination of the water supply, and therefore it is clear that Engineer A has an obligation to report the matter to her employer. Under the facts it appears that Engineer A has fulfilled this specific aspect of her obligation by reporting her concerns to City Administrator C and thereafter to certain members of the city council. However, under the facts of this case, we believe Engineer A had an ethical obligation under the Code to go considerably farther. As noted in the Code, where an engineer determines that a case may involve a danger to the public safety, the engineer has not merely an "ethical right" but has an "ethical obligation" to report the matter to the proper authorities and withdraw from further service on the project. We believe this is particularly clear when the engineer involved is a public servant (City Engineer and Director of Public Works).

---

[9]Code of Ethics Section I.l. "Engineers, in the fulfillment of their professional duties, shall hold paramount the safety, health and welfare of the public in the performance of their professional duties." Section II.l.a. "Engineers shall at all times recognize that their primary obligation is to protect the safety, health, property and welfare of the public. If their professional judgment is overruled under circumstances where the safety, health, property or welfare of the public are endangered, they shall notify their employer or client and such other authority as may be appropriate." Section II.4. "Engineers shall act in professional matters for each employer or client as faithful agents or trustees." Section III.2.b. "Engineers shall not complete, sign, or seal plans and/or specifications that are not of a design safe to the public health and welfare and in conformity with accepted engineering standards. If the client or employer insists on such unprofessional conduct, they shall notify the proper authorities and withdraw from further service on the project."

In the context of this case, we do not believe that Engineer A's act of reporting her concerns to City Administrator C or certain members of the city council constituted a reporting to the "proper authorities" as intended under the Code. Nor do we believe, Engineer A's decision to assume no responsibility for the plant and beds constitutes a "withdrawal from further service on the project." It is clear under the facts of this case that Engineer A was aware of a pattern of ongoing disregard for the law by her immediate superior as well as members of the city council. After several attempts to modify the views of her superiors, it is our view that Engineer A knew or should have known that the "proper authorities" were not the city officials, but more probably state officials (i.e., state water pollution control authority).

We cannot find it credible that a City Engineer/Director of Public Works for a medium-sized town would not be aware of this basic obligation. Engineer A's inaction permitted a serious violation of the law to continue and appeared to make Engineer A an "accessory" to the actions of City Administrator C and the others. It is difficult for us to say exactly at what point Engineer A should have reported her concerns to the "appropriate authorities." However, we would suggest that such reporting should have occurred at such time as Engineer A was reasonably certain that no action would be taken concerning her recommendations either by City Administrator C or the members of the city council and that, in her professional judgment, a probable danger to the public safety and health then existed.

In closing, we must acknowledge a basic reality that must confront all engineers faced with similar decisions. The engineer who makes the decision to "blow the whistle" will in many instances be faced with the loss of employment. While we recognize this sobering fact, we would be ignoring our obligation to the Code and hence to the engineering profession if, in matters of public health and safety, we were to decide otherwise. For an engineer to permit her professional obligations and duties to be compromised to the point of endangering the public safety and health does grave damage to the image and interests of all engineers.

Whereas one may agree with the opinion above rendered by the BER, the last paragraph raises more questions than answers. The board states that "we must acknowledge a basic reality that must confront all engineers faced with similar decisions. The engineer who makes the decision to 'blow the whistle' will in many instances be faced with the loss of employment." Although in an idealistic world it may be the right thing to do, blowing the whistle may not always be practical. The engineer may not be in a position where she can afford to loose her job. She would have to decide if the extent of danger to public safety and health is grave enough to outweigh the prospect of losing her job.

Finally, *conflicts of interest* arise in situations where an engineer may have a personal connection to a decision she is required to make during her course of employment. Personal connections can result in many benefits, including personal financial gain or financial gain to a relative or acquaintance. The personal connection may affect the ultimate decision the engineer makes but may not be in the best interest of her employer. Consider the example where an engineer has been contracted to purchase electronic timers for a current design project. In researching which companies would give the best product at the lowest price, she realizes that her uncle's company makes the timers needed. Although their products are not the best available, they are well within the required design parameters and are available for the lowest price. If the engineer purchases her uncle's timers, she would provide her company with an adequate timer at the lowest price. In addition, her decision would give her uncle's company a great financial boost. Conversely, she could recommend superior timers from an alternative company at a slightly higher price. Clearly, her decision is clouded by the conflicts of interest between loyalty to her employer and to her uncle.

In general engineers should strive to avoid conflicts of interest by recusing themselves from situations where the conflicts arise. In the above example, the engineer could pass on the task to another engineer on the team and let him make the decision. Alternatively, where the conflicts of interest cannot be avoided, the engineer should make them known to her supervisors and team members. For example, revisiting the timer example above, assume that the engineer's uncle's company made the best timers and offered them at the lowest price. Clearly it is in the companies best interest to purchase timers from him. By presenting these facts to her supervisors and team members, the engineer presents the rationale of her decision in the context of the conflicts of interest. Her supervisors and team members could then accept her decision (in the company's best interest) or pick an alternative (to avoid any *appearance* of conflicts of interest). Either way, the engineer has acted in an ethically responsible manner.

15.7 Tolerating disagreement and ambiguity

There may be several points of view because of the ambiguity in deciding the ethical/right course of action. Professional codes of conduct may not always furnish a direct answer either. Whereas one should be willing to tolerate disagreements, engineers should also search for possible areas of agreement and further clarity to try to reach consensus on an appropriate course of action. This tolerance involves being able to listen to others and accept alternative points of view.

15.8 Summary

This chapter has given a brief introduction to ethics as applied to the engineering profession: engineering ethics. The concepts illustrate the great responsibility that society places on the engineers, and accordingly the ethical behavior expected from them. For a more in-depth look at engineering ethics, visit the following sites

1. Penn State College of Engineering Ethics Web Site, *www.engr.psu.edu/ethics/*
2. Engineering Ethics at Texas A&M University, *ethics.tamu.edu*
3. Online Ethics Center for Engineering and Science at the CASE Western Reserve University, *onlineethics.org*

References

American Society of Mechanical Engineers, http://www.asme.org, viewed June 2003.

Bannes, L., Dorchester, E.D., Gregoritis, J.W., Pritzker. P.E., Simberg, R., Williamson, H., Wortley, C.A., NSBE Board of Engineering Review, Case 99-13.

Bannes, L., Dorchester, E.D., Gregoritis, J.W., Pritzker. P.E., Simberg, R., Williamson, H., Wortley, C.A., NSBE Board of Engineering Review, Case 99-4.

Baum, R., Ethics in Engineering, NY: Hastings Center, 1980.

Beard, F.W., Haefeli, R.J., James, E.C., Jarvis, R.W., Polk, J.L., Thompson, E.S., Roberts, J.K., NSBE Board of Engineering Review, Case 85-1.

Beard, F.W., Haefeli, R.J., James, E.C., Jarvis, R.W., Polk, J.L., Thompson, E.S., Roberts, J.K., NSBE Board of Engineering Review, Case 86-6.

Bechamps, E.N., Haefeli, R.J., Jarvis, R.W., Manning, L., Pritzker, P.E., Streeter, H., Koogle, H.G., NSBE Board of Engineering Review, Case 88-6.

Bechamps, E.N., Browne, J.F., Koogle, H.G., Manning, L., Pritzker, P.E., Streeter, H., and Haefeli, R.J., NSBE Board of Engineering Review, Case 89-2.

Cox, W.A., Middleton, W.W., Norris, W.E., Rauch, W.F., Smith, J.H., Tennant, O.A, and Nichols, R. L., NSBE Board of Engineering Review, Case 92-1.

Curd, M. *Professional Responsibility for Harmful Actions*, Dubuque: Kendall/Hunt Publishing, 1984.

Dym, C.L. and Little, P., *Engineering Design: A Project-Based Introduction*, New York: John-Wiley and Sons, 2003.

Fleddermann, C.B., *Engineering Ethics*, Upper Saddle River: Prentice Hall, 1999.

Harris, C., Pritchard, M. and Rabins, M., *Engineering Ethics: Concepts and Cases*, 2nd Edition, Belmont: Wadsworth Publishing Co., 2000.

James, E.C., Jones, L.E., Perrine, R.H., Polk, J.L., Roberts, J.K., Samborn, A.H., Beard, F.W., NSBE Board of Engineering Review, Case 82-2.

Martin, M., R. Schinzinger, *Ethics in Engineering*, New York: McGraw-Hill, 1989.

Nader, R., Petkas, P., and Blackwell, K., *Whistle blowing*, New York: Grossman, 1972.

Schinzinger, R. and Martin, M.W., *Introduction to Engineering Ethics*, New York: McGraw-Hill Higher Education, 2000.

Exercises

All the cases presented here are from the Board of Ethical Review of NSPE. The opinions of the board have been omitted making these cases ideal for class discussion or student assignments.

1. **Declining Employment After Acceptance (Bechamps et al., 1989).** The city of Orion began a recruitment process the first week of January for a city engineer/public works director. The recruitment was necessitated by the pending retirement of the former city engineer/public works director in May. The city wanted to have the new employee on board for orientation and training prior to the incumbent leaving. The city received a great number of applications and went through the laborious task of screening for finalists. During the screening period, Engineer A was in the area and requested an appointment to gather more information regarding the position. The appointment was granted and Engineer A was given information regarding the position, the city, housing, schools, etc. Engineer A expressed a strong interest in the position and stated he had friends living nearby. He also stated that he was familiar with the area. Engineer A was one of the four finalists interviewed for the position during the first week in March and was selected as the best qualified applicant. An offer of employment was extended to Engineer A on March 10, which was accepted. Engineer A agreed to start employment on or before April 10.

   During the period of March 15-April 10, several phone conversations were held with Engineer A during which he expressed some doubt as to his ability to start on April 10 due to obligations to his current employer and personal reasons. Engineer A was advised by the city that he would be permitted to return to his former home for meetings to satisfy his employment obligations. Engineer A was also advised by the city that if he was hesitant about employment due to personal reasons, the city could understand but that it would appreciate a decision so that it could begin a new recruitment process. Each time this was discussed, Engineer A stated that he wanted the position and would be there no later than April 10. On April 5th, Engineer A advised the city that he could not start on April 10th but that he could start on April 24th. Engineer A assured the city that this was a firm commitment. On April 23, Engineer A advised the city that he could not take the position. Was it ethical for Engineer A to deal with the city in the manner described?

2. **Credit for Engineering Work - Design Competitions (Cox et al., 1992).** Engineer A is retained by a city to design a bridge as part of an elevated highway system. Engineer A then retains the services of Engineer B, a structural engineer with expertise in horizontal geometry, superstructure design and elevations, to perform certain aspects of the design services. Engineer B designs the bridge's three curved welded plate girder spans which were critical elements of the bridge design. Several months following completion of the bridge, Engineer A enters the bridge design into a national organization's bridge design competition. The bridge design wins a prize. However, the entry fails to credit Engineer B for his part of the design. Was it ethical for Engineer A to fail to give credit to Engineer B for his part in the design?

3. **Copyright (Bannes et al., 1999).** Engineer A is employed by SPQ Engineering, an engineering firm in private practice involved in the design of bridges and other structures. As part of its services, SPQ Engineering uses a CAD software design product under a licensing agreement with a vendor. Although under the terms of the licensing agreement SPQ Engineering is not permitted to use the software at more than one workstation without paying a higher licensing fee, SPQ Engineering ignores this restriction and uses the software at a number of employee workstations. Engineer A becomes aware of this practice and calls a hotline publicized in a technical publication and reports his employers activities. Was it ethical for Engineer A to report his employers apparent violation of the licensing agreement on the hotline without first discussing his concerns with his employer?

4. **Providing a Design for Client's Competitor (Bannes et al., 1999).** Engineer A is hired by Developer X to perform design and construction-phase services for a subdivision for Developer X. Per the agreement with Developer X, Engineer A is paid 30% of his fee by Developer X. Engineer A submits the design drawings and plans to the county authorities, and permits are issued for the benefit of Developer X. Developer X cannot get financing for the project, and Developer X tells Engineer A that Engineer A should not disclose the contents of the drawings and plans to any unauthorized third party. Developer Y, a client of Engineer A and also a business competitor of Developer X, is interested in the subdivision project. Developer Y has secured financing for the project and approaches Engineer A, requesting that he perform the design on the project and provide the design documents for Developer Ys review. Since Engineer A was not paid his entire fee for his completed project design by Developer X, Engineer A agrees to provide the design drawings and plans to Developer Y and agrees to charge Developer Y only for the changes to the original subdivision design drawings and plans. The following questions are raised,

   a) Was it ethical for Engineer A to provide a copy of the design drawings and plans to Developer Y?
   b) Was it ethical for Engineer A to charge Developer Y for the changes to the original subdivision design drawings and plans?

Part V

Appendices

Appendix A

Creation of the PMWs in Excel

This appendix provides a step-by-step tutorial to create the **Excel-based** Project Management Workbooks. It assumes that the user has a basic knowledge of **Excel** and has read *Chapter 2 Management of the Design Process*

A.1 Task Enumeration: Work Breakdown Structures

Creating the Initial Workbook

1. Create a new blank **Excel** workbook.

2. Using Figure A.1 as a guide, enter the tasks and sub-tasks for your WBS (columns A and B only in the Figure).

3. Double-click on the current worksheet tab at the bottom of the **Excel** window and type the worksheet name, 'Work Breakdown Structure'. Press [Enter].

Automatic End Date Calculation

1. Add two extra columns to the current WBS worksheet: planned start date, planned end date (Figure A.2).

2. Select all the cells enclosed in the rectangle defined by the top most planned start date and the lowest actual end date (Figure A.2).

3. Format>Cells...

4. In the *Format Cells pop-up menu*, click on the **Number tab**.

5. Select **Date** under *Category* and make a format selection under *Type* that shows month, day and year (for example, 01-mar-03).

343

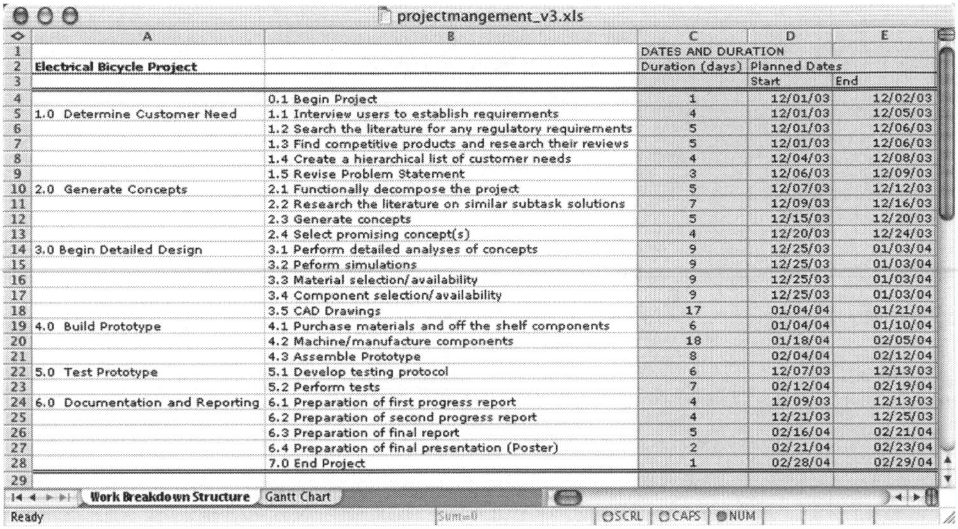

projectmangement_v3.xls

	A	B	C	D	E
1			DATES AND DURATION		
2	**Electrical Bicycle Project**		Duration (days)	Planned Dates	
3				Start	End
4		0.1 Begin Project	1	12/01/03	12/02/03
5	1.0 Determine Customer Need	1.1 Interview users to establish requirements	4	12/01/03	12/05/03
6		1.2 Search the literature for any regulatory requirements	5	12/01/03	12/06/03
7		1.3 Find competitive products and research their reviews	5	12/01/03	12/06/03
8		1.4 Create a hierarchical list of customer needs	4	12/04/03	12/08/03
9		1.5 Revise Problem Statement	3	12/06/03	12/09/03
10	2.0 Generate Concepts	2.1 Functionally decompose the project	5	12/07/03	12/12/03
11		2.2 Research the literature on similar subtask solutions	7	12/09/03	12/16/03
12		2.3 Generate concepts	5	12/15/03	12/20/03
13		2.4 Select promising concept(s)	4	12/20/03	12/24/03
14	3.0 Begin Detailed Design	3.1 Perform detailed analyses of concepts	9	12/25/03	01/03/04
15		3.2 Peform simulations	9	12/25/03	01/03/04
16		3.3 Material selection/availability	9	12/25/03	01/03/04
17		3.4 Component selection/availability	9	12/25/03	01/03/04
18		3.5 CAD Drawings	17	01/04/04	01/21/04
19	4.0 Build Prototype	4.1 Purchase materials and off the shelf components	6	01/04/04	01/10/04
20		4.2 Machine/manufacture components	18	01/18/04	02/05/04
21		4.3 Assemble Prototype	8	02/04/04	02/12/04
22	5.0 Test Prototype	5.1 Develop testing protocol	6	12/07/03	12/13/03
23		5.2 Perform tests	7	02/12/04	02/19/04
24	6.0 Documentation and Reporting	6.1 Preparation of first progress report	4	12/09/03	12/13/03
25		6.2 Preparation of second progress report	4	12/21/03	12/25/03
26		6.3 Preparation of final report	5	02/16/04	02/21/04
27		6.4 Preparation of final presentation (Poster)	2	02/21/04	02/23/04
28		7.0 End Project	1	02/28/04	02/29/04
29					

Work Breakdown Structure | Gantt Chart

Ready | Sum=0 | SCRL | CAPS | NUM

Figure A.1 – *Sample Work Breakdown Structure implemented in the* **project management workbook**

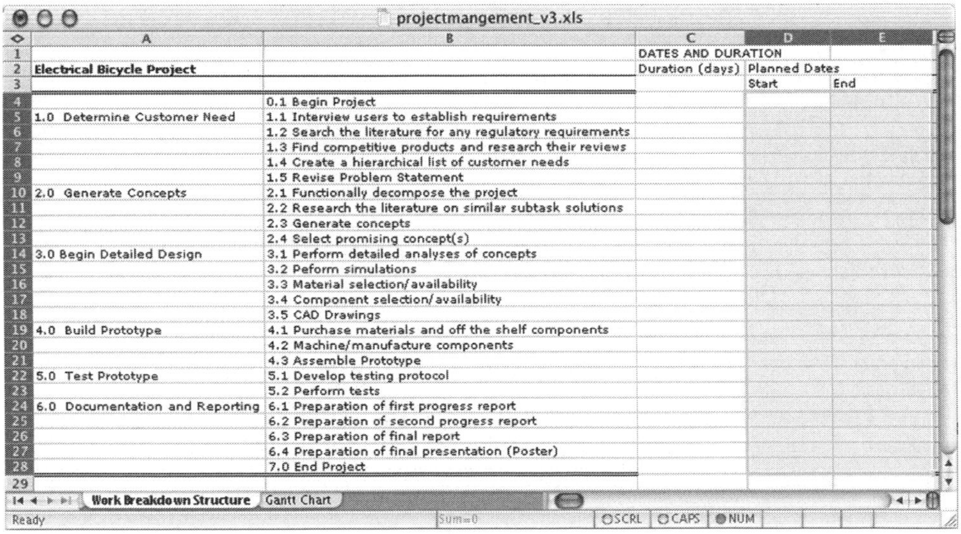

projectmangement_v3.xls

	A	B	C	D	E
1			DATES AND DURATION		
2	**Electrical Bicycle Project**		Duration (days)	Planned Dates	
3				Start	End
4		0.1 Begin Project			
5	1.0 Determine Customer Need	1.1 Interview users to establish requirements			
6		1.2 Search the literature for any regulatory requirements			
7		1.3 Find competitive products and research their reviews			
8		1.4 Create a hierarchical list of customer needs			
9		1.5 Revise Problem Statement			
10	2.0 Generate Concepts	2.1 Functionally decompose the project			
11		2.2 Research the literature on similar subtask solutions			
12		2.3 Generate concepts			
13		2.4 Select promising concept(s)			
14	3.0 Begin Detailed Design	3.1 Perform detailed analyses of concepts			
15		3.2 Peform simulations			
16		3.3 Material selection/availability			
17		3.4 Component selection/availability			
18		3.5 CAD Drawings			
19	4.0 Build Prototype	4.1 Purchase materials and off the shelf components			
20		4.2 Machine/manufacture components			
21		4.3 Assemble Prototype			
22	5.0 Test Prototype	5.1 Develop testing protocol			
23		5.2 Perform tests			
24	6.0 Documentation and Reporting	6.1 Preparation of first progress report			
25		6.2 Preparation of second progress report			
26		6.3 Preparation of final report			
27		6.4 Preparation of final presentation (Poster)			
28		7.0 End Project			
29					

Work Breakdown Structure | Gantt Chart

Ready | Sum=0 | SCRL | CAPS | NUM

Figure A.2 – *Addition of columns to represent planned and actual start and end dates into the* work breakdown strucutre *worksheet of the* **project management workbook**

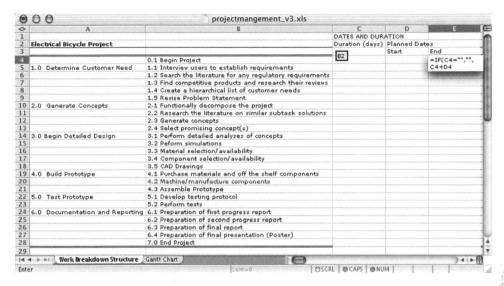

Figure A.3 – *Addition of formulas to automatically calculate planned end dates in the* work breakdown structure *worksheet*

6. Next you will enter in a formula to allow the automatic calculation of the planned end date based on values for the planned start date and corresponding task duration. Select the top *Planned end date* cell - $E4$ in this example (Figure A.3).

7. Enter the formula =IF(D4="","",C4+D4). Your actual cell references will vary depending on your specific spreadsheet. Essentially the formula leaves the current *end date* cell ($E4$) blank if the corresponding *start date* cell ($D4$) is blank. If a *start date* has been entered, the *end date* cell ($E4$) is populated with the sum of the *start date* ($D4$) and the *task duration* ($C4$).

8. In the *planned end date* column ($E4 : E28$ in this example) select the cells.

9. Edit>Fill>Down, to copy the end date formula into all the selected cells. Done! Each time a planned start is entered into your WBS, a planned end date will be automatically entered. Further if a task duration is changed, the corresponding end date will be changed accordingly.

10. Select the columns corresponding to 'Duration', 'Start Date' and 'End Date'. Change the background color, and add borders to the cells.

A.2 Incorporate the design structure matrix into the PMW

This section can be omitted to create a simpler PMW

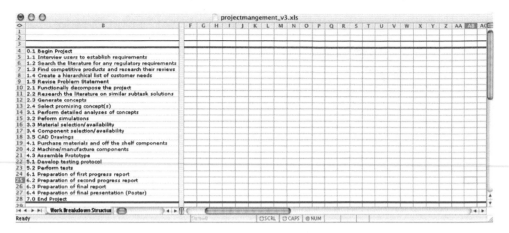

Figure A.4 – *Screen shot of split screen in* **project management workbook**

1. **Excel** allows you to split an active window into two screens (horizontally or vertically) enabling you to work on different non-adjacent portions of your worksheet. Split the window of the current WBS worksheet such that the right window only displays column B of the worksheet (the subtasks) and the left window displays the spreadsheet starting from the column *after* the 'Planned End Date' column (column F in this example) as shown in Figure A.4.

2. In the right screen, starting at column F, select the number of columns corresponding to the number of sub-tasks shown on the left screen.

3. Reduce the width of the columns to approximately, 3.5 mm (Format>Columns>Width...).

4. Change the background color of the cells to one different from that used for the WBS.

5. Add borders to all the cells.

6. Along the header row (row 3, in this example) type in all the sub-task numbers as headings for the column

7. Type in the same sub-task numbers along the diagonal, i.e, the inter-section cell between similar row and column headings. Your empty *design structure matrix* is complete.

8. Populate the DSM row by row as outlined in Section 2.2. A screenshot of the DSM for the work breakdown structure example is shown in Figure A.5.

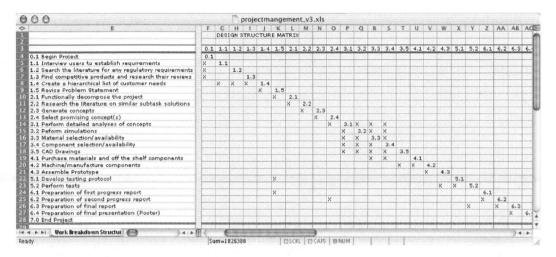

Figure A.5 – *Screen shot of the* design structure matrix *for the WBS example* *incorporated into the* **project management workbook**

A.3 Steps to Add an Activity Network

1. Fill out the design structure matrix and use it to sketch out a graphical activity network.

2. Split the WBS window into two screens as shown in Figure A.6. The left screen should display the task duration, planned start and end dates of the WBS. The left most column in the right screen should be the first column after the design structure matrix (column AF in this example). As shown in the figure, enter dummy task durations and task start dates. This enables checking of activity network formulas as they are entered.

3. For the number of paths, type in 'P1', 'P2', and so on in *every other column* as shown in Figure A.7.

4. Select all the columns that will form the activity network (columns $AF : AY$ in this example) and shade them a different color than the design structure matrix. Also add a border around each cell to make the cells visible.

5. The columns to the right of each path column will contain the task duration if that task forms part of the path. For path P1, enter the formula, =if(AF4="","",$C4) in the cell $AG4$. The formula adds the task duration in cell $AG4$ if the task is part of path 1 and is marked as such in cell $AF4$. Note that your actual cell references may be different. If cell $AF4$ is blank, i.e, the task is not part of the path, then the task duration cell, $AG4$, is also left blank. Test your formula entry by typing in an 'X' in cell $AF4$. The corresponding task duration should appear in cell $AG4$.

6. Select cells $AG4 : AG28$ (or from the cell corresponding to the first task to that corresponding to the last). Edit>Fill>Down, to copy the formula into the cells.

347

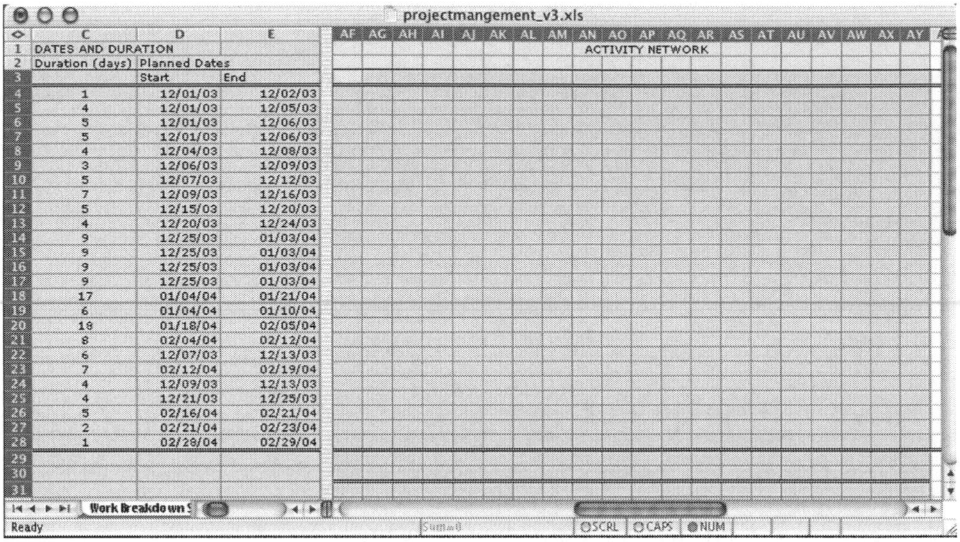

Figure A.6 – *First steps to add* activity network *into the* **project management workbook**

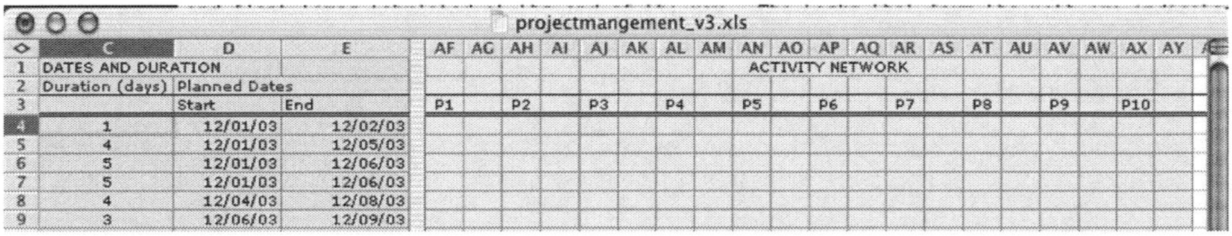

Figure A.7 – *Adding path labels for* activity network *in the* **project management workbook**

7. Cell $AG29$ will contain the sum of all the tasks in path 1. In cell $AG29$, enter the formula, =sum(AG4:AG28).

8. Cell $AG30$ will contain the float time for the path 1. In cell $AG30$, enter the formula, =max($AG29:$AY29)-AG29. Here AY corresponds to the last column of the activity network. Your actual cell references may vary. The function max() returns the maximum cell value (path duration) from which each individual path's duration is subtracted yielding each of the float times. By typing 'Xs' in column AF, assign some dummy tasks for path 1 to test your entries. Your current activity network should be similar to Figure A.8.

9. Finally, you will copy the formulas corresponding to path 1, to the other paths. Select cells $AG4 : AG30$. Edit>Copy.

10. Click in cell $AI4$, corresponding to path 2, Edit>Paste.

348

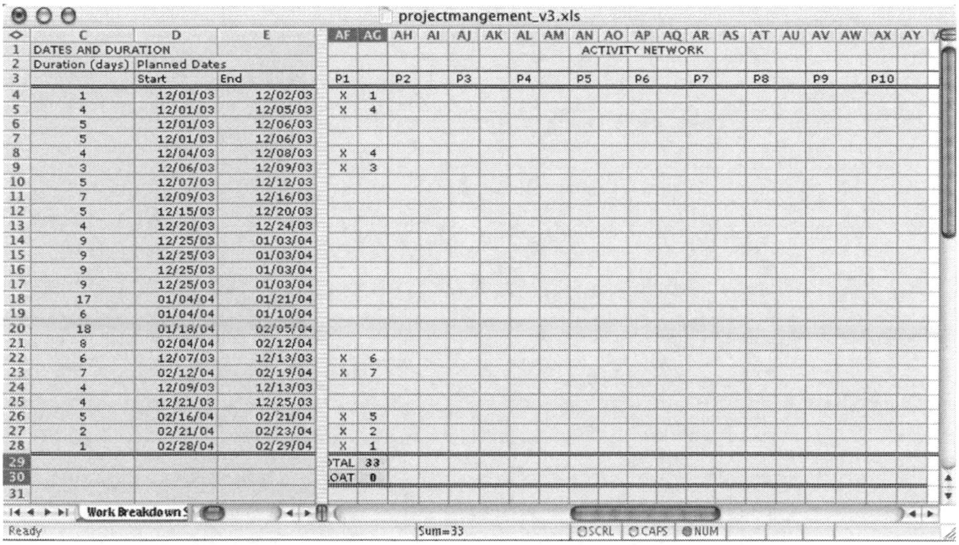

Row	Duration (days)	Planned Dates Start	End	P1			P2	P3	P4	P5	P6	P7	P8	P9	P10
1	DATES AND DURATION					ACTIVITY NETWORK									
4	1	12/01/03	12/02/03	X	1										
5	4	12/01/03	12/05/03	X	4										
6	5	12/01/03	12/06/03												
7	5	12/01/03	12/06/03												
8	4	12/04/03	12/08/03	X	4										
9	3	12/06/03	12/09/03	X	3										
10	5	12/07/03	12/12/03												
11	7	12/09/03	12/16/03												
12	5	12/15/03	12/20/03												
13	4	12/20/03	12/24/03												
14	9	12/25/03	01/03/04												
15	9	12/25/03	01/03/04												
16	9	12/25/03	01/03/04												
17	9	12/25/03	01/03/04												
18	17	01/04/04	01/21/04												
19	6	01/04/04	01/10/04												
20	18	01/18/04	02/05/04												
21	8	02/04/04	02/12/04												
22	6	12/07/03	12/13/03	X	6										
23	7	02/12/04	02/19/04	X	7										
24	4	12/09/03	12/13/03												
25	4	12/21/03	12/25/03												
26	5	02/16/04	02/21/04	X	5										
27	2	02/21/04	02/23/04	X	2										
28	1	02/28/04	02/29/04	X	1										
29				TOTAL	33										
30				OAT	0										

Figure A.8 – *Completion of path 1 formula entries for* activity network *in the* **project management workbook**

11. Repeat the above paste step for the remaining paths (for this example, that would be clicking and pasting into cells *AI*4, *AK*4, *AM*4, *A0*4, *AQ*4, *AS*4, *AU*4, *AW*4, and *AY*4).

12. Assign dummy tasks to each path to test the formulas. Your current activity network should be similar to Figure A.9.

13. Delete the dummy 'Xs' and the temporary task durations and start dates.

A.4 Steps to Add the Gantt Chart

1. Split the WBS window into two screens as shown in Figure A.10. The left screen should display the task duration, planned start and end dates of the WBS. The left most column in the right screen should be the first column after the activity network (column *AZ* in this example).

2. In the header row (row 3, in this example) type in 'Task', 'D1', 'D2' and 'D3' as shown in Figure A.10. The columns D1, D2 and D3 will contain the number of days from the start of the project to the initiation of the task, the number of days elapsed since the task was initiated, and the number of assigned days left to complete the tasks, respectively. The calculated values will be used to generate the Gantt chart.

3. Select all the columns that will form the Gantt chart calculation cells (columns *BA* : *BD* in this example) and shade them a color different than the activity network color.

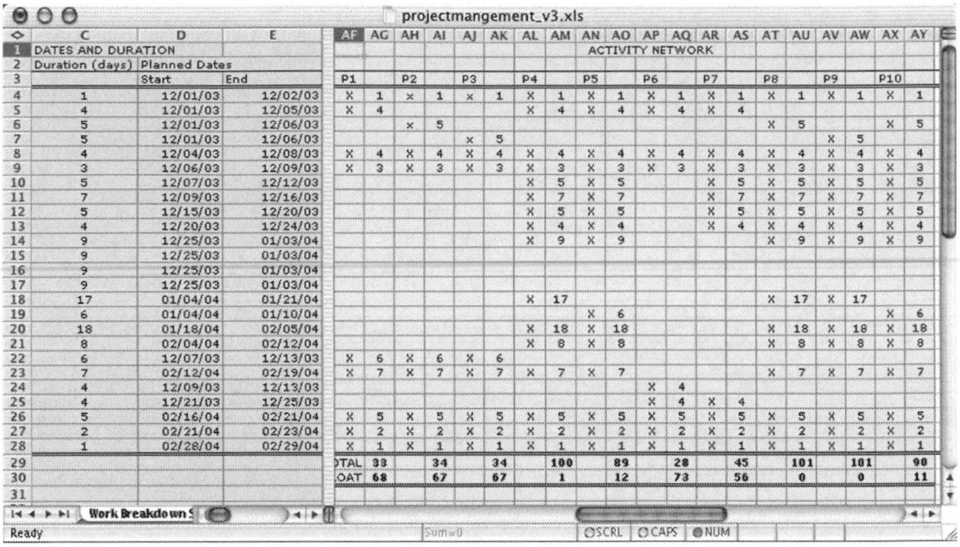

projectmangement_v3.xls

Figure A.9 – *Testing the formulas in the* activity network *in the* **project management workbook** *using dummy entries*

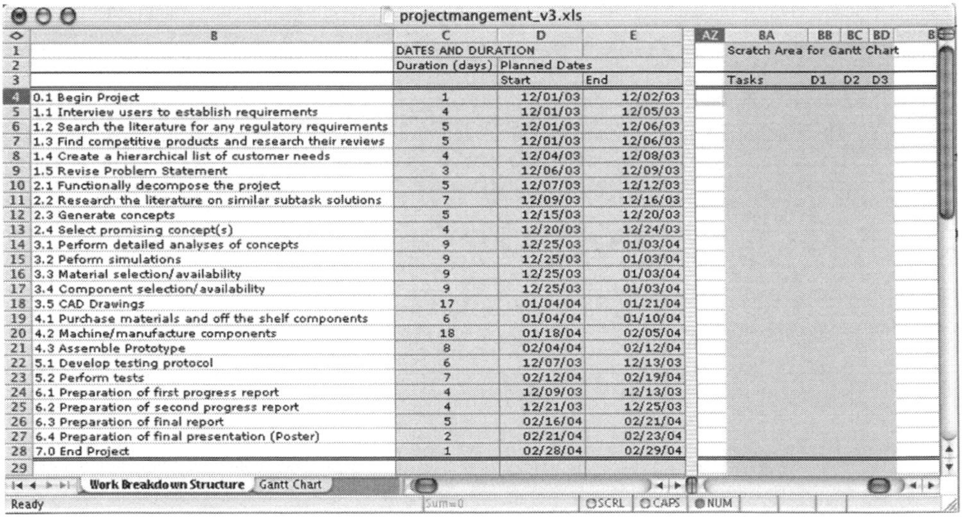

projectmangement_v3.xls

Task	Duration (days)	Planned Dates Start	End
0.1 Begin Project	1	12/01/03	12/02/03
1.1 Interview users to establish requirements	4	12/01/03	12/05/03
1.2 Search the literature for any regulatory requirements	5	12/01/03	12/06/03
1.3 Find competitive products and research their reviews	5	12/01/03	12/06/03
1.4 Create a hierarchical list of customer needs	4	12/04/03	12/08/03
1.5 Revise Problem Statement	3	12/06/03	12/09/03
2.1 Functionally decompose the project	5	12/07/03	12/12/03
2.2 Research the literature on similar subtask solutions	7	12/09/03	12/16/03
2.3 Generate concepts	5	12/15/03	12/20/03
2.4 Select promising concept(s)	4	12/20/03	12/24/03
3.1 Perform detailed analyses of concepts	9	12/25/03	01/03/04
3.2 Peform simulations	9	12/25/03	01/03/04
3.3 Material selection/availability	9	12/25/03	01/03/04
3.4 Component selection/availability	9	12/25/03	01/03/04
3.5 CAD Drawings	17	01/04/04	01/21/04
4.1 Purchase materials and off the shelf components	6	01/04/04	01/10/04
4.2 Machine/manufacture components	18	01/18/04	02/05/04
4.3 Assemble Prototype	8	02/04/04	02/12/04
5.1 Develop testing protocol	6	12/07/03	12/13/03
5.2 Perform tests	7	02/12/04	02/19/04
6.1 Preparation of first progress report	4	12/09/03	12/13/03
6.2 Preparation of second progress report	4	12/21/03	12/25/03
6.3 Preparation of final report	5	02/16/04	02/21/04
6.4 Preparation of final presentation (Poster)	2	02/21/04	02/23/04
7.0 End Project	1	02/28/04	02/29/04

Scratch Area for Gantt Chart — Tasks D1 D2 D3

Work Breakdown Structure / Gantt Chart

Figure A.10 – *Splitting the work breakdown structure worksheet window into two screens to facilitate the first steps in adding the Gantt chart to the* **project management workbook**

4. In the cell under 'Task' (cell $BA4$ in this example) enter the formula, =\$B4. This copies the corresponding subtask description from cell $B4$.

5. In the cell under the heading 'D1' (cell $BB4$ in this example) enter the formula, =D4-\$D\$4. This subtracts the current task date from the project initiation date in $D4$. Note that the '\$' symbols are for absolute addressing, i.e. even if the formula is copied to other cells, the second number will always reference cell \$D\$4.

6. In the cell under heading 'D2' (cell $BC4$ in this example) enter the formula, =IF(TODAY()>D4,IF((TODAY()-D4)>C4,C4,TODAY()-D4),0). These series of 'IF' statements provide the following logic:

```
If TODAY'S DATE is after the PROJECT START DATE then
    If (TODAY'S DATE-PROJECT START DATE)>TASK DURATION
        ELAPSED DAYS SINCE TASK INITIATED
                    = TASK DURATION (i.e., task complete)
    else
        ELAPSED DAYS SINCE TASK INITIATED
                    = TODAY'S DATE - PROJECT START DATE
    end
else
    ELAPSED DAYS SINCE TASK INITIATED = 0
        (i.e., task has not been started)
end
```

7. In the cell under the heading 'D3' (cell $BD4$ in this example) enter the formula, IF(BC4<C4,C4-BC4,0). The 'IF' statement provides the following logic:

```
If ELAPSED DAYS SINCE TASK INITIATED < TASK DURATION
                    (i.e., task not complete)
    # DAYS LEFT FOR TASK COMPLETION
                    = TASK DURATION - ELAPSED DAYS
else
    # DAYS LEFT FOR TASK COMPLETION = 0
                (i.e., task has been completed)
```

8. Select cells $BA4 : BD28$. Edit>Fill>Down to copy the formulas to all the tasks. Your calculation cells should be similar to Figure A.11.

9. With cells $BA4 : BD28$ still selected, click on the **Chart wizard button** or Insert>Chart...

10. In the *Chart Wizard pop-up menu*, select the *Stacked Bar* bar chart (Figure A.12).

11. Press the **Next button** twice to get to Step 3 of the Wizard (Figure A.13).

12. Under the (1) **Titles tab** enter 'Days' for the *Value (Y) axis*, (2) **Gridlines tab**, select both the Major and Minor Y-axis gridlines, and (3) **Legend tab**, unselect the *Show Legend* box.

13. Finish. In the final step of the wizard, select to have the chart created as a new worksheet called 'Gantt Chart' (Figure A.14).

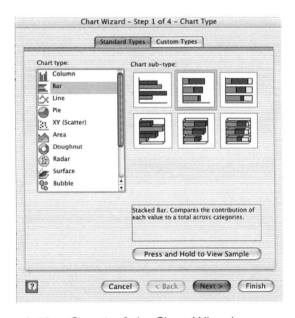

B	C	D	E	AZ	BA	BB	BC	BD	B
	DATES AND DURATION				Scratch Area for Gantt Chart				
	Duration (days)	Planned Dates							
		Start	End		Tasks	D1	D2	D3	
0.1 Begin Project	1	12/01/03	12/02/03		0.1 Begin Prc	1	1	0	
1.1 Interview users to establish requirements	4	12/01/03	12/05/03		1.1 Interview	0	4	0	
1.2 Search the literature for any regulatory requirements	5	12/01/03	12/06/03		1.2 Search th	0	5	0	
1.3 Find competitive products and research their reviews	5	12/01/03	12/06/03		1.3 Find com	0	5	0	
1.4 Create a hierarchical list of customer needs	4	12/04/03	12/08/03		1.4 Create a	3	4	0	
1.5 Revise Problem Statement	3	12/06/03	12/09/03		1.5 Revise Pr	5	3	0	
2.1 Functionally decompose the project	5	12/07/03	12/12/03		2.1 Functiona	6	5	0	
2.2 Research the literature on similar subtask solutions	7	12/09/03	12/16/03		2.2 Research	8	7	0	
2.3 Generate concepts	5	12/15/03	12/20/03		2.3 Generate	14	5	0	
2.4 Select promising concept(s)	4	12/20/03	12/24/03		2.4 Select pr	19	4	0	
3.1 Perform detailed analyses of concepts	9	12/25/03	01/03/04		3.1 Perform i	24	9	0	
3.2 Peform simulations	9	12/25/03	01/03/04		3.2 Peform s	24	9	0	
3.3 Material selection/availability	9	12/25/03	01/03/04		3.3 Material i	24	9	0	
3.4 Component selection/availability	9	12/25/03	01/03/04		3.4 Compone	24	9	0	
3.5 CAD Drawings	17	01/04/04	01/21/04		3.5 CAD Drav	34	13	4	
4.1 Purchase materials and off the shelf components	6	01/04/04	01/10/04		4.1 Purchase	34	6	0	
4.2 Machine/manufacture components	18	01/18/04	02/05/04		4.2 Machine/	48	0	18	
4.3 Assemble Prototype	8	02/04/04	02/12/04		4.3 Assemble	65	0	8	
5.1 Develop testing protocol	6	12/07/03	12/13/03		5.1 Develop	6	6	0	
5.2 Perform tests	7	02/12/04	02/19/04		5.2 Perform i	73	0	7	
6.1 Preparation of first progress report	4	12/09/03	12/13/03		6.1 Preparati	8	4	0	
6.2 Preparation of second progress report	4	12/21/03	12/25/03		6.2 Preparati	20	4	0	
6.3 Preparation of final report	5	02/16/04	02/21/04		6.3 Preparati	77	0	5	
6.4 Preparation of final presentation (Poster)	2	02/21/04	02/23/04		6.4 Preparati	82	0	2	
7.0 End Project	1	02/28/04	02/29/04		7.0 End Proje	89	0	1	

Figure A.11 – *Completed Gantt chart scratch area in the* **project management workbook**

Figure A.12 – *Step 1 of the* Chart Wizard pop-up menu

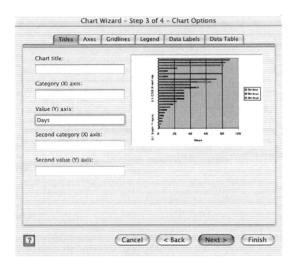

Figure A.13 – *Step 3 of the* Chart Wizard pop-up menu

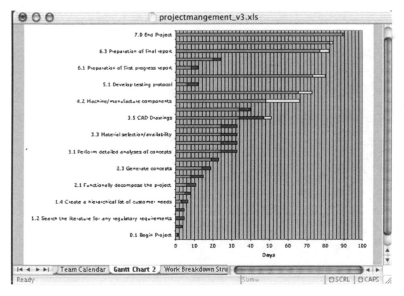

Figure A.14 – *Initial Gantt chart before formatting*

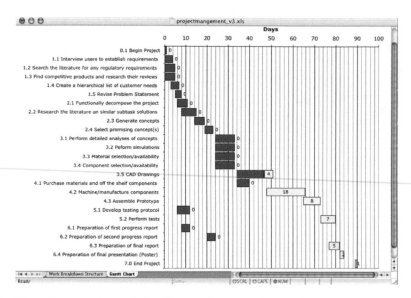

Figure A.15 – *Screen shot of the* Gantt chart *in the* **project management workbook**

14. In the chart double-click on any of the task names. This brings up the *Format Axis pop-up menu*. Under the **Scale tab**, change the *Number of Categories between tick mark labels* to '1'. Select the *Categories in Reverse Order* check-box.

15. Double-click on any of the bars corresponding to the first series (closest to the category axis). In the *Format Data Series pop-up menu* under the **Patterns tab** select 'none' for both *Border* and *Area*.

16. Double-click on any of the bars corresponding to the third data series. In the *Format Data Series pop-up menu* under the **Data labels** select the *Show value* radio button.

17. Figure A.15 shows a screen shot of the completed Gantt chart. Note that if your project has not begun, all the bars will be a single color.

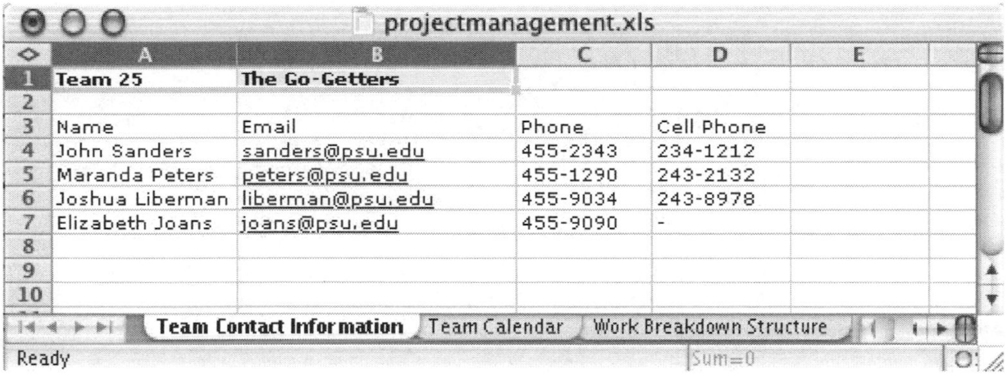

Figure A.16 – *Contact page in the* **Microsoft Excel** *project management workbook*

A.5   Design Team Contact Page

1. On a separate worksheet in the PMW create a team contact page as shown in Figure A.16.

2. Double-click on the worksheet tab at the bottom of the window name the worksheet, 'Contacts'.

A.6   Steps to Add a Team Calender

1. Double-click on the tab of an unused worksheet at the bottom of the **Excel** window. Type in 'Team Calendar' to change the name of the worksheet.

2. Using Figure A.17 as a guide, type in the headings 'Day', 'Date' and 'Calendar Entries'. Make them Boldface.

3. In cell $B2$ type in the project start date.

4. In cell $B3$ type in the formula =B2+1 to increment the date by one day.

5. Starting rom cell $B3$ select the number of cells in column B corresponding to the number of days minus one, available for the project. For example if the project runs for seven weeks, that corresponds to forty-nine days, you would select cells $B3 : B50$.

6. Edit>Fill>Down to copy the formula into all selected cells and produce a sequential list of dates from the project start to end dates.

7. Select all the populated cells in column B.

8. Format>Cells...

9. In the *Format Cells pop-up menu*, click on the **Number tab**.

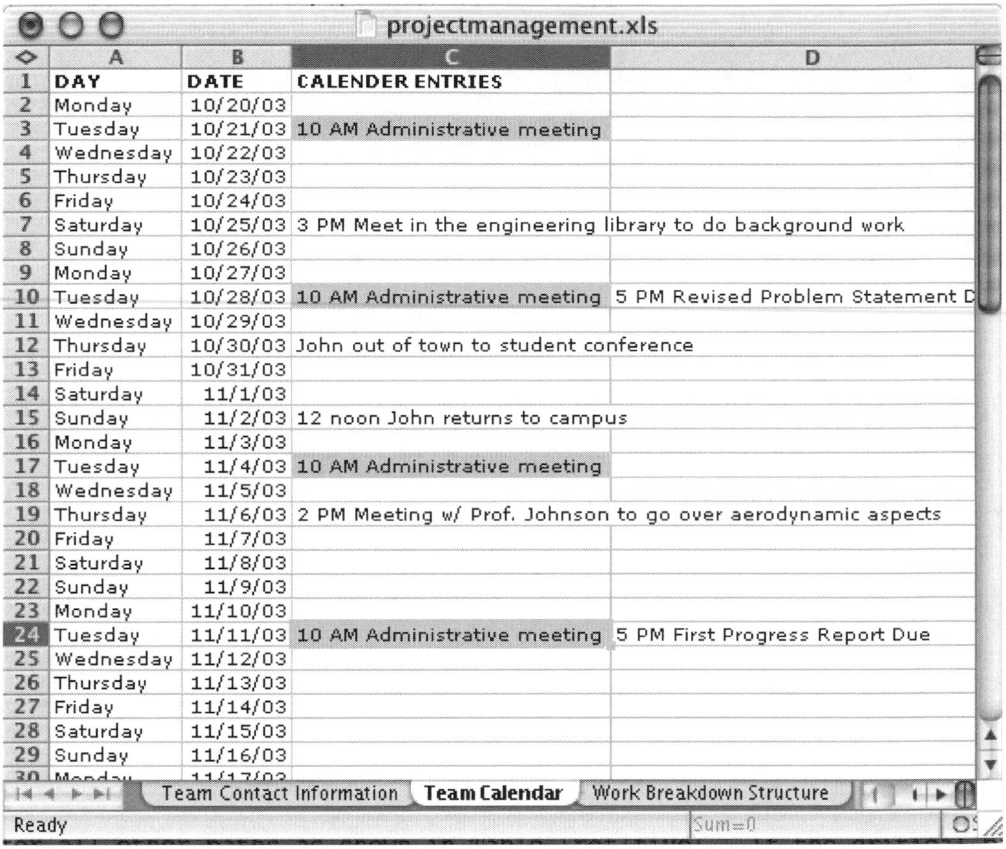
projectmanagement.xls

	A	B	C	D
1	DAY	DATE	CALENDER ENTRIES	
2	Monday	10/20/03		
3	Tuesday	10/21/03	10 AM Administrative meeting	
4	Wednesday	10/22/03		
5	Thursday	10/23/03		
6	Friday	10/24/03		
7	Saturday	10/25/03	3 PM Meet in the engineering library to do background work	
8	Sunday	10/26/03		
9	Monday	10/27/03		
10	Tuesday	10/28/03	10 AM Administrative meeting	5 PM Revised Problem Statement D
11	Wednesday	10/29/03		
12	Thursday	10/30/03	John out of town to student conference	
13	Friday	10/31/03		
14	Saturday	11/1/03		
15	Sunday	11/2/03	12 noon John returns to campus	
16	Monday	11/3/03		
17	Tuesday	11/4/03	10 AM Administrative meeting	
18	Wednesday	11/5/03		
19	Thursday	11/6/03	2 PM Meeting w/ Prof. Johnson to go over aerodynamic aspects	
20	Friday	11/7/03		
21	Saturday	11/8/03		
22	Sunday	11/9/03		
23	Monday	11/10/03		
24	Tuesday	11/11/03	10 AM Administrative meeting	5 PM First Progress Report Due
25	Wednesday	11/12/03		
26	Thursday	11/13/03		
27	Friday	11/14/03		
28	Saturday	11/15/03		
29	Sunday	11/16/03		

Figure A.17 – *Sample Team Calendar in the* **project management workbook**

10. Select **Date** under *Category* and make a format selection under *Type* that shows month, day and year (for example, 01-mar-03).

11. Next you will enter a formula in the column A to automatically determine the day corresponding to the date. In cell *A*2, enter the formula `=TEXT(B2,"dddd")`. The formula takes the date in cell *B*2 (stored as a number) and converts it to text. The "dddd" part of the formula requests that only the day portion of the date string be assigned to the cell.

12. From cell *A*2 select the number of cells in column A that correspond to the populated cells in column B.

13. `Edit>Fill>Down` to copy the formula into the selected cells. The day corresponding to the date in column B is inserted into column A.

14. Add entries to the calendar regularly to keep it up to date.

Appendix B

Illustrator 10 Tutorial

B.1 Coventions Used in the Text

To simplify the presentation of instructions, several conventions are used through out the text. First, where the instructions require you to depress a keyboard key, the name of the key will be enclosed in square brackets and written in Geneva font. For example, [Enter] means depress the 'Enter' key. Illustrator 10's interfaces for both Macintosh and Windows environments are very similar. Differences primarily occur in the use of keyboard shortcuts. For those instances in the text, both sets of keys will be given with the Macintosh version first, for example [Option/Alt] means press the 'Option' key if you working on a Macintosh, or the 'Alt' key for the Windows users.

Finally, hierarchal commands will be separated by a '>' symbol and written in Geneva font. For example the commands to change the font face to 'Arial Bold' illustrated in Figure B.1, would be written in the tutorial as Type>Font>Arial>Bold.

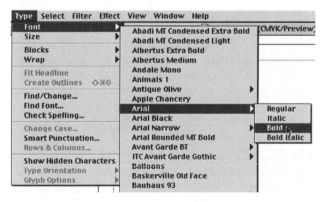

Figure B.1 – *Example of main menu hierarchal commands*

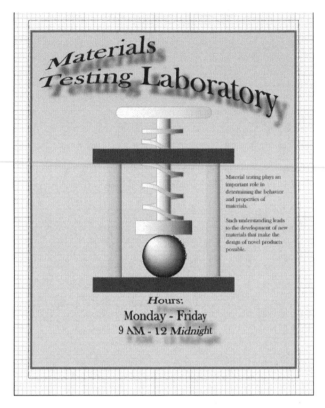

Figure B.2 – *Complete Illustration*

B.2 Introduction

Adobe Illustrator is useful for creating figures and schematics that can then be embedded in reports, presentations or Internet sites. Illustrator images are all in the *Scaleable Vector Graphics* format allowing resizing (magnification or reduction) without any loss in image quality. Once a figure is complete, Illustrator can export it into various other image formats including, JPEG, BMP or PDF. An explanation of these and other digital image formats follows in the next chapter.

The figure you are going to create during this tutorial is shown in Figure B.2. **Please remember to save your work often.**

Online Help

Additional information about how to use Illustrator can be found from it's extensive help files, accessible from Help>Illustrator Help (Figure B.3). The help files are displayed on the default browser installed on your computer.

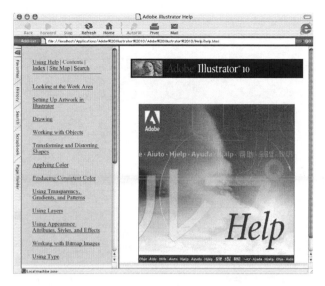

Figure B.3 – *Online help in* **Browser Window**

To find information using the help files index, click on the *Index link*. The left frame contains an alphabetized index. By clicking on the hyperlink next to an entry of interest, the corresponding information is displayed on the right frame (Figure B.4).

Information can also be found using the built-in direct search features by clicking on the *Search link*. By entering a few keywords into the find box and clicking on the **Search Button** a hyper-linked list of relevant pages is displayed in the left frame (Figure B.5). Clicking on the resulting linked relevant topics displays the complete pages in the right frame.

B.3 Basic Shapes and Shading

Creating a New Project

1. Start the Adobe Illustrator program.

2. Create a new project by selecting `File>New`.

3. In the *New Window pop-up menu* (see Figure B.6) enter 'Tutorial 1' for the file name. Set the *size* to 'letter', *units* to 'inches' and *color mode* to 'CMYK'. `[Enter]`.

Basic Shapes: The Rectangle Tool

1. `View>Show Grid`.

2. `View>Show Rulers`.

3. Select the **Rectangle Tool** (Figure B.7).

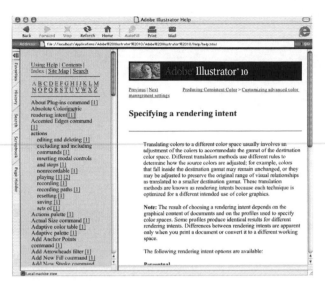

Figure B.4 – *Online help in* **Browser Window** *allowing you to find information by browsing the index*

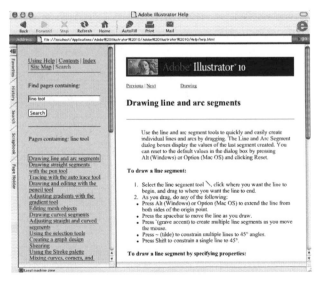

Figure B.5 – *Online help in* **Browser Window** *allowing you to find information using the search feature*

4. Click anywhere on the page.

5. In the *Rectangle pop-up menu* enter '5' for width, and '0.5' for height. [Enter].

6. Select the **Selection Tool** (Figure B.8).

7. Click on the center of the rectangle and drag it towards the bottom of the page (Figure B.9).

8. Select the **Rectangle Tool.**

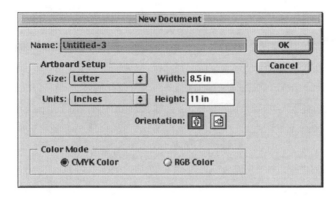

Figure B.6 – New Document *pop-up menu*

Figure B.7 – *Location of* **Rectangle Tool** *on main toolbar*

Figure B.8 – *Location of* **Selection Tool** *on main toolbar*

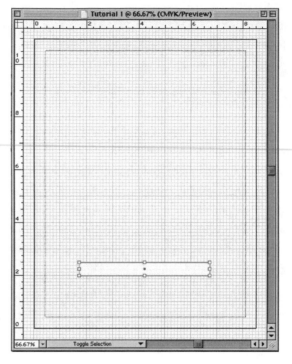

Figure B.9 – *Location of first rectangle*

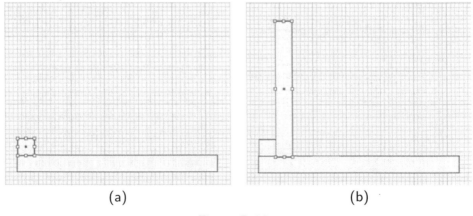

(a)   (b)

Figure B.10 –

9. Click anywhere on the page.

10. In the *Rectangle pop-up menu*, enter a width and height of '0.5' inches.

11. Using the **Selection Tool**, position the resulting square on the top left side of the base rectangle (Figure B.10(a)). The square will serve as a guide to accurately position the support columns onto the base.

12. Using the **Rectangle Tool**, create a rectangle of width of '0.50' and height of '4' inches.

13. Using the square as a guide, use the **Selection Tool** to place the rectangle created in step 12 as shown in Figure B.10(b). With the rectangle selected, you can use the arrow keys on the keyboard to nudge the object in all four directions.

Filling in Shapes with Gradient Colors: Linear Gradients

A gradient fill is a graduated blend between two or more colors.

1. Using the **Selection Tool**, select the upright rectangle.

2. Select the **Gradient Tool** (Figure B.11).

3. Check to see if the **Color Palette** is already in view (Figure B.12), click on the *Color* and *Gradient* tabs.

4. If the **Color Palette** is not in view: `Window>Color`.

5. In the **Gradient Palette**, select 'Linear' for Type; enter 0° for angle - the gradient lines will be vertical.

6. Drag the diamond-shaped slider to obtain a location of 87% (Step 6 in Figure B.12).

7. You will make the column vary in color between a light and a dark shade of green (viewing left to right). In the **Gradient Palette** click on the left small box below the slider (Step 7 in Figure B.12). The box represents the end color for the gradient.

8. The **Color Palette** changes to reflect the properties of the left box (Figure B.13).

9. Click on arrow on the right of the **Color Palette** (Step 9 in Figure B.13) to bring up more options.

Moving a selected object
A selected object can be nudged back and forth, up and down using the arrow keys on the keyboard.

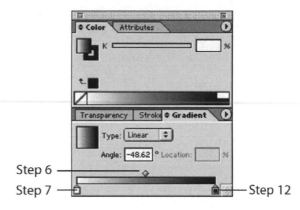

Step 6

Step 7 ———— Step 12

Figure B.12 – **Color** *and* **Gradient Palettes**

Figure B.11 – **Gradient Tool**

Step 9

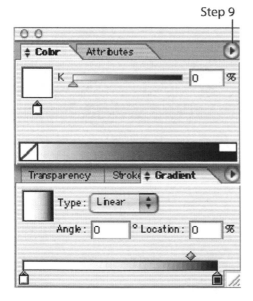

Figure B.13 – *New* **Color Palette** *showing properties of left gradient box*

Figure B.14 – *New menu options in* the **Color Palette**

364

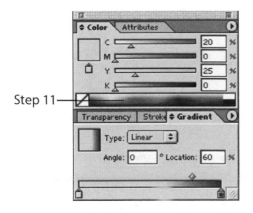

Figure B.15 – **Color Palette** *with new CMYK values*

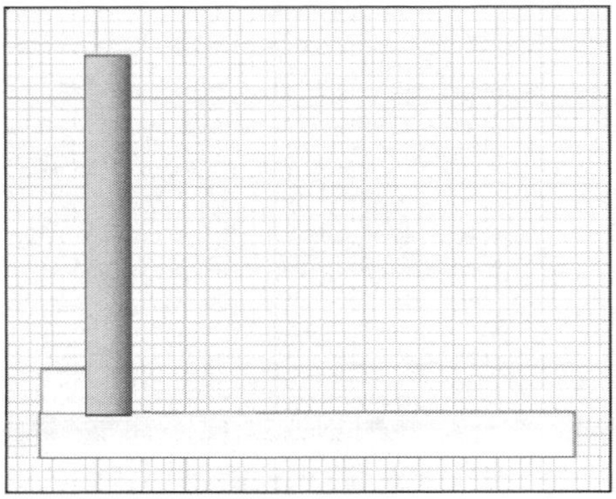

Figure B.16 –

10. In the resulting *pull-down menu*, select the CMYK color model (Figure B.14).

11. In the new **Color Palette**, click on a light green part of the CYMK spectrum. Enter '20'% for C, '0'% for M, '25'% for Y, and '0'% for K (Step 11 in Figure B.15).

12. In the **Gradient Palette** click on the right small box below the slider (Step 12 in Figure B.12)..

13. In the **Color Palette**, if the CMYK sliders are not visible, click on the arrow on the right and select the 'CYMK' from the *drop-down menu*.

14. To make the end color the desired dark shade of green, enter '90'% for C, '0'% for M, '90'% for Y, and '0'% for K.

15. Your picture should now be similar to Figure B.16.

Filling in Shapes with Solid Colors

1. With the **Selection Tool**, select the lower rectangle in your figure.

2. Click on the **Color Fill Tool** on the main toolbar (Figure B.17).

3. In the **Color Palette**, click on the brown portion of the CYMK spectrum (Figure B.18). The selected rectangle turns the same shade of brown.

4. Move the C, M, Y and K sliders until you get a 'dark' shade of brown. Your picture should now be similar to Figure B.19.

365

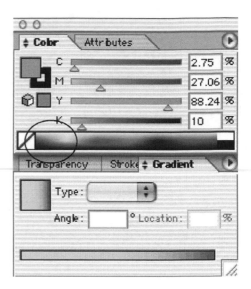

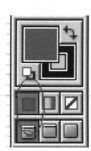

Figure B.17 – **Color Fill Tool**

Figure B.18 – *Brown region on CMYK spectrum*

Duplicating Objects

You will now duplicate the current base and left column to create the top support and right column, respectively.

1. Press and hold [Option/Alt].

2. Click on the small square and drag it to the extreme right of the base. Release the mouse button.

3. Release [Option/Alt]. A duplicate of the square is created on the right (Figure B.20). Nudge the square into position.

4. Repeat steps 1-3, except duplicate the column and locate as shown in Figure B.21.

5. Repeat steps 1-3, except duplicate the base and locate as shown in Figure B.22.

Selecting Multiple Objects

You are now going to delete both guide squares at the same time.

1. To delete both guide squares, use the **Selection Tool** to select the left guide square. [Shift]+Click on the right guide square. Both boxes are now selected.

2. [Delete].

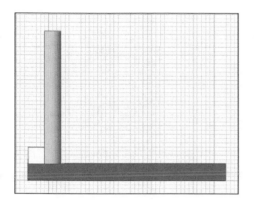

Figure B.19 –

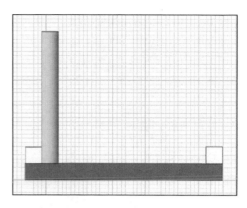

Figure B.20 – *Creation of the second guide square*

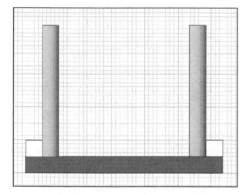

Figure B.21 – *Creation of the second column support*

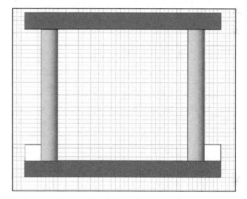

Figure B.22 – *Creation of the top support*

Grouping Objects Together

1. Click on the **Selection Tool.**

2. [Shift]+Click sequentially on all the objects drawn to this point to select them.

3. Object>Group. The four objects will now behave as one. They can always be ungrouped with the ungroup command (Object>Ungroup).

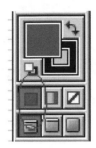

Figure B.23 – *Location of* **Pen Tool** *on main toolbar*

Figure B.24 – **Color Fill Tool**

Drawing Line Segments

You will begin by drawing a centerline about which all objects will now be placed.

1. Select the **Pen Tool** (Figure B.23).

2. Using the *top ruler* as a guide, click just below the 4.25 inch mark (measured from left to right) towards the top of the drawing (Step 2 in Figure B.25). This creates the first *anchor point* for your line.

3. Press and hold [Shift]. Again using the *top ruler* as a guide, click on the 4.25 inch mark (measured from left to right) towards the bottom of the drawing (Step 3 in Figure B.26). The second *anchor point* and the line are created. Pressing [Shift] ensures that your line is either, horizontal, vertical or at $45°$. If you do not press [Shift], you can create lines at any angle.

4. Click on the **Color Fill Tool** (Figure B.24).

5. In the **Color Palette**, click on the black part of the CYMK spectrum or enter CYMK=[0,0,0,100]. This changes the line color to black.

6. This line will serve as the center line for the drawing. Using the **Selection Tool**, click on the grouped objects to select them (Figure B.27).

7. Use the center squares on the grouped object's bounding box as a reference to center the drawing about the centerline. You can either drag the object with the mouse, or use the left/right arrow keys to nudge the object as necessary. Your picture should now be similar to Figure B.28.

Creating horizontal, vertical and $45°$ lines
Pressing [Shift] when placing anchor points using the **Pen Tool** will ensure that your lines are either horizontal , vertical or at 45^0.

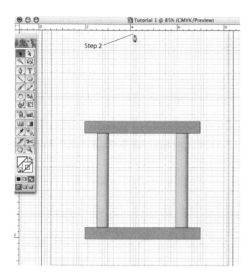

Figure B.25 – *Location of first* anchor point *using the* **Pen Tool**

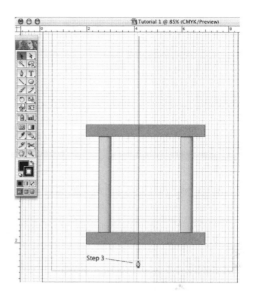

Figure B.26 – *Location of second* anchor point *using the* **Pen Tool**

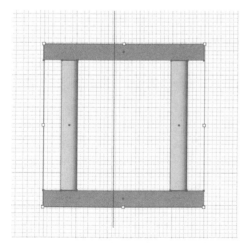

Figure B.27 – *Selection of grouped objects*

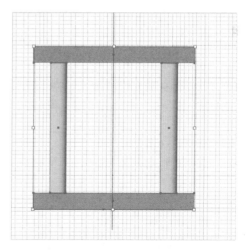

Figure B.28 – *Grouped objects aligned with center line*

Basic Shapes: The Ellipse Tool

1. Click on and hold on the **Rectangle Tool**.

2. The tool expands out revealing an expanded toolbar palette. Still holding down the mouse button, move the cursor over to the triangle on the far right of the expanded toolbar. This 'tears of' the toolbar.

3. On the 'torn off' toolbar, select the **Ellipse Tool** (Figure B.29).

4. Click anywhere on a blank part of the page.

5. In the *Ellipse pop-up menu*, enter a height and width of '1.5' inches.

6. Using the **Selection Tool** and the centerline as a guide, locate the circle as shown in Figure B.30.

7. On the 'torn off' toolbar, select the **Rectangle Tool**.

8. Click anywhere on a blank part of the page.

9. In the *Rectangle pop-up menu*, enter a height of '0.5' inch and width of '2' inches.

10. Using the **Selection Tool** and the centerline as a guide, locate the rectangle as shown in Figure B.31.

11. Re-select the **Rectangle Tool**.

12. Click anywhere on a blank part of the page.

13. In the *Rectangle pop-up menu*, enter a height of '2.4' inches and a width of '0.5' inches.

14. Using the **Selection Tool** and the center line as a guide, locate the rectangle as shown in Figure B.32.

'Tearing off' expanded toolbar palettes

Most tools in the main toolbar can be 'torn off' revealing an expanded toolbar palette. To 'tear off' an expanded palette, click on and hold the small black triangle on the bottom right of the tool. The tool expands. Still holding down the mouse button, move the cursor over the triangle on the far right of the expanded toolbar palette. This tears it off.

Figure B.29 – *Location of* **Ellipse Tool** *on expanded toolbar*

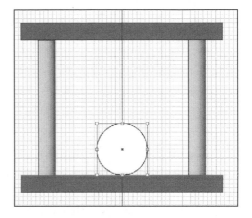

Figure B.30 –

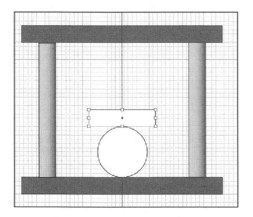

Figure B.31 –

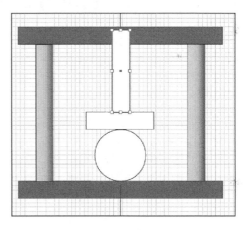

Figure B.32 –

Filling in Shapes with Gradient Colors: Radial Gradients

You are now going to shade the objects created in the previous section with the appropriate gradients.

1. Using the **Selection Tool**, select the circle.

2. Click on the **Gradient Tool** (Figure B.33).

3. In the **Gradient Palette**, click on the left small box below the slider (Step 3 in Figure B.34). In the **Color Palette**, click on the white part of the CMYK spectrum (extreme right) OR enter '0'% for C, M, Y and K. The left box (and the starting color) turn to white.

371

4. In the **Gradient Palette**, click on the right small box below the slider (Step 4 in Figure B.34). In the **Color Palette**, enter '100'% for C and M, and '0'% for Y and K. The left box (and the ending color) turn a deep purple.

5. Select 'Radial' for Type (Step 5 in Figure B.34).

6. Using the *Gradient Slider*, set a gradient location of 79%, OR you can click on the *Gradient Slider* and type in 79% directly into the *Location Box* (Step 6 in Figure B.34).

7. With the **Gradient Tool** still selected, click on the outer edge of the circle at approximately the "10 O'clock" position. The apparent direction of lighting shifts to this location. Your picture should now be similar to Figure B.35.

8. You will know apply linear gradients to the other two unshaded rectangles. Using the **Selection Tool** click on the lower unshaded rectangle to select it.

9. [Shift]+Click on the upper unshaded rectangle. Both rectangles are selected.

10. In the **Gradient Palette**, change the gradient 'Type' to linear.

11. In the **Gradient Palette**, click on the left small box below the slider. In the **Color Palette**, click on the white part of the CMYK spectrum (extreme right) OR enter '0'% for C, M, Y and K. The left box (and the starting color) turn to white.

12. In the **Gradient Palette**, click on the right small box below the slider. In the **Color Palette**, enter '0'% for C, M, and Y, and '75'% for K. The right box (and the ending color) turn a dark gray.

13. Using the *Gradient Slider*, set a gradient location of 75%, OR you can click on the *Gradient Slider* and type 75% directly into the *Location Box*. Your figure should be similar to Figure B.36.

14. Finally, click on the top support structure. `Object>Arrange>Bring to Front`. (Figure B.37)

Figure B.33 – **Gradient Tool**

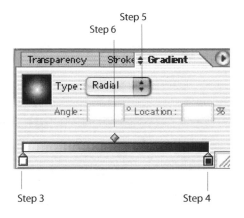

Figure B.34 – **Gradient Palette**

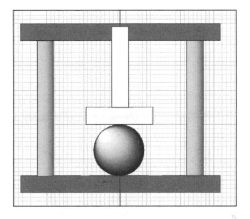

Figure B.35 –

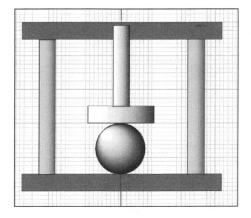

Figure B.36 –

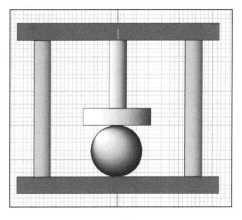

Figure B.37 –

Swatches: Saving Solid Colors and Gradients

Swatches allow you to save colors and gradients that you have created and would like to apply to other objects in your project. By using swatches you avoid having to go through all the original steps.

1. Window>Swatches to bring up the **Swatches Palette** (Figure B.38). If the palette is already visible it will appear checked in the Window menu. Selecting it again will close the palette.

2. Using the **Selection Tool**, click on the top brown rectangle. The settings in the **Color Window** change to reflect the rectangle color.

3. With reference to Figure B.38, click on the **New Swatch Button** on the **Swatch Palette**. A new swatch the same color as the selected rectangle appears in the **Swatch Palette** (Figure B.39).

4. Next you will create a swatch of the left light green gradient color of the columns. Using the **Selection Tool** click on either of the two green columns. The settings in both the **Color** and **Gradient Palettes** change to reflect the column's color properties.

5. In the **Gradient Palette**, click on the small left box below the slider.

6. Click on the **Color Fill Tool** (Figure B.40). The **Color Palette** settings change to reflect those of the left gradient color, and not of the entire gradient as before.

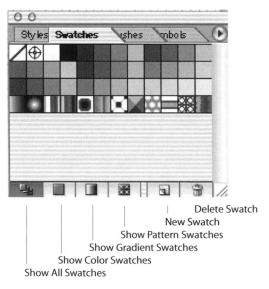

Figure B.38 – **Swatches Palette**

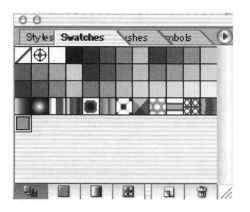

Figure B.39 – **Swatches Palette**

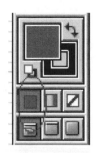

Figure B.40 – **Color Fill Tool**

7. With reference to Figure B.38, click on the **New Swatch Button** on the **Swatch Palette**. A new swatch the same color as the left gradient color appears in the **Swatch Palette**.

8. You can also save entire gradients with all corresponding settings as a swatch. Click on the center gray gradient column.

9. Click on the **New Swatch Button** on the **Swatch Palette**. A new swatch with the same gradient settings as the center column appears in the **Swatch Palette**.

B.4 Lines and Objects

Object Transformation: Changing Dimensions

1. Using the **Selection Tool**, select the center column.

2. Object>Arrange>Bring to Front, to bring the center column to the front (*See* Figure B.36 on Page 373).

3. Open the **Transformation Palette** by Window>Transform (Figure B.41). The X and Y boxes indicate the location of the object anchor on the page, where (0,0) is the bottom left corner of the page. The nine-square grid represents *anchor points* from which any change in size will be based. For example, by default the center square is selected, meaning that the object anchor is located in the center of the object. Any change in size or location will therefore be about the center.

4. We would like to increase the height of the center column from the top, keeping the current bottom position fixed. In the nine-square grid, click on the bottom, center square. This moves the anchor of the center column to the middle of base. Notice that the Y value changes accordingly (Figure B.42).

5. The W and H boxes indicate the current width and height of the column, respectively. Change the height from 2.4 to 4 inches by typing '4' into the H box.

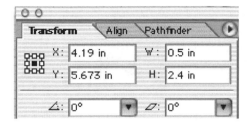

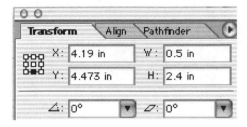

Figure B.41 – **Transformation Palette** *with default center* anchor point

Figure B.42 – **Transformation Palette** anchor point *changed to the bottom center position*

6. [Enter]. The center column increases in length.

Stroke vs. Fill/Using Swatches

In the previous section you used the **Color Fill Tool** to fill the inside of an object with a solid color. The outline of the object, however, was not affected and remained black. The **Stroke Tool** (Figure B.43) allows you to change the thickness and the color of an object outline. You will now change the outline color of all the objects in you current drawing.

1. Using the **Selection Tool**, click on the top brown support.

2. Object>Ungroup to ungroup the objects.

3. Click anywhere on a blank part of the page to de-select the selected objects.

4. Select the upper and the lower brown supports. Recall that you can use [Shift] to select multiple objects (*See* Page 367).

5. Click on the **Stroke Tool** (Figure B.43) in the main toolbar.

6. The **Color Palette** now shows the CYMK values of the outline. In the **Swatches Palette** click on the brown swatch you created in the previous chapter that corresponds to the supports' fill color. The stroke color changes from black to brown.

7. Select the left and right columns.

8. As the **Stroke Tool** is still selected the outline color, black CYMK=[0,0,0,100], is displayed in the **Color Palette**. In the **Swatches Palette** click on the light green swatch you created in the previous chapter that corresponds to the columns left gradient color. The stroke color changes from black to light green.

9. Select the two center rectangles (above the ball).

10. As the **Stroke Tool** is still selected the outline color, black CYMK=[0,0,0,100], is displayed in the **Color Palette**. Change the color to CYMK=[0,0,0,25] or light gray.

Arrangement of objects

When an object is drawn in Illustrator, it is placed in an imaginary plane above all pre-existing objects. For example with reference to the figures below, the rectangle was drawn after the circle and is therefore on a plane above the circle and covers any portion beneath it. Similarly, the line was drawn after the rectangle and therefore appears above it and the circle.

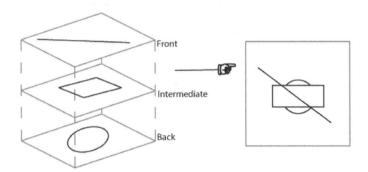

This order can be changed using the **Object>Arrange** commands. To change the relative location of an object:

- **Object>Arrange>Bring to Front** to move the selected object to the front. All objects previously above it drop down one step in the hierarchy.
- **Object>Arrange>Send to Back** to move the selected object to the back. All objects previously below it move up one step in the hierarchy.
- **Object>Arrange>Bring Forward** to move the selected object one step forward in the hierarchy.
- **Object>Arrange>Send Backward** to move the selected object one step backward in the hierarchy.

Fill

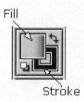

Stroke

Figure B.43 – *Stroke versus Fill*

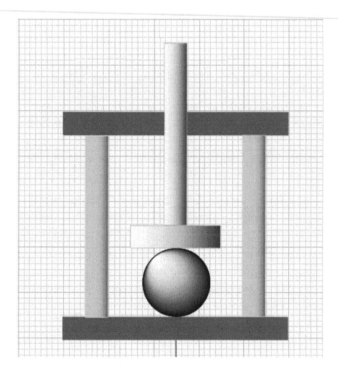

Figure B.44 – *Picture after completion of Stroke vs. Fill*

11. Select the purple ball.

12. As the **Stroke Tool** is still selected the outline color, black CYMK=[0,0,0,100], is displayed in the **Color Palette**. Change the color to CYMK=[100,100,0,0] or purple. Your picture should now be similar to Figure B.44

Creation of Curved Lines

Lines created in Illustrator are defined by a series of *anchor points* (see Figure B.45). Each anchor point, has a *Direction Line* and *Direction Handle* associated with it. A *Direction Line* defines the direction that the drawn line leaves the anchor point (Technically: The drawn line leaves the anchor point tangent to the *Direction Line*).

The length and orientation of the *Direction Line* can be changed using the *Direction Handle*. Changing the orientation of the *Direction Line*, changes the curvature of the line (Figure B.46(a) and B.46(b)), while increasing (or decreasing) the length of the *Direction Line* increases (or decreases) the depth of the curve (Figure B.46(c) and B.46(d)). You are now going to create the screw threads on the center column.

1. Click and hold on the **Pen Tool** .

2. The tool expands out revealing more tools. Still holding down the mouse button, move the pointer over to the triangle on the far right of the expanded toolbar. This 'tears of' the toolbar.

3. On the 'torn off' toolbar (Figure B.47), select the **Pen Tool**.

4. Click on a blank part of the page to place the line's first anchor point.

5. Holding down [Shift], click approximately 1 inch to the right of the first point - use the rulers as a guide. Recall that holding down [Shift] allows you to obtain perfectly horizontal or vertical lines.

6. Select the line using the **Selection Tool**.

7. In the **Color Palette**, click on the **Stroke Tool** (Figure B.48). Set the line stroke color to black (CYMK=[0,0,0,100]).

8. Click on the **Color Palette's Fill Button**. Set the line fill to 'None' by clicking on the **None Fill Button** (Figure B.49). This will prevent the curved portions of the line being filled in later on.

9. In the **Transformation Palette** (Figure B.50), enter '1.5' inches in the *W box*, changing the width of the line to exactly 1.5 inches.

10. In the same window, enter '-15°' in the *Orientation Box* (Step 8 in Figure B.50). Your line should be similar to Figure B.51(a).

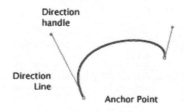

Figure B.45 – *Parameter terminology for defining curved lines*

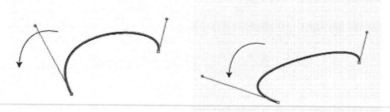

(a) Rotating the *Direction Line* using the *Direction Handle*

(b) Increased curvature from rotating the *Direction Line*

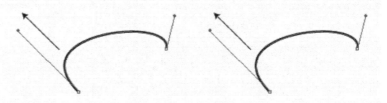

(c) Using the *Direction Handle* to lengthen the *Direction Line*

(d) Increasing the depth of the curve by lengthening the *Direction Line*

Figure B.46 – *Changing the curvature of a line using the* Direction Line *and the* Direction Handle

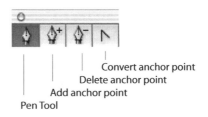

Convert anchor point
Delete anchor point
Add anchor point
Pen Tool

Figure B.47 – *'Tear away'* **Pen Tool** *toolbar*

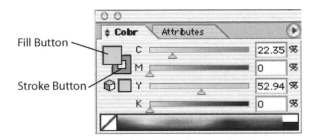

Figure B.48 – **Fill** *and* **Stroke Buttons** *in the* **Color Palette**

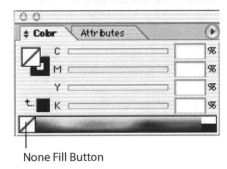

None Fill Button

Figure B.49 – **None Fill Button** *in* **Color Palette**

11. Using the **Convert Anchor Tool** (Figure B.47), click and hold the line's *left anchor point*.

12. Drag the *anchor point* to the *right*, creating a horizontal *direction handle* (and corresponding curve) approximately half the length of the line (Figure B.51(b)).

13. Click and hold on the lines right *anchor point*.

14. Drag the *anchor point* to the *right*, creating a horizontal *direction handle* (and corresponding curve) approximately half the length of the line (Figure B.51(c)).

15. You will now use the **Transformation window** to place an object at an exact location on the page. Select the curve using the **Selection Tool**.

16. From the **Transformation Palette**, write down on a piece of paper the X and Y *position* of the curve. Edit>Copy, to copy the curve. Edit>Paste, to paste a copy of the curve on to the drawing.

17. You would like the two curves to be exactly 0.15 inches apart. Recall that the (X,Y) positions of any object is measured from the bottom left corner of the page. On a piece of paper, subtract 0.15 inches from the noted Y position.

18. In the **Transformation Palette**, type in the new value into the *Y Box*. [Enter].

19. Type in the previously noted X value into the *X box*. [Enter]. The pasted curve is moved to a position 0.15 inches below the original curve (Figure B.51(d)).

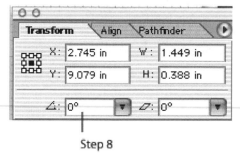

Step 8

Figure B.50 – *Transformation Palette*

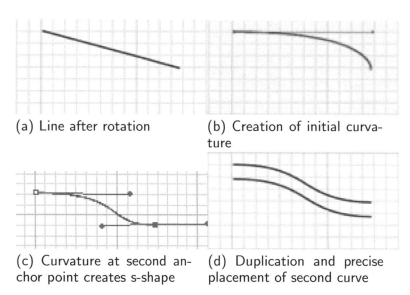

(a) Line after rotation

(b) Creation of initial curvature

(c) Curvature at second anchor point creates s-shape

(d) Duplication and precise placement of second curve

Figure B.51 – *Steps to create the screw threads*

Joining Line Segments and Curves to Form a Closed Object

The aim of this section is to join the two curves into a single closed object that can be shaded.

1. Click on the **Pen Tool**.

2. Move the cursor over the left *anchor point* in the top curve. When you are in the correct position, the 'x' symbol on the lower right of the pen cursor changes to a forward slash, '/' (Figure B.52). Click on the *anchor point* to select it.

3. Move the cursor over the left *anchor point* of the bottom curve. This time you will know you are in the correct position when the pen symbol changes to a square. This indicates that selection of this point will create a new **open object** consisting of the previous two unconnected curves and a new line segment between them. Click on the *anchor point* (Figure B.53(a))

4. Move the cursor over the right *anchor point* in the top curve. When you are in the correct position, the 'x' symbol on the lower right of the pen cursor changes to a forward slash, '/'. Click on the *anchor point* to select it.

5. Move the cursor over the right *anchor point* of the bottom curve. This time you will know you are in the correct position when the symbol changes to a circle. This indicates that selection of this point will create a new **closed object** consisting of the previous open object, and a new line segment between the two end points, thereby closing it. Click on the *anchor point* (Figure B.53(b)). The closed object now behaves as a single entity (similar to the rectangles and ellipses you created earlier), and not as the four lines that created it.

6. With the new closed object still selected, click on the gray gradient swatch you previously created in the **Swatches Palette**. The gradient fills the object (Figure B.53 (c)).

7. Using the **Stroke Tool**, change the outline color of the closed object to a light shade of gray defined by CYMK=[0,0,0,25]. (*See* steps on using the **Stroke Tool** on Page 376).

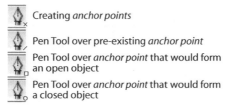

Figure B.52 – **Pen Tool** *symbols*

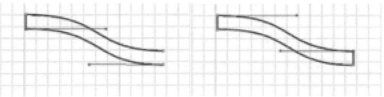

(a) Adding a line segment on the left, joining the two curves. The resultant shape is a stored as a single object

(b) Adding a second line segment to the left. The shape is considered a single closed entity

(c) Object after applying gradients

Figure B.53 – *Steps to create the screw threads*

Object Transformation: Reflection

Object transformation by reflection has the same effect as if the object were reflected in a mirror. Illustrator gives you the option to reflect about a horizontal or a vertical line.

1. Using the **Selection Tool**, select the centerline of your picture created in the previous tutorials, and bring it to the front: `Object> Arrange>Bring to Front`.

2. Select the thread you just created and position it as shown in Figure B.54. Align the thread's center with the picture's centerline.

3. Press and hold [`Option/Alt`], click on the thread and drag it slightly upwards to create a duplicate.

4. With the duplicate thread still selected: `Object>Transform> Reflect`. In the *pop-up menu* that appears, select 'horizontal' and click on the **OK Button**.

5. Click on the **Gradient Tool**. (*See* Page 363 for use of the **Gradient Tool**).

6. In the **Gradient Palette**, change the transition location to 50% by using the *Gradient Slider* or by clicking on the slider and then entering 50% in the *Location Box*.

7. Move the duplicate thread to the position shown in Figure B.55.

8. Select the center column and bring it to the front: `Object>Arrange >Bring to Front`.

9. Select the lower thread and bring it to the front: `Object>Arrange >Bring to Front`.

10. Press and hold `[Option/Alt]`. Click and drag the lower thread to the position shown in Figure B.56.

11. Press and hold `[Option/Alt]`. Click and drag the second lowest thread to the position shown in Figure B.57.

12. Counting from the bottom, click on thread 1, `[Shift]` and click on thread 3 to select them both.

13. Press and hold `[Option/Alt]`.

14. Click and drag thread 1 (creating a duplicate of thread 1 and 3) to the position shown in Figure B.58.

15. Counting from the bottom, click on thread 2, `[Shift]` and click on thread 4. This selects both.

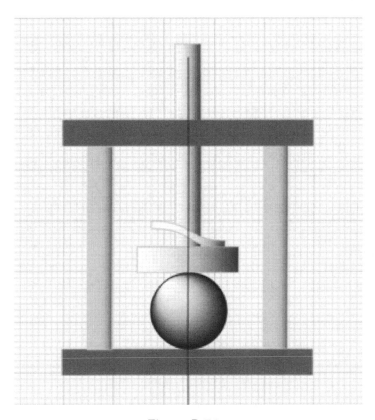

Figure B.54 –

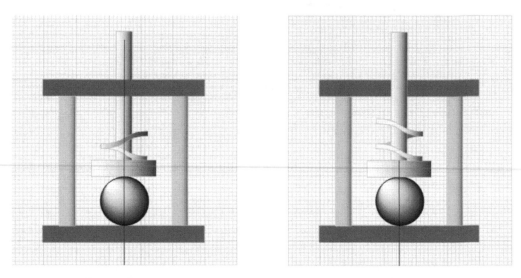

Figure B.55 –   Figure B.56 –

16. Press and hold [Option/Alt].

17. Click and drag thread 2 (creating a duplicate of thread 2 and 4) to the position shown in Figure B.59.

18. Select the **Rectangle Tool**. Click anywhere on a blank part of the page to create a rectangle of width 1.45 inches and height 0.15 inches.

19. Using the **Gradient Tool**, create a linear gradient from white on the left to gray on the right defined by CYMK=[0,0,0,75]. The gradient transition location should be 75%.

20. Use the **Stroke Tool** to change the outline color of the rectangle to a light gray defined by CYMK=[0,0,0,25] (*See* Page 376 for use of the **Stroke Tool**).

21. Place the rectangle on thread 8 as shown in Figure B.60.

22. If you have not done so, **save your work**.

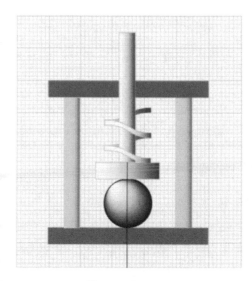

Figure B.57 –

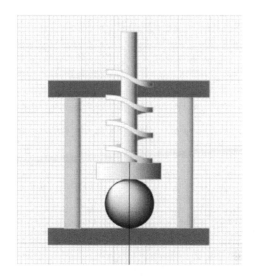

Figure B.58 –

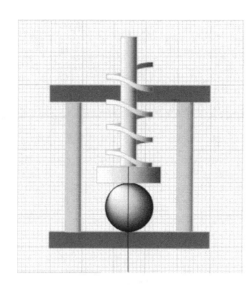

Figure B.59 –

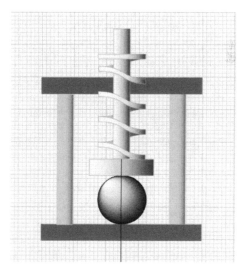

Figure B.60 – *Picture with all threads in place*

Merging Object Areas

1. On a blank part of the page, use the **Rectangle Tool** to create a rectangle of width 3 inches and height 0.4 inches.

2. Using the **Ellipse Tool**, create circle by specifying a width and height of 0.4 inches. (*See* Page 370 for use of the **Ellipse Tool**)

3. Using the **Selection Tool**, locate the circle on the left edge of the rectangle as shown in Figure B.61(a).

4. Make a duplicate of the circle and locate it on the right edge of the rectangle as shown in Figure B.61(b).

5. Select all three objects ([shift] and click on all three objects).

6. In the **Transform Palette**, click on the *Pathfinder Tab* OR Window>Pathfinder. This brings up the **Pathfinder Palette** (Figure B.62).

7. In the **Pathfinder Palette**, click on the **Add to Shape Area Button**. The three shapes, though separate now behave as a single area.

8. Click on the small arrow to the extreme right of the **Pathfinder Palette** (Figure B.62).

9. This brings up a *pull-down menu* (Figure B.63). Select the *Make Compound Shape* menu option to combine the three objects into a single entity.

10. Using the **Gradient Tool**, create a linear gradient from white to gray - CYMK=[0,0,0,7 - located at 75%.

11. Using the **Stroke Tool**, change the outline color to light gray, CYMK=[0,0,0,25]. Your figure should now look like Figure B.61(c).

Save your work often!
To avoid losing hours of work due to a system problem or loss of power to your computer, save your work often

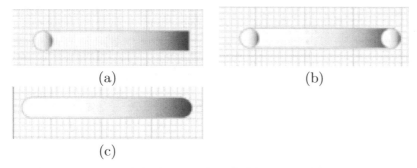

(a)

(b)

(c)

Figure B.61 – *Creating the handle using the* **Merge Tool**

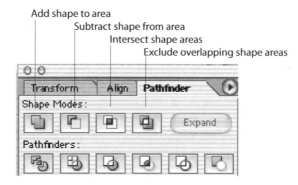

Figure B.62 – **Pathfinder Palette**

Figure B.63 – **Make Compound Shape** *menu option in* **Pathfinder Palette**

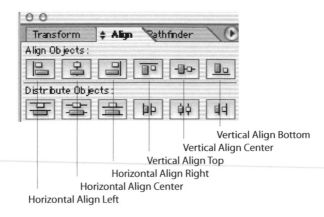

Figure B.64 – *The* **Align Palette**

Aligning Objects

1. The object you created in the previous section is the handle for the threaded screw in your drawing. Place the handle on top of the center column as shown in Figure B.65. *Align the top horizontal line of the handle slightly above the top horizontal line of the center column, i.e., create some overlap.*

2. To ensure that the two objects are aligned exactly, you can use the **the Alignment Tools**. Select both the handle and the center column.

3. Click on the *Alignment Tab* in the **Pathfinder Palette** OR `Window>Align` to bring up the **Align Palette** (Figure B.64).

4. Click on the **Vertical Align Top Tool** to exactly align the two objects using their top edges.

5. You will now re-group the top, bottom and column supports to act as a single entity. `[Shift]+Click` on the top, bottom, and both columns to select them all.

6. `Object>Group` to group them.

7. With the group still selected, `[Shift]+click` on all the threads, the rectangle on top of the ball, and the ball (Figure B.66)

8. Click on the **Horizontal Align Center Tool** to align all these objects about their center.

9. Deselect the objects by clicking anywhere on the page.

10. Click on the center line that you had used as a guide.

11. `[Delete]`, to delete the line.

12. Your current picture should now look like Figure B.67.

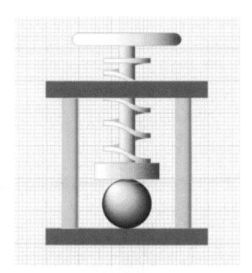

Figure B.65 – *Aligning the handle with the center column*

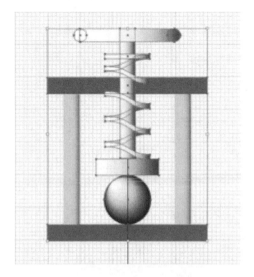

Figure B.66 – *Selection of objects using the* **Horizontal Align center Tool**

B.5 Working with Text

Selecting Multiple Objects II

In previous sections you learnt how to select multiple objects by [Shift]+Click using the **Selection Tool**. You can also select a block of objects using the **Selection Tool** to draw a marquee around them.

1. Click on the **Selection Tool**.

2. Using Figure B.68 as a guide, place the cursor at a point above and to the left of your drawing.

3. Click, hold and drag the mouse towards the bottom right of the drawing. A rectangular marquee is created around the objects (Figure B.69).

4. Release the mouse button. All the objects within the rectangular dashed box are selected (Figure B.70).

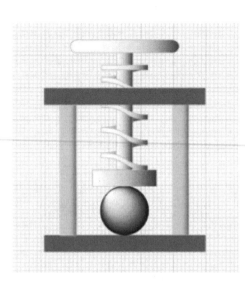

Figure B.67 – *Done!*

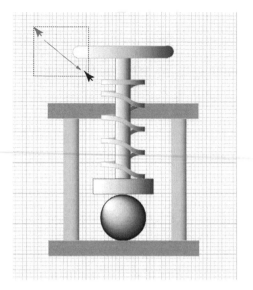

Figure B.68 – *Using the* **Selection Tool** *to select multiple objects*

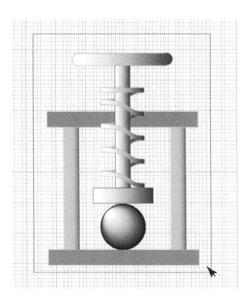

Figure B.69 – *Using the* **Selection Tool** *to select multiple objects*

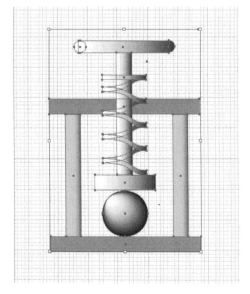

Figure B.70 – *Selected Objects*

Object Transformation: Scale

1. With all the objects selected from the previous section, `Object>Group`.

2. `Object>Transform>Scale`. In the *Scale pop-up menu* (Figure B.71), select 'Uniform', and enter a scale of '80'%.

3. Click on the **OK Button**. Your image is now 80% of it's original size.

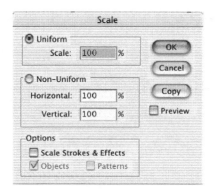

Figure B.71 – Scale pop-up menu

Uniform and Non-Uniform Scaling

In the *Scale pop-up menu* (Figure B.71) you can access two scaling options:

1. **Uniform** - the aspect ratio (relative size of height and width) of the selected object(s) is maintained. For example, if the object(s) is uniformly scaled by 80%, both the height and the width are scaled 80%.

2. **Non-Uniform** - the aspect ratio is not maintained. You can select different scaling percentages for both the Vertical (height) and Horizontal (width) dimensions.

Object Transformation: Exact Placement

1. In the **Transform Palette** (Figure B.72), click on the center square in the nine square *anchor point* array. This moves the anchor point of the grouped objects to the center. If the **Transform Palette** is not visible, `Window>Transform`.

2. Enter '4.25' for X and '5.5' for Y. This moves the center *anchor point* of the selected objects (and therefore the objects) to the middle of the page. Recall that the drawing origin (X=0, Y=0) is located at the bottom left corner of the page. X distances are measured horizontally and to the left of this position, and Y distances are measure vertically and upward from this position.

Creation of a New Layer

Layers act as transparent overlaid sheets on which you create your work. For complex artwork, layers allow the separation of different aspects of the image, making for easier editing and creation. You will create a new Layer in which you will add all the text for the figure.

1. `Window>Layers`.

2. In the **Layers Palette**, click on the **New Layer Button** (Figure B.73).

3. A new active layer is created - shaded in gray (Figure B.74).

Layers

Layers act as transparent overlaid sheets on which you create your work. For complex artwork, layers allow the separation of different aspects of the image, making for easier editing and creation. Layers assume a 'top down' line of sight, i.e. opaque items on any Layer will cover up any objects on layers directly below.

Clicking on the eye symbol adjacent to each Layer (Figure B.74) turn's the layer's visibility off and on, making work with complex art work easier. Also, clicking in the check box adjacent to each Layer (Figure B.74 toggles off and on the lock on the layer. When a Layer is locked, it cannot be edited. A locked Layer has a padlock symbol in the checkbox location.,

---

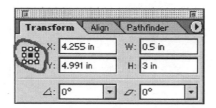

Figure B.72 – **Transform Palette**

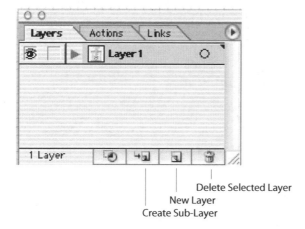

Delete Selected Layer
New Layer
Create Sub-Layer

Figure B.73 – **Layers Palette**

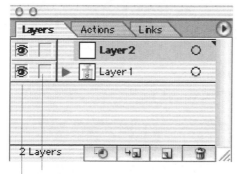

Toggle Lock Layer (Editable when blank)
Toggle Layer Visibility (Visible when eye symbol present)

Figure B.74 – *New created active layer*

Changing *Anchor Point* to top left position

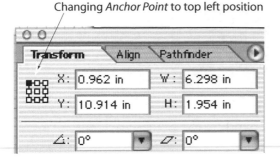
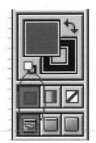

Figure B.75 – *Changing the location of the* anchor point *in the* **Transformation Palette**

Figure B.76 – **Color Fill Tool**

Text: Adding, Changing Font Face and Size

1. Select the **Text Tool**, **T**.

2. Click anywhere near the top of the page to insert the text cursor.

3. Type, 'Materials' [Enter] 'Testing Laboratory'.

4. Click on the **Selection Tool**.

5. Type>Font>Baskerville Old Face[1], to change the font face.

6. Type>Size>60, to change the font size.

7. You will now place the text box in the top left corner of your drawing using the **Transformation Palette**. In the **Transformation Palette**, change the object *anchor point* to the top left point (Figure B.75).

8. Enter '1' inch for X and '10' inch for Y in the **Transformation Palette** to move the text box to the top left corner (Figure B.77).

Changing Text Color

1. Select the 'Materials Testing Laboratory' text box.

2. Click on the **Color Fill Tool** (Figure B.76).

3. In the **Color Palette**, enter CYMK=[100,100,0,0] to change the text color to a dark shade of blue.

---

[1]If your computer does not have this font, substitute with any other font on your system.

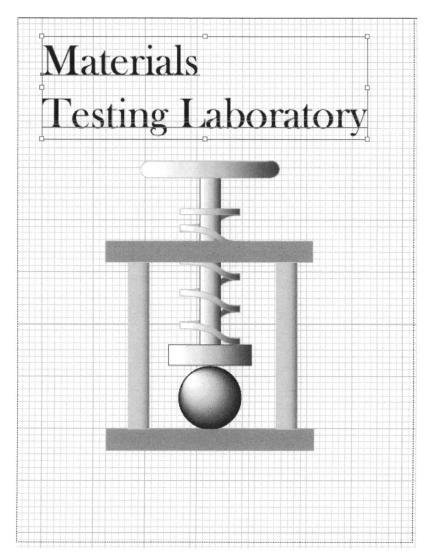

Figure B.77 – 'Materials Testing Laboratory' text box placed in the top left corner of the drawing

Distortion of Text: Warping Envelope

1. Object>Envelope Distort>Make with Warp. The *Warp Options pop-up menu* appears (Figure B.78).

2. Choose 'Arc' for style and check the 'Preview box'. Move the pop-up menu to a part of the screen that allows you to simultaneously view the text.

3. Change the *Bend, Horizontal* and *Vertical* settings, and view the effect they have on the text.

4. Once your curiosity has been satisfied, select 'Horizontal' warping; enter '16%' for *Bend*; '50%' and '10%' for *Horizontal* and *Vertical* distortion, respectively. Click on the **OK Button**. Your text should be similar to Figure B.80. Tables B.1 and B.2 illustrate the effect of other warp style options.

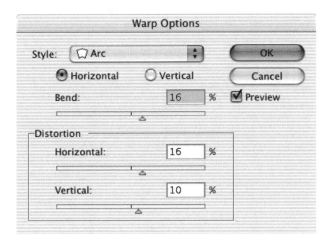

Figure B.78 – Warp Options pop-up menu

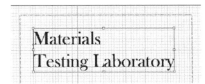

Figure B.79 – *'Material Testing Laboratory' text box*

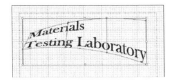

Figure B.80 – *'Material Testing Laboratory' text box after 'horizontal arc' warping*

Table B.1 – *Other possible warps. For all options, horizontal warp was selected, and both horizontal and vertical distortion set to 0. By varying the* **Warp Style** *and* **Bend** *options, numerous interesting shapes can be obtained.*

Warp Style	+40 Bend	-40 Bend
Arc		
Arc Lower		
Arch		
Bulge		
Shell Lower		
Shell Upper		
Flag		

Table B.2 – *Other possible warps continued... For all options, horizontal warp was selected, and both horizontal and vertical distortion set to 0. By varying the* **Warp Style** *and* **Bend** *options, numerous interesting shapes can be obtained.*

Warp Style	+40 Bend	-40 Bend
Wave		
Fish		
Rise		
Fisheye		
Inflate		
Squeeze		
Twist		

Typing Along a Path

Illustrator has the ability to make typed text follow a prescribed path.

1. First, you will create the path along which the typed text will follow. In the **Layer Palette** click on Layer 2 to select it.

2. Select the **Ellipse Tool** located on the expanded **Rectangle Tool** (Figure B.81).

3. Click anywhere on a blank part of the drawing.

4. Type in a height and width of '1.2' inches, [Enter].

5. Using the **Selection Tool**, move the circle until it overlaps the purple ball. Recall that you can use the arrow keys on the keyboard to nudge the circle in place.

6. In the **Layer Palette**, click on the eye adjacent to Layer 1 to make the Layer invisible. This leaves the outline of the circle you just drew.

7. Click on the small triangle on the lower right of the **Text Tool** to bring up the expanded **Text Tool Palette** (Figure B.82).

8. Still holding down the mouse button, move the cursor to the triangle on the far right of the expanded toolbar. This 'tears off' the toolbar.

9. From the expanded **Text Tool Palette**, select the **Path Type Tool**.

10. Click at approximately the 1 O'clock position on the ball outline to insert the text cursor.

11. Type>Size>6.

12. Type 'Drawn by *your name*', where you replace *your name* with your actual name. Note that the text follows the path of the circle (Figure B.83), and any stroke or fill color for the path is removed.

13. In the **Layer Palette**, click in the now empty check box where the eye symbol was adjacent to Layer 1. The eye symbol reappears and Layer 1 becomes visible again (Figure B.84).

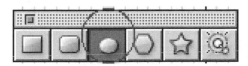

Figure B.81 – *Location of* **Ellipse Tool** *on expanded* **Rectangle Tool Palette**

Figure B.82 – *Location of* **Path Type Tool** *on expanded* **Text Tool Palette**

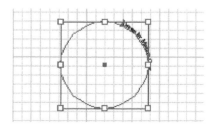

Figure B.83 – *Typed in text following path outline*

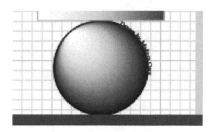

Figure B.84 – *Image after Layer 1 is reactivated*

Creating a Text Box and using the Spell Checking Tool

Text boxes allow you to insert a significant amount of text that can then be formatted in the same ways as in a word processing document.

1. Select the **Text Tool**.

2. Using Figure B.85 as a guide, click near the top of the right support column, hold and drag the cursor to create a text box as shown. The cursor is automatically placed at the top left of the text box.

3. Type>Font>Baskerville Old Face (Use alternate font face if necessary).

4. Type>Size>12.

5. Type, 'Material testing plays an important role in determinging the beha and properties of materials'. [Enter] [Enter]. 'Such an understanding leads to the development of new materials that make the design of novel products possible.' (Figure B.86). Note that '*determining*' is intentionally misspelled.

6. Type>Check Spelling. Using the *Check Spelling pop-up menu* (Figure B.87), correct any spelling mistakes. Note that Illustrator checks the spelling of all text in the drawing.

7. Click on the **Done Button**.

8. With the **Selection Tool**, click and select the text box.

9. In the **Color Palette** change the text color to an orange-brown color defined by CYMK=[2,75,90,0] (Figure B.88).

10. Using the **Text Tool**, click on a blank part of the page and type, 'Hours:' [Enter], 'Monday-Friday', [Enter], '9 AM - 12 MIDNIGHT'.

11. In the **Color Palette** change the text color to a dark blue defined by CYMK=[100,100,0

12. Window>Type>Character. To bring up the **Character Palette** (Figure B.89)

13. Click on the down arrow at the extreme right of the font row, and select 'Baskersville Old Face' font.

14. Change the font size to '24pt'.

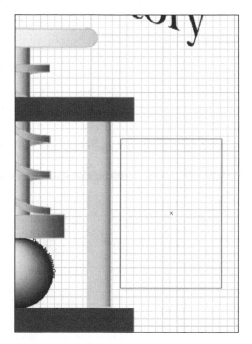

Figure B.85 – *Creation of a text box*

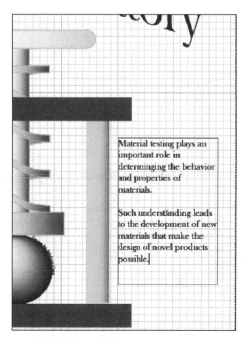

Figure B.86 – *Addition of text to text box*

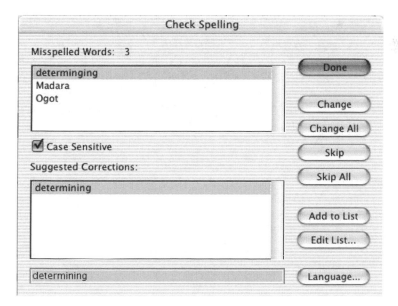

Figure B.87 – *Check Spelling window*

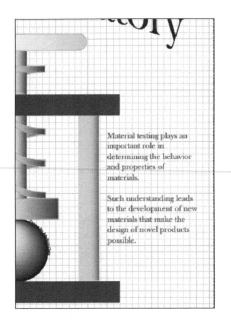

Figure B.88 – *Changing text color*

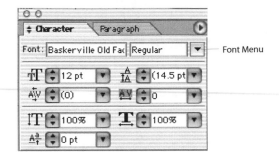

Figure B.89 – **Character Palette**

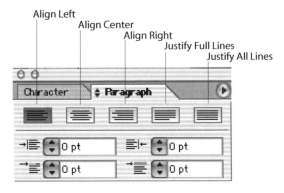

Figure B.90 – **Paragraph Window**

15. In the **Character Palette**, click on the *Paragraph* tab (Figure B.90).

16. Select the **Align Center Button**.

17. Move the text box below the diagram as shown in Figure B.91.

18. Object>Envelope Distort>Make with warp.

19. In the *Make with Warp pop-up menu*, choose 'Fisheye' for style, and check the preview box. Move the pop-up menu to a part of the screen that allows you to simultaneously view the text.

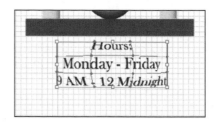

Figure B.91 – *Position of text*   Figure B.92 – *Text after warping*

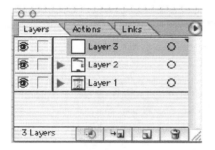

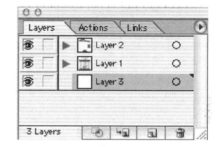

Figure B.93 – *Creation of new layer*

Figure B.94 – *Rearrangement of layers*

20. Change the *Bend*, *Horizontal* and *Vertical* settings, to view the effect they have on the text.

21. Once your curiosity has been satisfied, enter '30'% for *Bend*; '0'% and '0'% for *Horizontal* and *Vertical* distortion, respectively. Click on the **OK Button**. Your text should be similar to Figure B.92

Reordering and Locking Layers

1. In the **Layers Window**, create a new Layer by clicking on the **New Layer Button** ▣ (Figure B.93).

2. Click and hold Layer 3. Drag it *below* Layer 1. This reorders the layers as shown in Figure B.94. Note that if you drag it onto Layer 1, Layer 3 becomes a sub-Layer of Layer 1. You will now create and add a background image behind the existing drawing. With Layer 3 as the lowest layer, creation of the image here will accomplish that goal.

3. Make Layers 1 and 2 invisible by clicking on the 'eye' symbols adjacent to them.

4. Click on Layer 3 to make sure it is the active layer.

5. Using the **Rectangle Tool**, create a rectangle with a width and height of 7.5 and 10 inches, respectively (*See* Page 361 for a reminder on how to use the **Rectangle Tool**).

Step 10

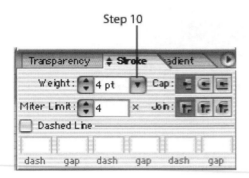

Figure B.95 – *Location of* Stroke Weight drop down menu

6. Use the **Transformation Palette** to place the *anchor point* of the rectangle in the middle and to move the *anchor point* to X=4.25 inches and Y=5.5 inches. This places the rectangle in the middle of the page (*See* Page 394 for a reminder on how to use the **Transformation Palette**).

7. Use the **Gradient Tool** to create a linear gradient fill from (left box) CYMK=[5,5,30,0] to (right box) CYMK=[5,5,65,0]. Locate the gradient slider at 80% and set a gradient angle of 90^0. (*See* Page 363 for a reminder on how to use the **Gradient Tool**)

8. Next you are going to change the color and the weight (thickness) of the box's stroke. In the **Color Palette**, change the stroke color to CYMK=[2,75,90,0]. (*See* Page 378 for a reminder on how to change the stroke color)

9. Click on the *Stroke tab* in the **Gradient Palette**.

10. Click on the down arrow (Step 10 in Figure B.95) to bring up the *Stroke Weight drop down menu*.

11. From the drop down menu select 4 pt.

12. In the **Layer Window** click on the check boxes where the 'eye' symbols adjacent to Layers 1 and 2 were to make the layers visible again.

13. Your illustration should now be similar to Figure B.96.

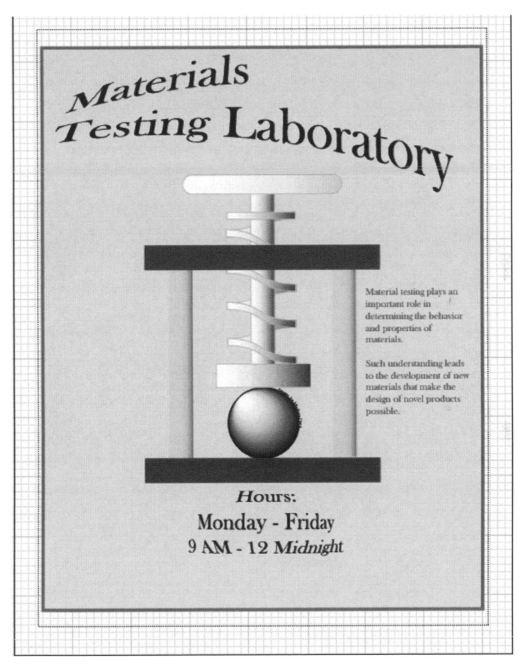

Figure B.96 – *Figure after the addition and transformation of text, and the addition of the background box*

Adding Drop Shadows

1. Using the **Selection Tool**, select the top and bottom text boxes.

2. `Effect>Stylize>Drop Shadow`.

3. In the *Drop Shadow pop-up menu* (Figure B.97), click in the Preview check box.

4. Set mode to 'Normal'; opacity to '80'% (*See* Page 409 for a description of opacity).

5. Set X offset to 0.25 in.; Y offset to 0.25 in.; and blur to 0.05 in.

6. Click on to select the 'Color' radio button.

7. Click on the square next to the color option. This brings up a *Color Picker pop-up menu* (Figure B.98) that allows you to change the drop shadow color.

8. Change the color to CYMK=[2,75,90,0]. Click on the **OK Button**.

9. In the *Drop Shadow pop-up menu*, click on the **OK Button**.

10. **Congratulations!** Your complete drawing should be similar to Figure B.99.

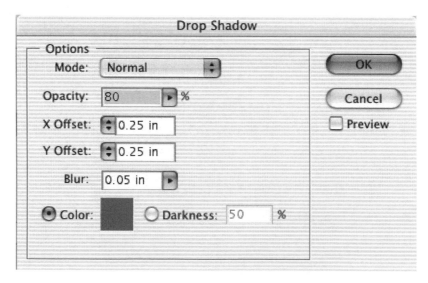

Figure B.97 – **Drop Shadow Window**

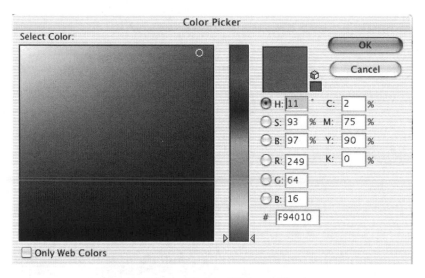

Figure B.98 – **Color Picker Window**

Opacity

Opacity or translucence refers to the extent to which you can see through a fill color. It ranges from 0% or totally transparent to 100% or opaque. As an example, the four figures below indicate the 'see through' effect or transparency of a gray Layer with opacities of 0%, 33%, 66% and 100% placed over an object below it.

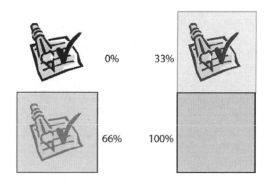

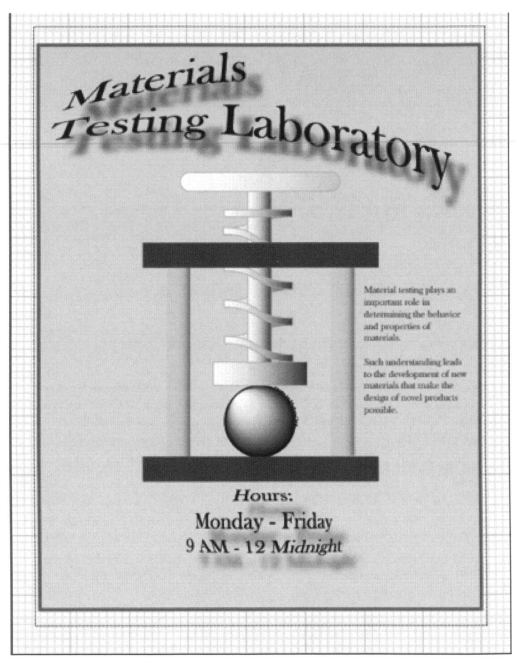

Figure B.99 – *Complete illustration after the addition of drop shadows*

The Navigator Window and Image Magnification

The **Navigator Window** allows you to easily move about different parts of your drawing, zooming in and out for easier detailed work (Figure B.100). The red rectangle in the window shows what portion of your drawing is currently visible (Compare Figure B.100 and B.101).

1. `Window>Navigator`. Note: If the Navigator menu item already has a check mark next to it, then the **Navigator Window** is active. Selecting it again will deactivate the window.

2. Zoom in to 200% magnification by either (a) clicking on the **Zoom In Button** (Step 2a in Figure B.100), (b) using the **Zoom Slider** (Step 2b in Figure B.100), or (c) by typing 200% directly in the *Magnification Box* (Step 2c in Figure B.100). Your image is enlarged to 200% (Figure B.102).

3. By clicking and holding the cursor in the middle of the red rectangle in the **Navigator Window** (Step 3 in Figure B.100) you can move the viewable area to anywhere of interest on the image. Move the rectangle until the lower text box is centered on the screen (Figure B.103).

4. Zoom out to 50% magnification by either (a) clicking on the **Zoom Out Button** (Step 4a in Figure B.104), (b) using the **Zoom Slider**, or (c) by typing 50% directly in the *Magnification Box* (Step 4c in Figure B.104). Your image is reduced in size to 50% (Figure B.105).

5. You can zoom in and out in pre-defined steps by using the *Magnification drop-down menu* at the bottom left of the Illustrator window. Click on the down arrow to the right of the drop-down menu (Step 5 in Figure B.105).

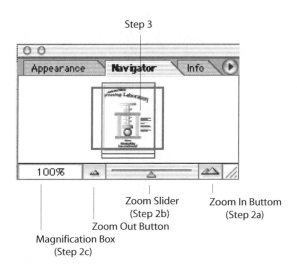

Figure B.100 – **Navigator Window** *showing steps to zoom in*

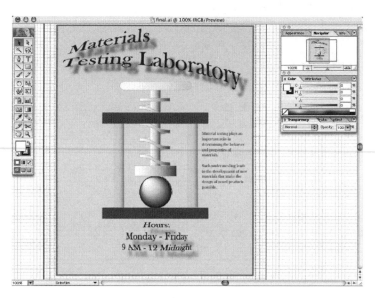

Figure B.101 – *Visible portion of image in Illustrator Window. Note similarity with rectangle in the* **Navigator Window**

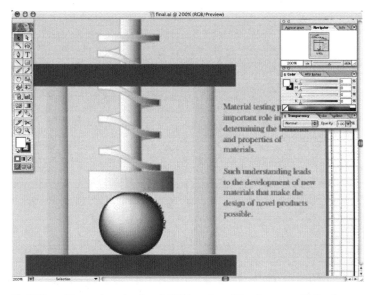

Figure B.102 – *Zooming in by 200% using* **Navigator Window** *controls*

412

Figure B.103 – *Using the red rectangle in the* **Navigator Window** *to move to an area of interest on the screen*

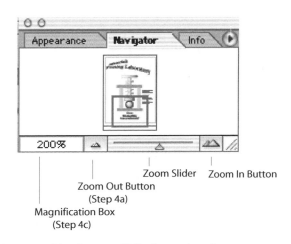

Figure B.104 – **Navigator Window** *showing steps to zoom out*

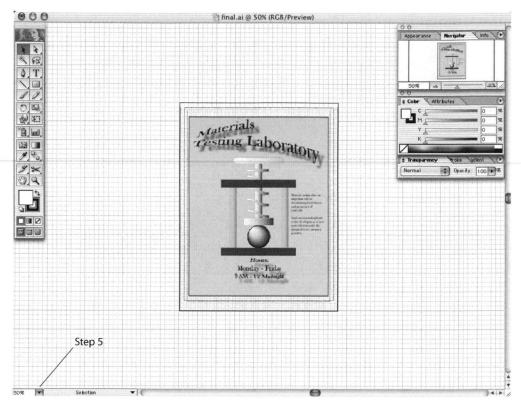

Figure B.105 – *Zooming out to 50% using* **Navigator Window** *controls*

6. In the *Magnification drop-down menu* (Figure B.106), select 'Fit on Screen'. Your image now fits exactly in the current window (Figure B.107)

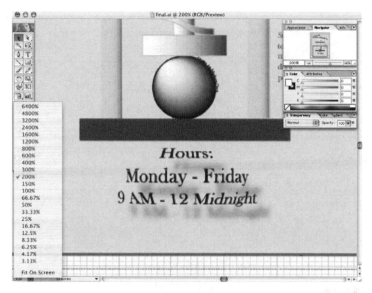

Figure B.106 – Magnification drop-down menu

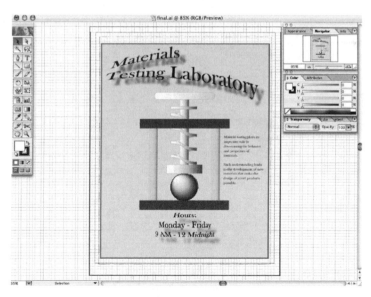

Figure B.107 – *Using the preset magnification values in the* Magnification pull-down menu

B.6 Exporting

Exporting Your File in Other Formats

Illustrator can export your final image in numerous formats. These include PDF, JPEG, TIFF and BMP. A description of each can be found on Page 85. In this tutorial, you will export your image in the PDF and JPEG formats.

1. File>Export. This brings up the *Export pop-up menu*.

2. From the *Format pull-down menu* (Figure B.108), choose the 'JPEG' format.

3. Enter a new file name and select a location to save your file.

4. Click on the **Export Button**.

5. In the *JPEG Options pop-up menu* (Figure B.109) select the 'Maximum' quality, and the 'CMYK' color model.

6. Click on the **OK Button**.

Saving Your Illustrations as a Portable Document Format (.PDF) file.

1. File>Save A Copy As.

2. In the *Save A Copy pop-up menu* (Figure B.110), enter a file name, select 'Adobe PDF (PDF)' as format.

3. Click on the **Save Button**.

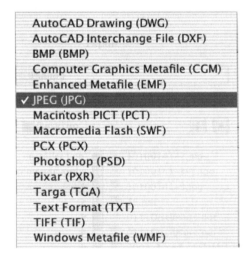

Figure B.108 – *Format pull-down menu*

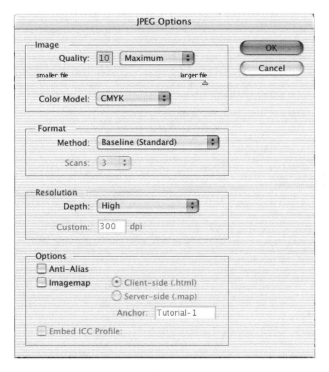

Figure B.109 – *JPEG format options*

4. This brings up the *Adobe PDF Format Options pop-up menu* (Figure B.111). Select Acrobat 5 compatibility, deselect the preserve editing capabilities options (if you are not going to edit the file directly, it results in a smaller file size); choose *Embed All Fonts* (the text in your drawing will not be dependent on the fonts available on the viewers computer. By embedding all fonts, your text will appear exactly as you intend).

5. Click on the **OK Button**. This saves your image as a pdf file. Your image has now been saved under three different formats: Illustrator (.ai), Portable Document Format (.pdf) and Joint Photographic Experts Group (.jpg)

417

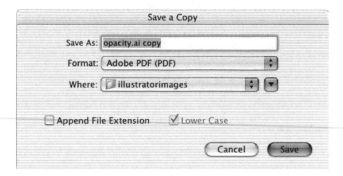

Figure B.110 – Save A Copy pop-up menu

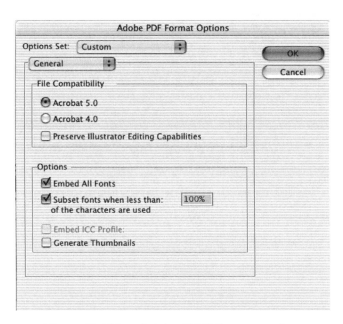

Figure B.111 – *Adobe PDF format options*

Saving for the Web

When creating images for the Internet you always need to take into account the time required for your image to download to a user's web browser. The larger the image file size, the longer it takes to download. As all web browsers display the text on a web page before the images are downloaded, a visitor to your web page may not wait to see an image if the download time is too long.

An image's file size can be reduced in two ways, (a) reducing it's dimensional size or (b) degrading it's quality. Illustrator's `Save for Web` command allows you to do both. You can see the effects of your decisions before you commit to them. Recall that for web site images, the dimensional size is measured in pixels.

Two of the most common formats used on the web are the JPEG and GIFF. A full description of each is given on Page 85.

1. `File>Open`. Select the Illustrator version of your picture.

2. `File>Save for Web`.

3. In the new window, click on the *4-up* tab (Figure B.112). The figure gives you four views of your image: the original Illustrator image (.ai extension) and three versions of a GIF image. Recall that the GIF format is best suited for flat, bold colors. Our image contains a large number of gradient and would look better with the JPEG format.

4. Select the JPEG format (Step 4 in the Figure B.112).

5. Set the 'Quality' setting to '100'.

6. Click on the arrow located by Step 6 in the figure.

7. In the resulting menu (Figure B.113), select `Repopulate Views`. The top right and the bottom two images change to reflect a JPEG version of your picture at different quality levels (Figure B.114). Under each figure, Illustrator lists the quality value, the file size and the approximate downloading time assuming a particular modem speed. Note that as the quality decreases so does the image clarity. These tools will help you decide on an appropriate compromise between image clarity and download time.

8. Click on the *Image Size* tab.

9. In the *Image Size Palette* (Figure B.115), under 'New Size', type in '50' for Percent. This will reduce the image size by 50%. Click on the **Apply Button**. The new dimensions appear in the 'Height' and 'Width' boxes. In addition each of the images in the four views are reduced in size. Beneath each figure, the corresponding file size and approximate download time are displayed (Figure B.116).

10. Click in the center of the lower left image. A black bounding box that surrounded the top right image, now surrounds the lower left image. The box indicates the current active figure.

11. Click on the 'Internet Explorer' symbol located by Step 11 in Figure B.116. NOTE: The symbol may be different depending on the default browser installed on your computer system. The active image is previewed on the browser (Figure B.117).

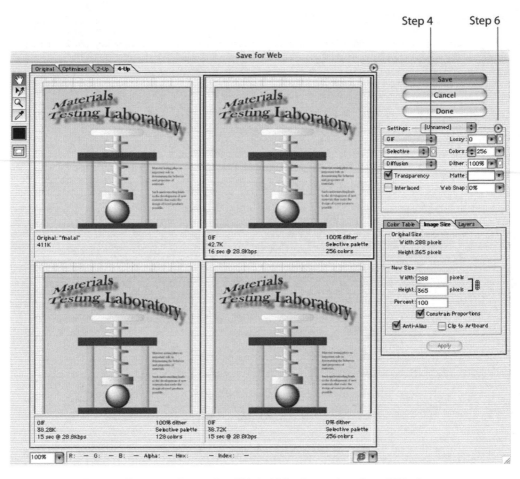

Figure B.112 – **Save for Web Window** *showing GIF views*

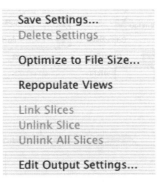

Figure B.113 –

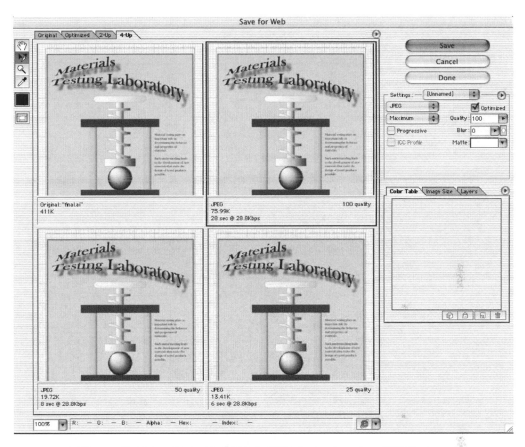

Figure B.114 – **Save for Web Window** *showing JPEG views*

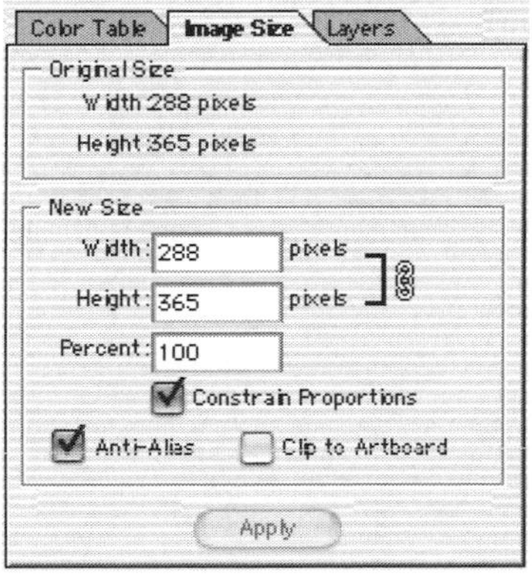

Figure B.115 – Image Size Palette

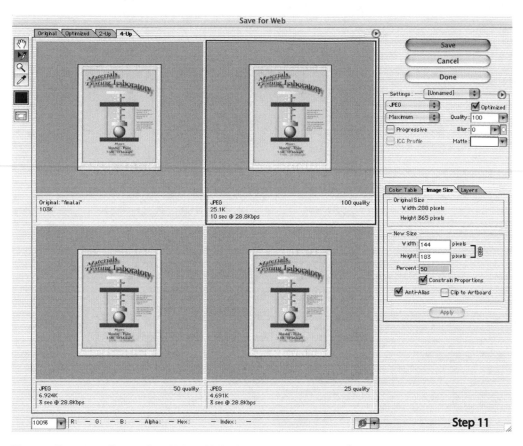

Figure B.116 – **Save for Web Window** *showing JPEG views with reduced image sizes*

12. In the **Save for Web Window**, click on the **Save Button**.

13. In the *Save pop-up menu* enter a filename and choose a location to save the active image. Click on the **Save Button**. Note that a copy of your image with the settings you selected is saved. Your original Illustrator image remains unaffected.

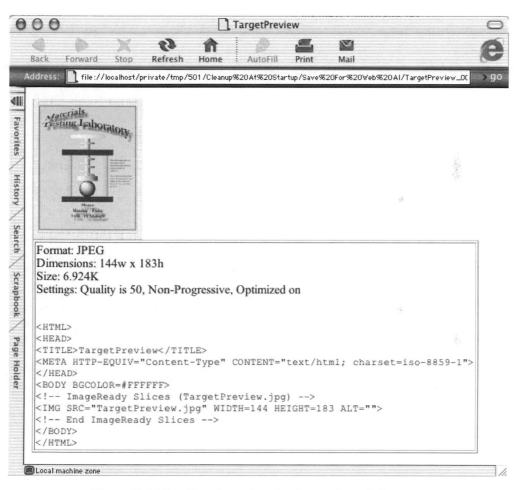

Figure B.117 – *Preview of active image in web browser*

Appendix C

TRIZ: Contradiction Matrices

Table C.1 – *TRIZ Contradiction Matrix: Improving (1-20) vs. Worsening (1-8)*

	Characteristic that is getting worse							
Characteristics to be improved	**1**	**2**	**3**	**4**	**5**	**6**	**7**	**8**
1	■	-	15 8 / 29 34	-	29 17 / 38 34	-	29 2 / 40 28	-
2	-	■	-	10 1 / 29 35	-	35 30 / 13 2	-	5 35 / 14 2
3	8 15 / 29 34	-	■	-	15 17 / 4	-	7 17 / 4 35	-
4	-	35 28 / 40 29	-	■	-	17 7 / 10 40	-	35 8 / 2 14
5	2 17 / 29 4	-	-	14 15 / 18 4	■	-	7 14 / 17 4	-
6	-	30 2 / 14 18	-	26 7 / 9 39	-	■	-	-
7	2 26 / 29 40	-	1 7 / 4 35	-	1 7 / 4 17	-	■	-
8	-	35 10 / 19 14	19 14	35 8 / 2 14	-	-	-	■
9	2 28 / 13 38	-	13 14 / 8	-	29 30 / 34	-	7 29 / 34	-
10	8 1 / 37 18	18 13 / 1 28	17 19 / 9 36	28 10	19 10 / 15	1 18 / 36 37	15 9 / 12 37	2 36 / 18 37
11	10 36 / 37 40	13 29 / 10 18	13 10 / 36	35 1 / 14 16	10 15 / 36 28	10 15 / 36 37	6 35 / 10	35 24.
12	8 10 / 29 40	15 10 / 26 3	29 34 / 5 4	13 14 / 10 7	5 34 / 4 10	-	14 4 / 15 22	7 2 / 35
13	21 35 / 2 39	26 39 / 1 40	13 15 / 1 28	37	2 11 / 13	39	28 10 / 19 39	34 28 / 35 40
14	1 8 / 40 15	40 26 / 27 1	1 15 / 8 35	15 14 / 28 26	3 34 / 40 29	9 40 / 28	10 15 / 14 7	9 14 / 17 15
15	19 5 / 34 31	-	2 19 / 9	-	3 17 / 19	-	10 2 / 19 30	-
16	-	6 27 / 19 16	-	1 40 / 35	-	-	-	35 34 / 38
17	36 22 / 6 38	22 35 / 32	15 19 / 9	15 19 / 9	3 35 / 39 18	35 38	34 39 / 40 18	35 6 / 4
18	19 1 / 32	2 35 / 32	19 32 / 16	-	19 32 / 26	-	2 13 / 10	-
19	12 18 / 28 31	-	12 28	-	15 19 / 25	-	35 13 / 18	-
20	-	19 9 / 6 27	-	-	-	-	-	-

Table C.2 – *TRIZ Contradiction Matrix: Improving (21-39) vs. Worsening (1-8)*

		Characteristic that is getting worse						
Characteristics to be improved		1	2	3	4	5	6	7	8
	21	8 36 / 38 31	19 26 / 17 27	1 10 / 35 37	-	19 38	17 32 / 13 38	35 6 / 38	30 6 / 25
	22	15 6 / 19 28	19 6 / 18 9	7 2 / 6 13	6 38 / 7	15 26 / 17 30	17 7 / 30 18	7 18 / 23	7
	23	35 6 / 23 40	35 6 / 22 32	14 29 / 10 39	10 28 / 24	35 2 / 10 31	10 18 / 39 31	1 29 / 30 36	3 39 / 18 31
	24	10 24 / 35	10 35 / 5	1 26	26	30 26	30 16	-	2 22
	25	10 20 / 37 35	10 20 / 26 5	15 2 / 29	30 24 / 14 5	26 4 / 5 16	10 35 / 17 4	2 5 / 34 10	35 16 / 32 18
	26	35 6 / 18 31	27 26 / 18 35	29 14 / 35 18	-	15 14 / 29	2 18 / 40 4	15 20 / 29	-
	27	3 8 / 10 40	3 10 / 8 28	15 9 / 14 4	15 29 / 28 11	17 10 / 14 16	32 35 / 40 4	3 10 / 14 24	2 35 / 24
	28	32 35 / 26 28	28 35 / 25 26	28 26 / 5 16	32 28 / 3 16	26 28 / 32 3	26 28 / 32 3	32 13 / 6	-
	29	28 32 / 13 18	28 35 / 27 9	10 28 / 29 37	2 32 / 10	28 33 / 29 32	2 29 / 18 36	32 28 / 2	25 10 / 35
	30	22 21 / 27 39	2 22 / 13 24	17 1 / 39 4	1 18	22 1 / 33 28	27 2 / 39 35	22 23 / 37 35	34 39 / 19 27
	31	19 22 / 15 39	35 22 / 1 39	17 15 / 16 22	-	17 2 / 18 39	22 1 / 40	17 2 / 40	30 18 / 35 4
	32	28 29 / 15 16	1 27 / 36 13	1 29 / 13 17	15 17 / 27	13 1 / 26 12	16 40	13 29 / 1 40	35
	33	25 2 / 13 15	6 13 / 1 25	1 17 / 13 12	-	1 17 / 13 16	18 16 / 15 39	1 16 / 35 15	4 18 / 39 31
	34	2 27 / 35 11	2 27 / 35 11	1 28 / 10 25	3 18 / 31	15 13 / 32	16 25	25 2 / 35 11	1
	35	1 6 / 15 8	19 15 / 29 16	35 1 / 29 2	1 35 / 16	35 30 / 29 7	15 16	15 35 / 29	-
	36	26 30 / 34 36	2 26 / 35 39	1 19 / 26 24	26	14 1 / 13 16	6 36	34 26 / 6	1 16
	37	27 26 / 28 13	6 13 / 28 1	16 17 / 26 24	26	2 13 / 18 17	2 39 / 30 16	29 1 / 4 16	2 18 / 26 31
	38	28 26 / 18 35	28 26 / 35 10	14 13 / 17 28	23	17 14 / 13	-	35 13 / 16	-
	39	35 26 / 24 37	28 27 / 15 3	18 4 / 28 38	30 7 / 14 26	10 26 / 34 31	10 35 / 17 7	2 6 / 34 10	35 37 / 10 2

Table C.3 – *TRIZ Contradiction Matrix: Improving (1-20) vs. Worsening (9-16)*

		Characteristic that is getting worse							
		9	10	11	12	13	14	15	16
Characteristics to be improved	1	2 8 / 15 38	8 10 / 18 37	10 36 / 37 40	10 14 / 35 40	1 35 / 19 39	28 27 / 18 40	5 34 / 31 35	-
	2	-	8 10 / 19 35	13 29 / 10 18	13 10 / 29 14	26 39 / 1 40	28 2 / 10 27	-	2 27 / 19 6
	3	13 4 / 8	17 10 / 4	1 8 / 35	1 8 / 10 29	1 8 / 15 34	8 35 / 29 34	19	-
	4	-	28 10	1 14 / 35	13 14 / 15 7	39 37 / 35	15 14 / 28 26	-	1 40 / 35
	5	29 30 / 4 34	19 30 / 35 2	10 15 / 36 28	5 34 / 29 4	11 2 / 13 39	3 15 / 40 14	6 3	-
	6	-	1 18 / 35 36	10 15 / 36 37	-	2 38	40	-	2 10 / 19 30
	7	29 4 / 38 34	15 35 / 36 37	6 35 / 36 37	1 15 / 29 4	28 10 / 1 39	9 14 / 15 7	6 35 / 4	-
	8	-	2 18 / 37	24 35	7 2 / 35	34 28 / 35 40	9 14 / 17 15	-	35 34 / 38
	9	■	13 28 / 15 19	6 18 / 38 40	35 15 / 18 34	28 33 / 1 18	8 3 / 26 14	3 19 / 35 5	-
	10	13 28 / 15 12	■	18 21 / 11	10 35 / 40 34	35 10 / 21	35 10 / 14 27	19 2	-
	11	6 35 / 36	36 35 / 21	■	35 4 / 15 10	35 33 / 2 40	9 18 / 3 40	19 3 / 27	-
	12	35 15 / 34 18	35 10 / 37 40	34 15 / 10 14	■	33 1 / 18 4	30 14 / 10 40	14 26 / 9 25	-
	13	33 15 / 28 18	10 35 / 21 16	2 35 / 40	22 1 / 18 4	■	17 9 / 15	13 27 / 10 35	39 3 / 35 23
	14	8 13 / 26 14	10 18 / 3 14	10 3 / 18 40	10 30 / 35 40	13 17 / 35	■	27 3 / 26	-
	15	3 35 / 5	19 2 / 16	19 3 / 27	14 26 / 28 25	13 3 / 35	27 3 / 10	■	-
	16	-	-	-	-	39 3 / 35 23	-	-	■
	17	2 28 / 36 30	35 10 / 3 21	35 39 / 19 2	14 22 / 19 32	1 35 / 32	10 30 / 22 40	19 13 / 39	19 18 / 36 40
	18	10 13 / 19	26 19 / 6	-	32 30	32 3 / 27	35 19	2 19 / 6	-
	19	8 35	16 26 / 21 2	23 14 / 25	12 2 / 29	19 13 / 17 24	5 19 / 9 35	28 35 / 6 18	-
	20	-	36 37	-	-	27 4 / 29 18	35	-	

Table C.4 – *TRIZ Contradiction Matrix: Improving (21-39) vs. Worsening (9-16)*

		Characteristic that is getting worse							
		9	10	11	12	13	14	15	16
Characteristics to be improved	21	15 35 2	26 2 36 35	22 10 35	29 14 2 40	35 32 15 31	26 10 28	19 35 10 38	16
	22	16 35 38	36 38	-	-	14 2 39 6	26	-	-
	23	10 13 28 38	14 15 18 40	3 36 37 10	29 35 3 5	2 14 30 40	35 28 31 40	28 27 3 18	27 16 18 38
	24	26 32	-	-	-	-	-	10	10
	25	-	10 37 36 5	37 36 4	4 10 34 17	35 3 22 5	29 3 28 18	20 10 28 18	28 20 10 16
	26	35 29 34 28	35 14 3	10 36 14 3	35 14	15 2 17 40	14 35 34 10	3 35 10 40	3 35 31
	27	21 35 11 28	8 28 10 3	10 24 35 19	35 1 16 11	-	11 28	2 35 3 25	34 27 6 40
	28	28 13 32 24	32 2	6 28 32	6 28 32	32 35 13	26 6 32	28 6 32	10 26 24
	29	10 28 32	28 19 34 36	3 35	32 30 40	30 18	3 27	3 27 40	-
	30	21 22 35 28	13 35 39 18	22 2 37	22 1 3 35	35 24 30 18	18 35 37 1	22 15 33 28	17 1 40 33
	31	35 28 3 23	35 28 1 40	2 33 27 18	35 1	35 40 27 39	15 35 22 2	15 22 33 31	21 39 16 22
	32	32 13 8 1	35 12	35 19 1 37	1 28 13 27	11 13 1	1 3 10 32	27 1 4	35 16
	33	18 13 34	28 13 35	2 32 12	15 34 29 28	32 35 30	32 40 3 28	29 3 8 25	1 16 25
	34	34 9	1 11 10	13	1 13 2 4	2 35	11 1 2 9	11 29 28 27	1
	35	35 10 14	15 17 20	35 16	15 37 1 8	35 30 14	35 3 32 6	13 1 35	2 16
	36	34 10 28	26 16	19 1 35	29 13 28 15	2 22 17 19	2 13 28	10 4 28 15	-
	37	3 4 16 35	36 28 40 19	35 36 37 32	27 13 1 39	11 22 39 30	27 3 15 28	19 29 39 25	25 34 6 35
	38	28 10	2 35	13 35	15 32 11 13	18 1	25 13	6 9	-
	39	-	28 15 10 36	10 37 14	14 10 34 40	35 3 22 39	29 28 10 18	35 10 2 18	20 10 16 38

Table C.5 – *TRIZ Contradiction Matrix: Improving (1-20) vs. Worsening (17-24)*

		Characteristic that is getting worse							
		17	18	19	20	21	2	23	24
Characteristics to be improved	1	6 29 / 4 38	19 1 / 32	35 12 / 34 31	-	12 36 / 18 31	6 2 / 34 19	5 35 / 3 31	10 24 / 35
	2	28 19 / 32 22	19 32 / 35	-	18 19 / 28 1	15 19 / 18 22	18 19 / 28 15	5 8 / 13 30	10 15 / 35
	3	10 15 / 19	32	8 35 / 24	-	1 35	7 2 / 35 39	4 29 / 23 10	1 24
	4	3 35 / 38 18	3 25	-	-	12 8	6 28	10 28 / 24 35	24 26
	5	2 15 / 16	15 32 / 19 13	19 32	-	19 10 / 32 18	15 17 / 30 26	10 35 / 2 39	30 26
	6	35 39 / 38	-	-	-	17 32.	17 7 / 30	10 14 / 18 39	30 16
	7	34 39 / 10 18	2 13 / 10	35	-	35 6 / 13 18	7 15 / 13 16	36 39 / 34 10	2 22
	8	35 6 / 4	-	-	-	30 6	-	10 39 / 35 34	-
	9	28 30 / 36 2	10 13 / 19	8 15 / 35 38	-	19 35 / 38 2	14 20 / 19 35	10 13 / 28 38	13 26
	10	35 10 / 21	-	19 17 / 10	1 16 / 36 37	19 35 / 18 37	14 15.	8 35 / 40 5	-
	11	35 39 / 19 2	-	14 24 / 10 37	-	10 35 / 14	2 36 / 25	10 36 / 3 37	
	12	22 14 / 19 32	13 15 / 32	2 6 / 34 14	-	4 6 / 2	14	35 29 / 3 5	-
	13	35 1 / 32	32 3 / 27 15	13 19	27 4 / 29 18	32 35 / 27 31	14 2 / 39 6	2 14 / 30 40	-
	14	30 10 / 40	35 19	19 35 / 10	35	10 26 / 35 28	35	35 28 / 31 40	-
	15	19 35 / 39	2 19 / 4 35	28 6 / 35 18	-	19 10 / 35 28		28 27 / 3 18	10
	16	19 18 / 36 40	-	-	-	16	-	27 16 / 18 38	10
	17		32 30 / 21 16	19 15 / 3 17	-	2 14 / 17 25	21 17 / 35 38	21 36 / 29 31	-
	18	32 35 / 19		32 1 / 19	32 35 / 1 15	32	13 16 / 1 6	13 1	1 6
	19	19 24 / 3 14	2 15 / 19		-	6 19 / 37 18	12 22 / 15 24	35 24 / 18 5	-
	20	-	19 2 / 35 32	-		-	-	28 27 / 18 31	-

Table C.6 – *TRIZ Contradiction Matrix: Improving (21-39) vs. Worsening (17-24)*

		Characteristic that is getting worse							
		17	18	19	20	21	22	23	24
Characteristics to be improved	21	2 14 / 17 25	16 6 / 19	16 6 / 19 37	-	■	10 35 / 38	28 27 / 18 38	10 19
	22	19 38 / 7	1 13 / 32 15	-	-	3 38	■	35 27 / 2 37	19 10
	23	21 36 / 39 31	1 6 / 13	35 18 / 24 5	28 27 / 12 31	28 27 / 18 38	35 27 / 2 31	■	-
	24	-	19	-	-	10 19	19 10	-	■
	25	35 29 / 21 18	1 19 / 26 17	35 38 / 19 18	1	35 20 / 10 6	10 5 / 18 32	35 18 / 10 39	24 26 / 28 32
	26	3 17 / 39	-	34 29 / 16 18	3 35 / 31	35	7 18 / 25	6 3 / 10 24	24 28 / 35
	27	3 35 / 10	11 32 / 13	21 11 / 27 19	36 23	21 11 / 26 31	10 11 / 35	10 35 / 29 39	10 28
	28	6 19 / 28 24	6 1 / 32	3 6 / 32	-	3 6 / 32	26 32 / 27	10 16 / 31 28	-
	29	19 26	3 32	32 2	-	32 2	13 32 / 2	35 31 / 10 24	-
	30	22 33 / 35 2	1 19 / 32 13	1 24 / 6 27	10 2 / 22 37	19 22 / 31 2	21 22 / 35 2	33 22 / 19 40	22 10 / 2
	31	22 35 / 2 24	19 24 / 39 32	2 35 / 6	19 22 / 18	2 35 / 18	21 35 / 2 22	10 1 / 34	10 21 / 29
	32	27 26 / 18	28 24 / 27 1	28 26 / 27 1	1 4	27 1 / 12 24	19 35	15 34 / 33	35 24 / 18 16
	33	26 27 / 13	13 17 / 1 24	1 13 / 24	-	35 34 / 2 10	2 19 / 13	28 32 / 2 24	4 10 / 27 22
	34	4 10	15 1 / 13	15 1 / 28 16	-	15 10 / 32 2	15 1 / 32 19	2 35 / 34 27	-
	35	27 2 / 3 35	6 22 / 26 1	19 35 / 29 13	-	19 1 / 29	18 15 / 1	15 10 / 2 13	-
	36	2 17 / 13	24 17 / 13	27 2 / 29 28	-	20 19 / 30 34	10 35 / 13 2	35 10 / 28 29	-
	37	3 27 / 35 16	2 24 / 26	35 38	19 35 / 16	19 1 / 16 10	35 3 / 15 19	1 18 / 10 24	35 33 / 27 22
	38	26 2 / 19	8 32 / 19	2 32 / 13	-	28 2 / 27	23 28	35 10 / 18 5	35 33
	39	35 21 / 28 10	26 17 / 19 1	35 10 / 38 19	1	35 20 / 10	28 10 / 29 35	28 10 / 35 23	13 15 / 23

431

Table C.7 – *TRIZ Contradiction Matrix: Improving (1-20) vs. Worsening (25-32)*

		Characteristic that is getting worse						
		25	26	27	28	29	30	31	32
Characteristics to be improved	1	10 35 / 20 28	3 26 / 18 31	3 11 / 1 27	28 27 / 35 26	28 35 / 26 18	22 21 / 18 27	22 35 / 31 39	27 28 / 1 36
	2	10 20 / 35 26	19 6 / 18 26	10 28 / 8 3	18 26 / 28	10 1 / 35 17	2 19 / 22 37	35 22 / 1 39	28 1 / 9
	3	15 2 / 29	29 35	10 14 / 29 40	28 32 / 4	10 28 / 29 37	1 15 / 17 24	17 15	1 29 / 17
	4	30 29 / 14	-	15 29 / 28	32 28 / 3	2 32 / 10	1 18	-	15 17 / 27
	5	26 4	29 30 / 6 13	29 9	26 28 / 32 3	2 32	22 33 / 28 1	17 2 / 18 39	13 1 / 26 24
	6	10 35 / 4 18	2 18 / 40 4	32 35 / 40 4	26 28 / 32 3	2 29 / 18 36	27 2 / 39 35	22 1 / 40	40 16
	7	2 6 / 34 10	29 30 / 7	14 1 / 40 11	26 28	25 28 / 2 16	22 21 / 27 35	17 2 / 40 1	29 1 / 40
	8	35 16 / 32 18	35 3	2 35 / 16	-	35 10 / 25	34 39 / 19 27	30 18 / 35 4	35
	9	-	10 19 / 29 38	11 35 / 27 28	28 32 / 1 24	10 28 / 32 25	1 28 / 35 23	2 24 / 35 21	35 13 / 8 1
	10	10 37 / 36	14 29 / 18 36	3 35 / 13 21	35 10 / 23 24	28 29 / 37 36	1 35. / 40 18	13 3 / 36 24	15 37 / 18 1
	11	37 36 / 4	10 14 / 36	10 13 / 19 35	6 28 / 25	3 35	22 2 / 37	2 33 / 27 18	1 35 / 16
	12	14 10 / 34 17	36 22	10 40 / 16	28 32 / 1	32 30 / 40	22 1 / 2 35	35 1	1 32 / 17 28
	13	35 27	15 32 / 35	-	13	18	35 24 / 30 18	35 40 / 27 39	35 19
	14	29 3 / 28 10	29 10 / 27	11 3	3 27 / 16	3 27	18 35 / 37 1	15 35 / 22 2	11 3 / 10 32
	15	20 10 / 28 18	3 35 / 10 40	11 2 / 13	3	3 27 / 16 40	22 15 / 33 28	21 39 / 16 22	27 1 / 4
	16	28 20 / 10 16	3 35 / 31	34 27 / 6 40	10 26 / 24	-	17 1 / 40 33	22	35 10
	17	35 28 / 21 18	3 17 / 30 39	19 35 / 3 10	32 19 / 24	24	22 33 / 35 2	22 35 / 2 24	26 27
	18	19 1 / 26 17	1 19	-	11 15 / 32	3 32	15 19	35 19 / 32 39	19 35 / 28 26
	19	35 38 / 19 18	34 23 / 16 18	19 21 / 11 27	3 1 / 32	-	1 35 / 6 27	2 35 / 6	28 26 / 30
	20	-	3 35 / 31	10 36 / 23	-	-	10 2 / 22 37	19 22 / 18	1 4

Table C.8 – *TRIZ Contradiction Matrix: Improving (21-39) vs. Worsening (25-32)*

		Characteristic that is getting worse							
		25	26	27	28	29	30	31	32
Characteristics to be improved	21	35 20 / 10 6	4 34 / 19	19 24 / 26 31	32 15 / 2	32 2	19 22 / 31 2	2 35 / 18	26 10 / 34
	22	10 18 / 32 7	7 18 / 25	11 10 / 35	32	-	21 22 / 35 2	21 35 / 2 22	-
	23	15 18 / 35 10	6 3 / 10 24	10 29 / 39 35	16 34 / 31 28	35 10 / 24 31	33 22 / 30 40	10 1 / 34 29	15 34 / 33
	24	24 26 / 28 32	24 28 / 35	10 28 / 23	-	-	22 10 / 1	10 21 / 22	32
	25		35 38 / 18 16	10 30 / 4	24 34 / 28 32	24 26 / 28 18	35 18 / 34	35 22 / 18 39	35 28 / 34 4
	26	35 38 / 18 16		18 3 / 28 40	13 2 / 28	33 30	35 33 / 29 31	3 35 / 40 39	29 1 / 35 27
	27	10 30 / 4	21 28 / 40 3		32 3 / 11 23	11 32 / 1	27 35 / 2 40	35 2 / 40 26	-
	28	24 34 / 28 32	2 6 / 32	5 11 / 1 23		-	28 24 / 22 26	3 33 / 39 10	6 35 / 25 18
	29	32 26 / 28 18	32 30	11 32 / 1	-		26 28 / 10 36	4 17 / 34 26	-
	30	35 18 / 34	35 33 / 29 31	27 24 / 2 40	28 33 / 23 26	26 28 / 10 18		-	24 35 / 2
	31	1 22	3 24 / 39 1	24 2 / 40 39	3 33 / 26	4 17 / 34 26	-		-
	32	35 28 / 34 4	35 23 / 1 24	-	1 35 / 12 18	-	24 2	-	
	33	4 28 / 10 34	12 35	17 27 / 8 40	25 13 / 2 34	1 32 / 35 23	2 25 / 28 39	-	2 5 / 12
	34	32 1 / 10 25	2 28 / 10 25	11 10 / 1 16	10 2 / 13	25 10	35 10 / 2 16	-	1 35 / 11 10
	35	35 28	3 35 / 15	35 13 / 8 24	35 5 / 1 10	-	35 11 / 32 31	-	1 13 / 31
	36	6 29	13 3 / 27 10	13 35 / 1	2 26 / 10 34	26 24 / 32	22 19 / 29 40	19 1	27 26 / 1 13
	37	18 28 / 32 9	3 27 / 29 18	27 40 / 28 8	26 24 / 32 28	-	22 19 / 29 28	2 21	5 28 / 11 29
	38	24 28 / 35 30	35 13	11 27 / 32	28 26 / 10 34	28 26 / 18 23	2 33	2	1 26 / 13
	39	-	35 38	1 35 / 10 38	1 10 / 34 28	18 10 / 32 1	22 35 / 13 24	35 22 / 18 39	35 28 / 2 24

Table C.9 – *TRIZ Contradiction Matrix: Improving (1-20) vs. Worsening (33-39)*

		Characteristic that is getting worse						
		33	34	35	36	37	38	39
Characteristics to be improved	1	35 3 2 24	2 27 28 11	29 5 15 8	26 30 36 34	28 29 26 32	26 35 18 19	35 3 24 37
	2	6 13 1 32	2 27 28 11	19 15 29	1 10 26 39	25 28 17 15	2 26 35	1 28 15 35
	3	15 29 35 4 7	1 28 10	14 15 1 16	1 19 26 24	35 1 26 24	17 24 26 16	14 4 28 29
	4	2 25	3	1 35	1 26	26	-	30 14 7 26
	5	15 17 13 16	15 13 10 1	15 30	14 1 13	2 36 26 18	14 30 28 23	10 26 34 2
	6	16 4	16	15 16	1 18 36	2 35 30 18	23	10 15 17 7
	7	15 13 30 12	10	15 29	26 1	29 26 4	35 34 16 24	10 6 2 34
	8	-	1	-	1 31	2 17 26	-	35 37 10 2
	9	32 28 13 12	34 2 28 27	15 10 26	10 28 4 34	3 34 27 16	10 18	-
	10	1 28 3 25	15 1 11	15 17 18 20	26 35 10 18	36 37 10 19	2 35	3 28 35 37
	11	11	2	35	19 1 35	2 36 37	35 24	10 14 35 37
	12	32 15 26	2 13 1	1 15 29	16 29 1 28	15 13 39	15 1 32	17 26 34 10
	13	32 35 30	2 35 10 16	35 30 34 2	2 35 22 26	35 22 39 23	1 8 35	23 35 40 3
	14	32 40 28 2	27 11 3	15 3 32	2 13 25 28	27 3 15 40	15	29 35 10 14
	15	12 27	29 10 27	1 35 13	10 4 29 15	19 29 39 35	6 10	35 17 14 19
	16	1	1	2	-	25 34 6 35	1	20 10 16 38
	17	26 27	4 10 16	2 18 27	2 17 16	3 27 35 31	26 2 19 16	15 28 35
	18	28 26 19	15 17 13 16	15 1 19	6 32 13	32 15	2 26 10	2 25 16
	19	19 35	1 15 17 28	15 17 13 16	2 29 27 28	35 28	32 2	12 28 35
	20	-	-	-	-	19 35 16 25	-	1 6

Table C.10 – *TRIZ Contradiction Matrix: Improving (21-39) vs. Worsening (33-39)*

		Characteristic that is getting worse						
		33	**34**	**35**	**36**	**37**	**38**	**39**
Characteristics to be improved	**21**	26 35 / 10	35 2 / 10 34	19 17 / 34	20 19 / 30 34	19 35 / 16	28 2 / 17	28 35 / 34
	22	35 32 / 1	2 19	-	7 23	35 3 / 15 23	2	28 10 / 29 35
	23	32 28 / 2 24	2 35 / 34 27	15 10 / 2	35 10 / 28 24	35 18 / 10 13	35 10 / 18	28 35 / 10 23
	24	27 22	-	-	-	35 33	35	13 23 / 15
	25	4 28 / 10 34	32 1 / 10	35 28.	6 29	18 28 / 32 10	24 28 / 35 30	-
	26	35 29 / 25 10	2 32 / 10 25	15 3 / 29	3 13 / 27 10	3 27 / 29 18	8 35	13 29 / 3 27
	27	27 17 / 40	1 11	13 35 / 8 24	13 35 / 1	27 40 / 28	11 13 / 27	1 35 / 29 38
	28	1 13 / 17 34	1 32 / 13 11	13 35 / 2	27 35 / 10 34	26 24 / 32 28	28 2 / 10 34	10 34 / 28 32
	29	1 32 / 35 23	25 10	-	26 2 / 18	-	26 28 / 18 23	10 18 / 32 39
	30	2 25 / 28 39	35 10 / 2	35 11 / 22 31	22 19 / 29 40	22 19 / 29 40	33 3 / 34	22 35 / 13 24
	31	-	-	-	19 1 / 31	2 21 / 27 1	2	22 35 / 18 39
	32	2 5 / 13 16	35 1 25 / 11 9	2 13 / 15	27 26 / 1	6 28 / 11 1	8 28 / 1	35 1 / 10 28
	33		12 26 / 1 32	15 34 / 1 16	32 26 / 12 17	-	1 34 / 12 3	15 1 / 28
	34	1 12 / 26 15		7 1 / 4 16	35 1 25 / 13 11	-	34 35 / 7 13	1 32 / 10
	351	15 34 / 1 16 7	1 16 / 7 4		15 29 / 37 28	1	27 34 / 35	35 28 / 6 37
	36	27 9 / 26 24	1 13	29 15 / 28 37		15 10 / 37 28	15 1 / 24	12 17 / 28
	37	2 5	12 26	1 15	15 10 / 37 28		34 21	35 18
	38	1 12 / 34 3	1 35 / 13	27 4 / 1 35	15 24 / 10	34 27 / 25		5 12 / 35 26
	39	1 28 / 7 19	1 32 / 10 25	1 35 / 28 37	12 17 / 28 24	35 18 / 27 2	5 12 / 35 26	

Appendix D

NSPE: Code of Ethics for Engineers

 NSPE Code of Ethics is available on their website (www.nspe.org) and is reprinted here with permission - include version and the additional information at the end of the code.

Preamble

Engineering is an important and learned profession. As members of this profession, engineers are expected to exhibit the highest standards of honesty and integrity. Engineering has a direct and vital impact on the quality of life for all people. Accordingly, the services provided by engineers require honesty, impartiality, fairness, and equity, and must be dedicated to the protection of the public health, safety, and welfare. Engineers must perform under a standard of professional behavior that requires adherence to the highest principles of ethical conduct.

D.1 Fundamental Canons

Engineers, in the fulfillment of their professional duties, shall:

1. Hold paramount the safety, health and welfare of the public.

2. Perform services only in areas of their competence.

3. Issue public statements only in an objective and truthful manner.

4. Act for each employer or client as faithful agents or trustees.

5. Avoid deceptive acts.

6. Conduct themselves honorably, responsibly, ethically, and lawfully so as to enhance the honor, reputation, and usefulness of the profession.

D.2 Rules of Practice

1. Engineers shall hold paramount the safety, health, and welfare of the public.

   a) If engineers' judgment is overruled under circumstances that endanger life or property, they shall notify their employer or client and such other authority as may be appropriate.
   b) Engineers shall approve only those engineering documents that are in conformity with applicable standards.
   c) Engineers shall not reveal facts, data, or information without the prior consent of the client or employer except as authorized or required by law or this Code.
   d) Engineers shall not permit the use of their name or associate in business ventures with any person or firm that they believe are engaged in fraudulent or dishonest enterprise.
   e) Engineers shall not aid or abet the unlawful practice of engineering by a person or firm.
   f) Engineers having knowledge of any alleged violation of this Code shall report thereon to appropriate professional bodies and, when relevant, also to public authorities, and cooperate with the proper authorities in furnishing such information or assistance as may be required.

2. Engineers shall perform services only in the areas of their competence.

   a) Engineers shall undertake assignments only when qualified by education or experience in the specific technical fields involved.
   b) Engineers shall not affix their signatures to any plans or documents dealing with subject matter in which they lack competence, nor to any plan or document not prepared under their direction and control.
   c) Engineers may accept assignments and assume responsibility for coordination of an entire project and sign and seal the engineering documents for the entire project, provided that each technical segment is signed and sealed only by the qualified engineers who prepared the segment.

3. Engineers shall issue public statements only in an objective and truthful manner.

   a) Engineers shall be objective and truthful in professional reports, statements, or testimony. They shall include all relevant and pertinent information in such reports, statements, or testimony, which should bear the date indicating when it was current.
   b) Engineers may express publicly technical opinions that are founded upon knowledge of the facts and competence in the subject matter.
   c) Engineers shall issue no statements, criticisms, or arguments on technical matters that are inspired or paid for by interested parties, unless they have prefaced their comments by explicitly identifying the interested parties on whose behalf they are speaking, and by revealing the existence of any interest the engineers may have in the matters.

4. Engineers shall act for each employer or client as faithful agents or trustees.

   a) Engineers shall disclose all known or potential conflicts of interest that could influence or appear to influence their judgment or the quality of their services.

   b) Engineers shall not accept compensation, financial or otherwise, from more than one party for services on the same project, or for services pertaining to the same project, unless the circumstances are fully disclosed and agreed to by all interested parties.

   c) Engineers shall not solicit or accept financial or other valuable consideration, directly or indirectly, from outside agents in connection with the work for which they are responsible.

   d) Engineers in public service as members, advisors, or employees of a governmental or quasi-governmental body or department shall not participate in decisions with respect to services solicited or provided by them or their organizations in private or public engineering practice.

   e) Engineers shall not solicit or accept a contract from a governmental body on which a principal or officer of their organization serves as a member.

5. Engineers shall avoid deceptive acts.

   a) Engineers shall not falsify their qualifications or permit misrepresentation of their or their associates' qualifications. They shall not misrepresent or exaggerate their responsibility in or for the subject matter of prior assignments. Brochures or other presentations incident to the solicitation of employment shall not misrepresent pertinent facts concerning employers, employees, associates, joint venturers, or past accomplishments.

   b) Engineers shall not offer, give, solicit or receive, either directly or indirectly, any contribution to influence the award of a contract by public authority, or which may be reasonably construed by the public as having the effect of intent to influencing the awarding of a contract. They shall not offer any gift or other valuable consideration in order to secure work. They shall not pay a commission, percentage, or brokerage fee in order to secure work, except to a bona fide employee or bona fide established commercial or marketing agencies retained by them.

D.3   Professional Obligations

1. Engineers shall be guided in all their relations by the highest standards of honesty and integrity.

   a) Engineers shall acknowledge their errors and shall not distort or alter the facts.

   b) Engineers shall advise their clients or employers when they believe a project will not be successful.

c) Engineers shall not accept outside employment to the detriment of their regular work or interest. Before accepting any outside engineering employment they will notify their employers.

d) Engineers shall not attempt to attract an engineer from another employer by false or misleading pretenses.

e) Engineers shall not promote their own interest at the expense of the dignity and integrity of the profession.

2. Engineers shall at all times strive to serve the public interest.

a) Engineers shall seek opportunities to participate in civic affairs; career guidance for youths; and work for the advancement of the safety, health, and well-being of their community.

b) Engineers shall not complete, sign, or seal plans and/or specifications that are not in conformity with applicable engineering standards. If the client or employer insists on such unprofessional conduct, they shall notify the proper authorities and withdraw from further service on the project.

c) Engineers shall endeavor to extend public knowledge and appreciation of engineering and its achievements.

3. Engineers shall avoid all conduct or practice that deceives the public.

a) Engineers shall avoid the use of statements containing a material misrepresentation of fact or omitting a material fact.

b) Consistent with the foregoing, engineers may advertise for recruitment of personnel.

c) Consistent with the foregoing, engineers may prepare articles for the lay or technical press, but such articles shall not imply credit to the author for work performed by others.

4. Engineers shall not disclose, without consent, confidential information concerning the business affairs or technical processes of any present or former client or employer, or public body on which they serve.

a) Engineers shall not, without the consent of all interested parties, promote or arrange for new employment or practice in connection with a specific project for which the engineer has gained particular and specialized knowledge.

b) Engineers shall not, without the consent of all interested parties, participate in or represent an adversary interest in connection with a specific project or proceeding in which the engineer has gained particular specialized knowledge on behalf of a former client or employer.

5. Engineers shall not be influenced in their professional duties by conflicting interests.

a) Engineers shall not accept financial or other considerations, including free engineering designs, from material or equipment suppliers for specifying their product.

b) Engineers shall not accept commissions or allowances, directly or indirectly, from contractors or other parties dealing with clients or employers of the engineer in connection with work for which the engineer is responsible.

6. Engineers shall not attempt to obtain employment or advancement or professional engagements by untruthfully criticizing other engineers, or by other improper or questionable methods.

a) Engineers shall not request, propose, or accept a commission on a contingent basis under circumstances in which their judgment may be compromised.

b) Engineers in salaried positions shall accept part-time engineering work only to the extent consistent with policies of the employer and in accordance with ethical considerations.

c) Engineers shall not, without consent, use equipment, supplies, laboratory, or office facilities of an employer to carry on outside private practice.

7. Engineers shall not attempt to injure, maliciously or falsely, directly or indirectly, the professional reputation, prospects, practice, or employment of other engineers. Engineers who believe others are guilty of unethical or illegal practice shall present such information to the proper authority for action.

a) Engineers in private practice shall not review the work of another engineer for the same client, except with the knowledge of such engineer, or unless the connection of such engineer with the work has been terminated.

b) Engineers in governmental, industrial, or educational employ are entitled to review and evaluate the work of other engineers when so required by their employment duties.

c) Engineers in sales or industrial employ are entitled to make engineering comparisons of represented products with products of other suppliers.

8. Engineers shall accept personal responsibility for their professional activities, provided, however, that engineers may seek indemnification for services arising out of their practice for other than gross negligence, where the engineer's interests cannot otherwise be protected.

a) Engineers shall conform with state registration laws in the practice of engineering.

b) Engineers shall not use association with a nonengineer, a corporation, or partnership as a "cloak" for unethical acts.

9. Engineers shall give credit for engineering work to those to whom credit is due, and will recognize the proprietary interests of others.

a) Engineers shall, whenever possible, name the person or persons who may be individually responsible for designs, inventions, writings, or other accomplishments.

b) Engineers using designs supplied by a client recognize that the designs remain the property of the client and may not be duplicated by the engineer for others without express permission.

c) Engineers, before undertaking work for others in connection with which the engineer may make improvements, plans, designs, inventions, or other records that may justify copyrights or patents, should enter into a positive agreement regarding ownership.

d) Engineers' designs, data, records, and notes referring exclusively to an employer's work are the employer's property. The employer should indemnify the engineer for use of the information for any purpose other than the original purpose.

e) Engineers shall continue their professional development throughout their careers and should keep current in their specialty fields by engaging in professional practice, participating in continuing education courses, reading in the technical literature, and attending professional meetings and seminars.

Appendix E

Component Tables

Table E.1 – *American Standard Unified and American National Coarse Threads*

Size (Nom. Diam., in.)	Threads per in.	Tap drill	Tensile Stress Area, in^2
0 (0.060)			
1 (0.073)	64	No. 53	0.00263
2 (0.086)	56	No. 50	0.00370
3 (0.099)	48	No. 47	0.00487
4 (0.112)	40	No. 43	0.00604
5 (0.125)	40	No. 38	0.00796
6 (0.138)	32	No. 36	0.00909
8 (0.164)	32	No. 29	0.0140
10(0.190)	24	No. 25	0.0175
12(0.216)	24	No. 16	0.0242
1/4	20	No. 7	0.0318
5/16	18	Letter F	0.0524
3/8	16	5/16	0.0775
7/16	14	Letter U	0.1063
1/2	13	27/64	0.1419
9/16	12	31/64	0.182
5/8	11	17/32	0.226
3/4	10	21/32	0.334
7/8	9	49/64	0.462
1	8	7/8	0.606

Table E.2 – *American Standard Unified and American National Fine Threads*

Size (Nom. Diam., in.)	Threads per in.	Tap drill	Tensile Stress Area, in^2
0 (0.060)	80	3/64	0.00180
1 (0.073)	72	No. 53	0.00278
2 (0.086)	64	No. 50	0.00394
3 (0.099)	56	No. 45	0.00523
4 (0.112)	48	No. 42	0.00661
5 (0.125)	44	No. 37	0.00830
6 (0.138)	40	No. 33	0.01015
8 (0.164)	36	No. 29	0.01474
10(0.190)	32	No. 21	0.0200
12(0.216)	28	No. 14	0.0258
1/4	28	No. 3	0.0364
5/16	24	Letter I	0.0580
3/8	24	Letter Q	0.0878
7/16	20	25/64	0.1187
1/2	20	29/64	0.1599
9/16	18	33/64	0.203
5/8	18	37/64	0.256
3/4	16	11/16	0.373
7/8	14	13/16	0.509
1	12	59/64	0.663

Table E.3 – *Common soldering alloys and their applications (Compiled from Mc-Master(2002))*

Name	Composition	Applications, Approx. Melting Temp, °F (°C)
Lead-Silver	Pb(96%)-Ag(4%)	High temperature joints, 580 (305)
Lead-Silver-Tin	Pb(97.5%)-Ag(1.5%)-Sn(1%)	Due to it's high melting point, these solders find use with commutators, armatures, and initial solders-its high melting point prevents remelting when soldering successive joints. 565-574 (296-301)
Lead-Tin-Silver	Pb(54%)-Sn(45%)-Ag(1%)	Provides exceptional strength. It contains silver for a nondulling, bright, long-lasting finish and is ideal for stainless steel. 350-410 (177-210)
Tin-Lead	Sn(95%)-Pb(5%) Sn(63%)-Pb(37%) Sn(60%)-Pb(40%) Sn(50%)-Pb(50%) Sn(40%)-Pb(60%) Sn(30%)-Pb(70%)	Plumbing and heating. 460 (238) Has the lowest melting point of the tin/lead solders, with exceptional tinning and wetting properties. It is well suited for stainless steel and electronic work. 361 (183) The most popular dipping solder for electronic applications. 370 (188) General Purpose and well suited for standard tinning and sheet metal work. 390 (199) Automobile Radiators. 405 (207) Ideal for machine and torch soldering. 361-491
Tin	Sn(100%)	Has good strength and is nondulling. Finds applications on food processing equipment, as well as for alloying and tinning wire. 410 (210)
Tin-Antimony	Sn(95%)-Sb(5%)	Provides high strength and a bright finish, it melts and flows like a 50/50 tin/lead solder. Primary applications include air-conditioning, refrigeration, food containers and high-temperature applications. It is not recommended for use with electronics. 450-464 (232-240)
Tin-Zinc	Sn(91%)-Zn(9%)	Aluminum joining. 390 (199)
Tin-Silver	Sn(96%)-Ag(4%)	Often referred to as silver-bearing solder, tin/silver solder is used on food service equipment, refrigeration, plumbing, heating, and air conditioning. 430-444 (221-229)

Table E.4 – *Relationship between pipe size, outer diameter, NPT threads, inner diameter, wall thickness and schedules 10 and 40 sizes - all units inches*

Pipe Size	Pipe OD	NPT Threads per inch	Schedule 10 Pipe ID	Thickness	Schedule 40 Pipe ID	Thicknes
1/8	0.405	27	0.307	0.049	0.269	0.068
1/4	0.540	18	0.410	0.065	0.364	0.088
3/8	0.675	18	0.545	0.083	0.493	0.091
1/2	0.840	14	0.674	0.083	0.622	0.109
3/4	1.050	14	0.884	0.109	0.824	0.113
1	1.315	11 1/2	1.097	0.109	1.049	0.133
1 1/4	1.660	11 1/2	1.442	0.109	1.380	0.140
1 1/2	1.900	11 1/2	1.682	0.109	1.610	0.145
2	2.375	11 1/2	2.157	0.109	2.067	0.154
2 1/2	2.875	8	2.635	0.120	2.469	0.203
3	3.500	8	3.260	0.120	3.068	0.216
4	4.500	8	4.260	0.120	4.026	0.237
5	5.563	8	5.295	0.134	5.047	0.258
6	6.625	8	6.357	0.134	6.065	0.280
8	8.625	8	8.329	0.148	7.981	0.322

Table E.5 – *Relationship between pipe size, outer diameter, NPT threads, inner diameter, wall thickness and schedules 80 and 160 sizes - all units inches*

Pipe Size	Pipe OD	NPT Threads per inch	Schedule 80 Pipe ID	Thickness	Schedule 160 Pipe ID	Thickness
1/8	0.405	27	0.215	0.095	–	–
1/4	0.540	18	0.302	0.119	–	–
3/8	0.675	18	0.423	.126	–	–
1/2	0.840	14	0.546	0.147	0.466	0.187
3/4	1.050	14	0.742	0.154	0.614	0.218
1	1.315	11 1/2	0.957	0.179	0.815	0.250
1 1/4	1.660	11 1/2	1.278	0.191	1.160	0.250
1 1/2	1.900	11 1/2	1.500	0.200	1.338	0.281
2	2.375	11 1/2	1.939	0.218	1.689	0.343
2 1/2	2.875	8	2.323	0.276	2.125	0.375
3	3.500	8	2.900	0.300	2.626	0.437
4	4.500	8	3.826	0.337	3.438	0.531
5	5.563	8	4.813	0.375	4.313	0.625
6	6.625	8	5.761	0.432	5.189	0.718
8	8.625	8	7.625	0.500	6.813	0.906

Appendix F

Common Unit Conversions

Converting from	To	Multiply by
Length		
foot (ft)	$meter(m)$	0.3048
inch $(in.)$	$meter(m)$	0.0254
yard	feet (ft)	3.0
mile $(mi.)$, US statue	meter (m)	1,609.3
mile $(mi.)$, International nautical	meter (m)	1,852
Area		
foot^2 (ft^2)	meter^2 (m^2)	0.092903
inch^2 $(in.^2)$	meter^2 (m^2)	0.00064516
Volume		
inch^3 $(in.^3)$	meter^3 (m^3)	0.016387
foot^3 (ft^3)	inch^3 $(in.^3)$	1,728
foot^3 (ft^3)	meter^3 (m^3)	0.028317
Density		
pound mass/inch^3 (lbm/in^3)	kilogram/meter^3 (kg/m^3)	27,680
Force		
pound force (lb)	newton(N)	4.4482
Mass		
pound mass (lb)	kilogram kg	0.4536
Pressure		
pound/inch^2 $(lb/in^2$ or $psi)$	newton/meter^2 $(N/m^2$ or $Pa)$	27,680

Glossary

Activity networks. Graphical or text method that displays the parallel and sequential relationships between tasks.

Aesthetics. Related to art and beauty.

Annealing. The heating of a metal to a suitable temperature, holding the temperature fixed for a period of time, followed by controlled slow cooling. It is use to reduce hardness, recrystalize strain hardened metals, and to reduce residual stresses from other processes.

Assembly drawing. A drawing that illustrates how all the parts of a particular design are assembled. The parts are typically drawn in their operating positions. Also included in the drawing is a parts list or bill of materials.

Automation. Replacing human operation (manual) with machinery.

Benchmarking. Comparing similar products or functions using the same set of evaluation criteria. Results are typically presented in tabular form.

Bill of materials. Listing of parts or materials in an assembly. Typically included as part of an assembly drawing.

Bioaccumulation. Increase in the concentration of a chemical over time in a biological organism compared to the chemical's concentration in the environment.

Biomimicry. The study of nature's models to imitate or take inspiration from these designs and processes to solve human problems.

Bonding. The process of permanently combining two (or more) parts. Examples of bonding include brazing, gluing, welding, or soldering.

Brainstorming. A method to generate a large number of concepts by freely suggesting ideas with no evaluation or criticism.

Brittle. Brittle materials exhibit little or no yielding before fracture.

CAD. An acronym that could refer to Computer-Aided Design, Computer-Aided Drafting or Computer-Aided Design/Drafting depending on the context.

CAM. Computer-Aided Manufacturing, and is the use of computers to plan and control the production process. Examples include, robotics and numerically controlled machines.

Casting-metals. Metal casting is a process in which a metal or metal alloy is poured into a mold, solidifying in the shape of the mold cavity.

449

Cold Working. Manufacturing processes performed on or around room temperature.

Composites. Combination of materials typically consisting of a structural element and a resin binder. A common example is fibreglass that uses glass strands as the structural element.

Concurrent engineering. Teams from different disciplines all work closely together during the entire design process, rather than serially where each discipline team waits for other disciplines to complete their part of the design before beginning theirs.

Creep. Plastic deformation under a sustained load.

Critical Path. Longest task completion path form the beginning to the end of a project. It defines the minimum time required to complete a project.

Design for manufacturability (DFM). The incorporation of manufacturing processes directly into the design process, resulting in higher quality products at lower cost.

Ductility. Ability of a material to *plastically* strain without fracture.

Ecological Footprint. Indicator and measure of the effective land area necessary to support human activity.

Elastic Limit. Maximum stress beyond which permanent deformation occurs.

Elastomers. Polymers that undergo large *elastic* deformation when subjected to relatively low loads. Commonly referred to as rubbers.

Factor of Safety (Dimensionless). The ratio of Ultimate stress to allowable stress of a member. The allowable stress is the maximum operating stress for that member. The factor of safety must always be greater than one.

Fatigue. Failure of a material due to cyclical (on and off) loading stress values below the ultimate strength. Failure depends on value of loading and number of cycles.

Hardness. Resistance to permanent indentation, i.e., good hardness means the material is resistant to scratch and wear. Several tests are available to measure hardness, the most common ones being the Brinell Hardness Test (used for testing metals and nonmetals of low to medium hardness), the Rockwell Hardness Test and the Vickers Hardness Test.

Hot Working. Manufacturing processes performed at the materials recrystallization temperature.

Life Cycle Assessment. Assessment of a product's "life" from the time it is created from raw materials, through the product's use, and eventually ends with disposal.

Malleable. The extent to which a material can be permanently compressed without rupturing.

Plasticity. Permanent (non-elastic) deformation of a material under load.

Quenching. Rapid cooling of a material to increase its hardness.

Recrystallization temperature. Approximately half the material melting point.

Rolling. Rolling involves two opposing rolls drawing material into the gap between them, thereby reducing the material's thickness.

Stiffness. Ability of a material to resist deformation due to applied loading.

Strength. Ability of a material to resist yielding.

Strain. Deformation per unit length of a material (object) under an applied load. It is a dimensionless quantity.

Stress. Measure of intensity of a force acting on a material.

Sustainability. The ability to meet the needs of the present without compromising the ability of future generations to meet their own needs.

Thermoplastics. Polymers that do not undergo an irreversible curing process during heating.

Thermosets. Polymers that undergo an irreversible curing process during heating.

Viscoelasticity. Process where a molten polymer expands when imposed stresses are removed.

Ultimate Strength. Maximum stress level that the material can resist before breaking.

Work Hardening. when a material becomes stronger as the strain increases during the plastic region of the stress strain curve. It is a property that metals exhibit to a greater or lessor degree, and is an important factor during certain manufacturing processes, for example metal forming

Yield Strength (Units of Stress) Maximum stress levels that the material can resist before yielding.

Young's Modulus of Elasticity, E Proportionality constant between stress and strain of a material within the linear elastic region. The higher the value of E, the more ductile the material. The lower the value of E, the more brittle the material.

Index

ISBN 141203850-2

Edwards Brothers Malloy
Oxnard, CA USA
June 2, 2014